Concrete Structures Subjected to Impact and Blast Loadings and Their Combinations

Concrete Structures Subjected to Impact and Blast Loadings and Their Combinations

Chunwei Zhang and Gholamreza Gholipour

CRC Press

Taylor & Francis Group

Boca Raton London New York

CRC Press is an imprint of the
Taylor & Francis Group, an **informa** business

Cover image: Chunwei Zhang, Gholamreza Gholipour

First published 2022
by CRC Press/Balkema
Schipholweg 107C, 2316 XC Leiden, The Netherlands
e-mail: enquiries@taylorandfrancis.com
www.routledge.com – www.taylorandfrancis.com

CRC Press/Balkema is an imprint of the Taylor & Francis Group, an informa business

Library of Congress Cataloging-in-Publication Data
Names: Zhang, Chunwei, (Civil Engineer), author. | Gholipour, Gholamreza, author.
Title: Concrete structures subjected to impact and blast loadings and their combinations / Chunwei Zhang & Gholamreza Gholipour.
Description: Boca Raton : CRC Press, 2022. | Includes bibliographical references and index.
Identifiers: LCCN 2021055051 (print) | LCCN 2021055052 (ebook)
Subjects: LCSH: Buildings, Reinforced concrete. | Concrete walls--Blast effect. | Impact.
Classification: LCC TA683.2 .Z43 2022 (print) | LCC TA683.2 (ebook) | DDC 624.1/8341--dc23/eng/20220217
LC record available at https://lccn.loc.gov/2021055051
LC ebook record available at https://lccn.loc.gov/2021055052

ISBN: 978-1-032-20051-4 (hbk)
ISBN: 978-1-032-20127-6 (pbk)
ISBN: 978-1-003-26234-3 (ebk)

DOI: 10.1201/9781003262343

Typeset in Times New Roman
by KnowledgeWorks Global Ltd.

Contents

Foreword

The world has witnessed the catastrophic consequences of structural failure caused by terrorist bombing attacks and accidental explosions. One of the recent examples is the accidental explosion in Beirut, which resulted in over 6,000 injures, 200 fatalities, and destroyed a major part of the city with an estimate of over US$10 billion in infrastructure damage affecting over 300,000 individuals. Some RC structures during their service life may experience extreme loadings from terrorist attacks and accident explosions. For protection of life and economy, they need to be designed to resist such loadings effectively. One very likely scenario is the combined loadings from impact and blast, for example, the explosion occurs after vehicle or ship that carries explosives impacts on structures. There are many studies of impact and blast loading effects on structures independently. The study of the combined impact and blast loading effects is limited, though it is a realistic loading scenario. Prof. Chunwei Zhang, who is an expert on structural engineering, has devoted many years researching on the combined impact and blast loading effects on RC structures. This book summarizes the research results of the author in this area, and it is the first book that presents a comprehensive and systematic analysis and design methodologies of structures against combined impact and blast loads.

Combined loading scenarios may vary with respect to the various loading parameters including the loading sequence and the time lag between the onsets of the loads. Impact and blast loads can occur simultaneously or sequentially. In the case of simultaneous-combined loadings, the time lag between the initiations of the impact and blast loads is very short (milliseconds), which can include considerable temperature and strain rate coupled effects. However, in the case of sequential-combined loadings, the time lag between the onsets of the impact and blast loads is relatively long. Owing to the substantial discrepancies between the loading mechanisms of impact and blast loads, the accumulative impulses of their combinations are extremely sensitive to the loading sequence and the time lag, which significantly affect the structural responses and could change the failure modes of the structure. Therefore, a comprehensive understanding of the dynamic responses and failure behaviors of civil structures under impact and blast loads, and their combinations, is essential for effective design of structures to resist such loadings.

This book is suitable for students (undergraduate or postgraduate), engineers, and researchers who are interested in analysis and design of structures against blast and impact loads. It introduces the analysis methods for concrete structures when subjected to impact and blast loads. The theory of the combinations of impact and blast loads is presented that can provide primary insights for readers to develop new ideas on impact and blast engineering. This book also provides a comprehensive state-of-the-art review on the previous studies and structural analysis of structures under extreme loads that can be very helpful for postgraduate students and researchers in Civil Engineering.

Hong Hao
BS, MSc, PhD
DistFIAPS, FTSE, FASCE, FISEAM
ARC Laureate Fellow, John Curtin Distinguished Professor
Department of Civil Engineering
Curtin University, Australia

Preface

With increasing the number of terrorist attacks on strategic structures such as high-rise buildings and bridges in recent years, recognizing the dynamic responses and failure behaviors of such structures under extreme loads such as impact and blast loads, and their combinations, is the topic of importance. Many research works in the literature are focused on investigating the responses of reinforced concrete (RC) structures and isolated members under sole impact or blast loadings. However, evaluating the responses and behaviors of RC structures under the combination of impact and blast loadings is the gap study of previous research works. The combined actions of impact and blast loadings may be applied to RC structures during accidental or intentional (i.e., terrorist attacks) collision of vessels, vehicles, etc., carrying explosive materials. Under such loading scenarios, the vulnerability of RC structures can be assessed with the variability of several key parameters including the loading sequence and the time lag between the initiations of impact and blast loadings. Therefore, a comprehensive investigation of the dynamic responses and failure behaviors of various structural members, such as RC beams, columns, and bridge piers, is conducted using finite element (FE) simulations in LS-DYNA software under combined actions of impact and blast loadings varying in various loading- and structural-related parameters.

Firstly, comprehensive state-of-the-art reviews on the responses and failure behaviors of concrete structures subjected to lateral impact and blast loads are presented in Chapters 2 and 3 based on various approaches of structural analysis and existing provisions and recommendations in the current design codes and guidelines.

The failure behaviors of RC columns and bridge piers under ship collisions are investigated in Chapters 4 and 5 considering the key dynamic factors, such as the strain rate effects of materials and the inertia of the pier and superstructure, which have not been well and accurately studied in the literature. The majority of current design codes propose equivalent static approaches or simplified analytical methods to analyze and design the structures under extreme loads. However, these approaches are not only able to predict the brittle failure behaviors of RC structures, but also they neglect the effectiveness of aforementioned dynamic factors that can lead to underestimated and unconservative structural responses in the design of RC structures. Hence, it is attempted to investigate the effects of these dynamic factors on the failure behaviors and progressive damage process of RC bridge piers under ship collisions in Chapters 4 and 5.

The primary objective of this book is to explore the dynamic responses of RC structures under combined actions of impact and blast loads. In Chapter 6, the methodology for applying impact and blast loadings is proposed in the manners of simultaneously and subsequently on a simply-supported RC beam varying some key loading- and structural-related parameters. Also, a motion-based damage index based on the residual flexural capacity of the RC beam is proposed. Different failure modes are observed by varying the loading sequence and the time lag parameters between the initiations of applied loads. In line with this study, the sensitivity levels of these key loading parameters (i.e., the loading sequence and the time lag) are assessed to the impact loading rate (i.e., impact velocity) on simply-supported RC beams using a parametric study and defining two damage indices based on the residual shear force and flexural moment capacities of the beam in Chapter 7.

In Chapter 8, the vulnerability of axially-loaded RC columns subjected to the combinations of lateral impact and blast loadings is investigated by proposing a damage index based on the residual axial load capacity of columns from a multistep loading process numerically in LS-DYNA. Furthermore, the sensitivity levels of the column failure to various loading-related parameters are assessed in comparison with those with under sole impact and blast loadings. To study the applicability of the proposed combined loading scenarios in the real world, the combined actions of impact and blast loadings arising from the collision of two typical vessels, including 2,000-DWT barge and 5,000-DWT container ship carrying 200 and 500 kg TNT (as the explosive materials), respectively, are applied on a typical girder bridge pier in Chapter 9. In organizing the combined loading scenarios, it is attempted to investigate the vulnerability of the bridge pier to several loading-related parameters, including the loading locations, impact velocity, and the time lag between the initiations of the applied loads with considering the existing limitations for the draft depth and the velocity of such vessels in inland waterways according to the current design codes.

Chunwei Zhang
B.E., MSc, PhD (HIT), FISEAM
Distinguished Professor and Founding Director
Center for Multi-disciplinary Infrastructure Engineering
Shenyang University of Technology
Shenyang City, China

Founding Director, Structural Vibration Control Group
Qingdao University of Technology
Qingdao City, China

Acknowledgments

The authors would like to acknowledge the following financial support by the Ministry of Science and Technology of China (Grant No. 2019YFE0112400), the Department of Science and Technology of Shandong Province (Grant No. 2021CXGC011204), the State Key Laboratory of Precision Blasting of Jianghan University, the Key Research and Development Program of Liaoning Province, the Taishan Scholar Priority Discipline Talent Group program funded by the Shandong Province, and the first-class discipline project funded by the Education Department of Shandong Province, to the completion of the book and relevant research investigations. These immense supports have facilitated the conceptualization and growth of knowledge concerning blast and impact loading effects onto civil structures, and especially their combination effects.

Chapter 1

Introduction

1.1 Background

Since the construction of reinforced concrete (RC) structures has been widely increased in recent decades, the need for recognizing different behavioral aspects of such structures from micro (materials) to macro scales (structural responses) under various types of loadings from static to dynamic and impulsive loadings is the topic of importance nowadays. As reported by Buth et al. [1], the collapse of bridge superstructures in several cases, such as due to the collision of trucks with bridge piers, from 1965 to 2008 was observed in the United States. Harik et al. [2] classified the bridge failures that occurred due to various causes in the United States during the period of 1951–1988. It was reported that 42 out of 79 (i.e., about 53%) bridges collapsed due to collisions in which 19 cases (≈24%) caused by ships, 11 cases (≈14%) by trucks, and 6 cases (≈8%) by trains. Moreover, in 4 cases (≈5%), the bridges collapsed due to exploding or burning of fuel-tanker trucks. Based on a study done by Wardhana and Hadipriono [3], during 1989–2000 in the United States, 12% of the total bridge failures occurred due to lateral impact forces (arising from the collision of trucks, barges, ships, and trains) and 3% of the failures caused by fire and explosions. For the period of 2000–2008, Hersi [4] revealed that 24% of bridge failures in the United States were occurred due to collision and lateral impact loads, whereas fire and explosions caused only 1% of bridge failures during a similar time period. In addition, Cook [5] reported that 12% of collapsed bridges in the United States during 1987–2011 were occurred due to collision, and 3% of them failed due to explosions and the combination of fire and collision. According to a study by Lee et al. [6] during 1980–2012 in the United States, 15.3% and 2.8% of failures were occurred due to collisions, and fire and explosions, respectively. Comparing to bridge failure studies in the United States, Scheer [7] classified the causes of bridge failures that occurred all over the world. It was found that 13% of 440 failure cases were due to ship collisions and 5% of them failed due to fire or explosion.

The effects of extreme loads such as impact and blast loads on RC structures and infrastructures have been notably studied since the 1960s. In 1964, the dynamic principles and analyses for structures under impulsive load were proposed by Biggs [8]. A comprehensive treatment of explosion hazard mitigation free field by Baker [9] and Clough and Penzien [10] adopted a more systematic and advanced theory for structural dynamics. Subsequently, Hetherington and Smith [11] presented the design of structures to resist weapon effects. The Structural Engineering Institute (SEI) of the American Society of Civil Engineers (ASCE) published a state-of-the-practice report

DOI: 10.1201/9781003262343-1

to provide guidance to structural engineers in the design of civil structures to resist the effects of terrorist bombings. More recently, Krauthammer [12] addressed a broad range of scientific and technical issues involved in mitigating the extreme loading effect associated with blast, shock, and impact. In addition, a few design manuals can be referred to, e.g., the Tri-service manual TM-5-1300 [13], Army Technical Manual TM-5-855-1 (US Department of the Army, 1986) [14], ASCE Manual 42 [15], ASCE guidelines for blast-resistant buildings in petrochemical facilities [9], UFC 3-340-02 [16], GSA [17] criteria, DOE/TIC-11268 [18], and FEMA Reference Manual [19].

The effects of blast loads are directly related to stress-wave propagation in the structure which are generally applied to the surface of target structures. On the basis of the interactions and relationships between blast loading pressure (P) and its impulse (I) response on target structure, known as P-I diagrams, blast loads are divided into three classes, including (i) impulsive (in which loading duration, t_d, is very short relative to structure period time, T), (ii) dynamic ($t_d \approx T$), and (iii) quasi-static ($t_d > T$). Another common criterion to classify the intensity of blast loadings based on Hopkinson-Cranz law [20] is known as the scaled distance (Z) of detonation from the target defining a relationship between equivalent weight (W) and standoff distance (R) of the explosive charge. According to this criterion, blast loads can be categorized into near-field and far-field detonations. There exist different definitions in classifying blast loads on scaled distance. According to the American Society of Civil Engineers (ASCE SEI 59-11) [21], blast loads with scaled distances less than 1.2 m/kg$^{1/3}$ are identified as close-in detonations. Gel'fandetal et al. [22] defined another criterion based on the dimension of the charge r_0. According to this criterion, blast loads with standoff distance R_n between the ranges 0 and $20r_0$ ($0 < R_n < 20r_0$) are recognized as close-in detonations. In addition, UFC 3-340-02 [23] considered a scaled distance equal to 0.4 m/kg$^{1/3}$ as the sensitive level for scaling of blast loads. As such, blast loads with a scaled distance less than this value were considered close-in explosions.

The analysis of structures under explosions has been widely carried analytically, numerically, and experimentally in the literature. Krauthammer et al. [24, 25] proposed an analytical method to analyze RC beams under idealized blast loadings uniformly distributed on the theoretical Timoshenko beam. Most of the design codes and guidelines listed previously employ simplified approaches to idealized blast loads and predict the resistance of structures by considering dynamic increase factors. The occurrence of concrete spallation is the most probable failure mode in RC structures under explosions. The basic theory of spallation of RC structures associated with the concept of stress-wave propagation under blast loadings was proposed by McVay [26] and the intensity damage states of RC structures based on the spallation were classified. Although the global failure behaviors of structures on displacement-based damage criterion can be predicted by simplified methods, they are unable to capture brittle shear failures and spallation. Moreover, these approaches cannot take into account the effects of structural dimensions, material dynamic properties, stress-wave propagation, and interlock resistance forces in the cross section. Therefore, many research works utilized experimental and numerical finite element (FE) methods to realistically analyze the failure behaviors and dynamic responses of structures under explosions. There exist several approaches to conduct experimental field and laboratory tests of blast loads on structures using different facilities such as (i) free-air bursts using explosive charges [27], (ii) shock tube facilities [28], (iii) the University of California, San Diego (UCSD) simulator [28], (iv) Gas Blast Simulator (GBS) [29–33]. Owing to expensive costs and security limitations and difficulties of experimental tests, numerous research studies focused on the simulation of blast loads on structures using high-fidelity FE software codes such as ABAQUS [34], LS-DYNA [35], and AUTODYN [36].

Impact loading is another type of dynamic and extreme loadings in which its duration may reach 1,000 times shorter than earthquakes. Impact loads can be characterized into three types based on their intensity and duration (t_d). These types are (i) quasi-static loading in which the structure reaches its maximum response before ending of the impact duration, (ii) dynamic loading in which the structure reaches its maximum response almost at the same time with the ending of the impact duration, and (iii) impulsive loading in which impact duration ends before reaching the structure its maximum response. Accordingly, structural components can demonstrate different behaviors under concentrate lateral impact loads, including localized and overall responses. When a structural member is subjected to high-rate impact loading with very short duration relative to the structure natural period (T), the stress-wave propagation and inertia resistance of the structure are predominant on the responses. Under this loading condition, it is more expected to observe localized failure in the structure. However, when this ratio (t_d/T) is large, the structural responses and failure modes are dependent on the stiffness of the structure and the structure tends to fail in overall modes [37]. Another classification of impact loadings based on their dissipative mechanism was proposed by Eurocode [38], including (i) hard impact in which the initial kinetic energy was dissipated by striking objects such as colliding of vessels and vehicles with deformable bows with concrete structures and (ii) soft impact in which the major part of the initial kinetic energy was dissipated by the impacted structure such as impacting the rocks and rigid objects on concrete structures. However, concrete structures might suffer localized failure modes and damages such as brittle spalling, scabbing, perforation, and punching shear failure [39, 40] under high-rate impact loads, or overall failure modes under rather low-rate impact loads that cannot capture using simplified approaches.

Several current design codes such as the American Association of State Highway and Transportation Officials (AASHTO) [41], European Committee for Standardization (CEN) [38, 42], China Ministry of Railways (CMR) [43], Standards Australia (SA/SNZ, 2002) [44], AS 1170.1 [45], UK's Highways Agency [46] proposed equivalent static approaches to estimate the impact forces. Besides, there exist some design manuals and documents in the literature to investigate the impact forces and dynamic responses of various structures based on a series of experimental collision tests of vehicles [1, 47, 48] and vessels [49–56].

For vehicle collisions, AASHTO [57] recommended to employ an equivalent static impact force of 2,668 kN laterally applied to the height of 1.5 m from the base of columns or bridge piers based on a series of experimental collision tests of a heavy truck with rigid steel columns carried out by Buth et al. [1, 47]. Unlike AASHTO that does not take into account the structural characteristics of impacted columns and the initial impact loading conditions, Eurocode 1 (CEN) [42] estimates the vehicle collision loads using a simplified equation by considering the deformations suffered by striking vehicle and stuck columns, and the initial kinetic energy of the vehicle. However, the effects of key dynamic factors arising from high-rate collisions such as inertia, strain rate effects of the materials, and stress-wave propagations on the estimation of impact loads and structural responses have not been considered that may lead to unconservative approach in designing the structures. According to several vehicle-column collision investigations in the literature [58–61], the inability of these simplified methods in the prediction of dynamic responses and failure behaviors of RC structures especially those associated with the brittle shear failure modes were found. Therefore, numerous experimental [47] and numerical [61–64] studies have been carried out to explore the dynamic responses and failure behaviors of RC structures under vehicle collisions.

The experimental tests of ship-ship impacts conducted by Minorsky [50] in 1959 are known as the pioneer of vessel collision studies. An empirical formula was proposed based on the dissipation of the initial kinetic energy due to the deformations of the ship bow. Afterwards, Woisin [51] modified Minorky's formula for high-energy vessel collisions. Based on the empirical formulas proposed by Meir-Dorenberg [49] by conducting a series of pendulum hammer impact tests on the scaled models of barges in 1983, AASHTO [65] provided a static equation for computing barge collision loads using the absorbed energy due to the deformations of barge bow. Like the existing provisions for vehicle collisions, the equivalent static equations proposed by AASHTO for vessel collisions are not able to predict the dynamic responses and failure modes of impacted structures since the dynamic factors and the structural properties of the impacted structure are not taken into account. Similarly, Eurocode 1 [66] assumes that not only the impacted structures are rigid and fixed but also the behaviors of impacting vessels are elastic. Despite some experimental tests of barge collision with lock gates [67] and lock walls [68] carried out by the US Army Corps of Engineers, the first full-scale barge collision tests with two piers of St. George Island Causeway Bridge were conducted by Consolazio et al. [53–55] in 2004 with the purpose of filling the gap of knowledge in the contribution and effects of structural dynamic factors and geometrical characteristics of the impacted structure on the impact responses. The force-deformation results from vessel collisions with bridge piers were analytically [69] and numerically [70–72] studied by numerous research works in the literature. In addition, the structural response of the bridge piers subjected to vessel collisions was investigated using various simplified analytical methods [73–76] and FE high-resolution techniques [77–80]. However, the influences of the superstructure inertia and the strain rate effects of materials on the impact responses and failure behaviors of bridge piers and columns were not carefully investigated in these studies.

1.2 Challenges in relevant research areas

With increasing the risk of applying extreme loads such as impact and blast loads to concrete structures in recent years, the structural analysis and protective design of civil structures and infrastructures are the topic of importance. Under more critical circumstances, such structures may be subjected to the combined actions of impact and blast loads during the intentional terrorist attacks or accidental events. Such combined loading scenarios can occur during the impact of heavy objects with relatively low velocities such as vehicles, trucks, and vessels, or rather lighter objects with higher velocities such as missiles and projectiles that carry the explosive materials could be led to the subsequent blast loading after the onset of impact loadings on civil structures such as bridges, framed buildings, etc.

Previous studies have attempted to capture some insights into the dynamic responses of concrete structures under sole impact or blast loads. However, investigating the performance of concrete structures under synergistic effects of impact and blast loads has not yet been addressed in the available technical literature. Besides, the current design codes and guidelines suffer from the lack of practical information and deficiencies in the protective and resilient design of concrete structures against sole impact and blast loads and their combinations. RC structures, retrofitted concrete, and resilient structures may undergo substantially different dynamic responses and failure behaviors under combined actions of impact and blast loads compared to those under sole impact and blast loads. Accordingly, existing specifications in the current design codes may not be effective and sufficient in the protective design of resilient structures under combined loads. Therefore, it is worth to comprehensively investigate the performance and structural failure behaviors of concrete structures under the synergistic effects of impact and blast loads using experimental tests and numerical techniques.

The major challenges existing in the analysis and design of concrete structures under combined actions of impact and blast loads can be drawn through the following questions:

- How do the sequential blast/impact loads affect the impact/blast-induced behaviors of concrete structures?
- How do the critical damage mechanisms and failure modes of concrete structures take place under simultaneous- and subsequent-combined loadings?
- Do the current design guidelines provide efficient and sufficient resilient solutions that resist to the synergistic effects of impact and blast loads?
- What factors should be considered in the design of concrete structures against combined loadings?
- What design recommendations and solutions are required to improve the performance of concrete structures under combined loads?

The main objective of this book is to investigate the vulnerability of RC structures subjected to impact and blast loads, and their combinations varying in terms of two specific loading parameters, including the loading sequence and the time lag between the initiations of the applied loads. The specific objectives of this book are:

1 To investigate the vulnerability of axially loaded RC columns and bridge piers under lateral impact loads and especially ship collisions using comprehensive FE simulations and loading scenarios varying in various loading- and structural-related parameters.
2 To formulate the strain rate effects of the concrete materials on the impact responses of RC structures by proposing an analytical simplified model with two-degree-of-freedom (2-DOF).
3 To discover the sensitivity thresholds of the impact responses, internal forces, and the location of plastic hinges formed in RC columns to the variability of the axial load ratio (ALR), and the location impact load by proposing a series of empirical equations between them through a parametric study.
4 To evaluate the dynamic responses and failure behaviors of RC structures subjected to combined actions of impact and blast loadings varying in terms of various parameters, including the loading sequence, the time lag between the onsets of the applied loads, the ALR, the loading location, and the impact velocity.
5 To propose different damage indices based on the residual capacities of RC structures using multistep loading approaches.

6 To study the sensitivity thresholds of the proposed damages indices of RC structures to key loading parameters under combined loadings in comparison with
   those resulted under sole impact and blast loadings.
7 To analyze the nonlinear failure behaviors of RC bridge piers subjected to vessel impact combined with blast loads to propose the essential protective design
   recommendations.

1.3 Significance of the book

Under more critical circumstances, RC structures may be subjected to the combined actions of impact and blast loads during the intentional terrorist attacks
or accidental events arising from the collision of vehicles, vessels, etc., carrying
explosive materials, which may lead to the sequent blast loading after the onset of
impact loadings on civil structures such as bridges, framed buildings, and so on.
Combined loading scenarios may vary in terms of two key loading parameters,
including the loading sequence and the time lag between the onsets of the loads.
In addition, impact and blast loads can apply simultaneously or subsequently to
structures under such loading combinations. In the case of simultaneous-combined
loadings, the time lag between the initiations of the impact and blast loads is very
short (milliseconds) which can include considerable temperature and strain-rate-
coupled effects. However, in the case of sequential-combined loadings, the time lag
between the onsets of the impact and blast loads is relatively long. Owing to substantial discrepancies between the loading mechanisms of impact (which are commonly
applied as the concentrated loading) and blast loads (which are commonly applied
as distributed loading), accumulative impulses of their combinations are extremely
sensitive to the loading sequence and the time lag parameters that can play a very
important role in the determination of failure modes. Therefore, a comprehensive
study on the vulnerability of RC structures under the combined actions of impact
and blast loads is required.

   The study reported in this book includes two parts. In the first part, the influences
of various parameters, including the axial load, the strain rate effects of materials,
and loading rate on the dynamic responses and failure behaviors of axially loaded
RC bridge piers and columns under lateral impact loads and ship collisions, were
numerically investigated using FE simulations in LS-DYNA. In addition, an analytical simplified model with 2-DOF is proposed to formulate the strain rate effects of
the concrete materials as the dynamic increase factors in the global responses of the
impacted pier.

   In the second part, the vulnerabilities of structural components, including RC
beams, columns, and bridge piers, were numerically assessed subjected to combined
actions of impact and blast loads varying in terms of various parameters, including
the loading sequence, the time lag between the onsets of the applied loads, the ALR,
the loading location, and the impact velocity. To do this, different damage indices
regarding the applicability of the structural member were proposed based on their
residual capacities calculated using multistep loading methodologies. Due to the
lack of experimental tests to investigate RC structures under the combined actions
of impact and blast loads in the literature, the experimental results and data from
previous impact and blast tests are used to separately validate the FE models of RC
structures developed in LS-DYNA.

1.4 Structure of the book

This book consists of ten chapters and their contents are summarized as follows:

Chapter 1: This chapter contains the research background associated with the effects and actions of impact and blast loadings on RC structures, research objectives, and the outline of the book.

Chapter 2: This chapter presents a comprehensive state-of-the-art review on the responses and failure behaviors of various types of concrete structures subjected to lateral impact loads based on the analytical, numerical, and experimental studies. In addition, the influences of various structural- and loading-related parameters on the impact resistance and failure behaviors of different concrete structures under lateral impact loads are reviewed.

Chapter 3: This chapter presents a comprehensive review on existing analytical, numerical, and experimental studies in the literature investigating the loading mechanisms, dynamic responses, and failure behaviors of various concrete structures subjected to blast loads.

Chapter 4: This chapter evaluates the impact responses, internal forces, failure modes, and the location of plastic hinges formed in RC columns and bridge piers under lateral impact loads varying in terms of several loading-related parameters, including the ALR, the location of impact loads, and the impact velocity through a parametric study.

Chapter 5: This chapter presents a study on the progressive damage behaviors and nonlinear failure modes of a cable-stayed bridge pier subjected to ship collisions using FE simulations in LS-DYNA. In addition, an analytical 2-DOF system is proposed to formulate the strain rate effects of materials for the ship-bridge collision system.

Chapter 6: This chapter numerically investigates the nonlinear dynamic responses and failure behaviors of simply-supported RC beams subjected to the combination of impact and blast loads with the variations in structural-related parameters, including the beam depth, the span length and the configuration of the reinforcements, and loading-related parameters, including the loading sequence, and the time lag between the onsets of the applied loads. A damage index based on the residual flexural moments of the beams is proposed through a multistep loading method.

Chapter 7: The influences of impact loading rate (i.e., impact velocity) on residual capacities and damage states of RC beams with simple supports subjected to combined actions of impact and blast loads are assessed in this chapter. Two different damage indices on the basis of the residual shear and flexural capacities of the RC beams are proposed through multistep loading procedures. In addition, the sensitivities of the damage indices were assessed to the impact velocity, the loading sequence, and the time lag between the onsets of the applied loads.

Chapter 8: In this chapter, the failure behaviors and dynamic responses of a typical RC column commonly used in medium-rise buildings are numerically evaluated under combined actions of impact and explosion loadings varying in several loading-related parameters, including the loading sequence, the time lag (t_L) between the onsets of the applied loads, the ALR, the loading location, and the impact velocity (V_{impact}) using a damage index (*DI*) on the basis of the column residual axial load carrying capacity.

Chapter 9: This chapter studies the vulnerability of an RC girder bridge pier, which is numerically assessed under the combination of vessel collisions and blast loadings varying in terms of several loading-related parameters, including the vessel type, the location of applied loads, the impact velocity, and the time lag between the initiations of loadings. Some design recommendations are suggested to enhance the resistance of bridge piers against combined loadings

Chapter 10: The summary, particularly conclusions, and main findings of this book are presented in this chapter. Also, several recommendations are suggested for future works.

References

1. Buth CE, Williams WF, Brackin MS, Lord D, Geedipally SR, Abu-Odeh AY. Analysis of large truck collisions with bridge piers: phase 1. Report of guidelines for designing bridge piers and abutments for vehicle collisions (FHWA/TX-10/9-4973-1). College Station, TX, 2010.
2. Harik IE, Shaaban AM, Gesund H, Valli GYS, Wang ST. United States bridge failures, 1951–1988. *J Perform Constr Fac* 1990;4(4):272–277.
3. Wardhana K, Hadipriono FC. Analysis of recent bridge failures in the United States. *J Perform Constr Fac* 2003;17(3):144–150.
4. Hersi M. Analysis of bridge failure in United States (2000–2008). M.Sc. Thesis, The Ohio State University, Columbus, OH, 2009.
5. Cook W. Bridge failure rates, consequences, and predictive trends. Ph.D. Dissertation, Department of Civil and Environmental Engineering, Utah State University, Logan, UT, 2014.
6. Lee GC, Mohan S, Huang C, Fard BN. A study of US bridge failures (1980–2012). Report No. MCEER-13-0008, University at Buffalo, Buffalo, NY, 2013.
7. Scheer J. *Failed bridges: case studies, causes and consequences.* Ernst & Sohn-Wiley, Germany, 2010.
8. Biggs JM. *Introduction to structural dynamics.* McGraw Hill, New York, NY, 1964.
9. Baker WE. *Explosion hazards and evaluation.* Elsevier Scientific Publishing Company, Amsterdam, NY, 1983.
10. Clough RW, Penzien J. *Dynamics of structures*, 2nd edn. McGraw-Hill, Tokyo, 1993.
11. Hetherington JG, Smith PD. *Blast and ballistic loading of structures.* CRC Press, Boca Raton, FL, 2014.
12. Krauthammer T. *Modern protective structures.* CRC Press, Boca Raton, FL, 2008.
13. TM-5-1300. Design of structures to resist the effects of accidental explosions. Technical Manual, US DoA, Washington, DC, 1990.
14. TM-5-855-1. Fundamentals of protective design for conventional weapons, Technical Manual, US DoA, Washington, DC, 1986.
15. ASCE. Design of structures to resist nuclear weapons effects. Manual 42, American Society of Civil Engineers, Reston, VA, 1985.
16. US Department of Defense. Structures to resist the effects of accidental explosions. Report no. UFC 3-340-02, United Facilities Criteria, Washington, DC, 2008.
17. US General Services Administration (GSA). ISC security design criteria for new federal office buildings and major modernization projects. Washington, DC, 2003.
18. Baker WE, Kulesz JJ, Westine PS, Cox PA, Wilbeck JS. A manual for the prediction of blast and fragment loading on structures. DOE/TIC-11268, US Department of the Army, US Department of Energy, Washington, DC, 1992.

19. FEMA 427. Primer for design of commercial buildings to mitigate terrorist attacks, Federal Emergency Management Agency, Washington, DC, 2003.
20. Kennedy WD. *Explosions and explosives in air.* NDRC, Washington, DC, 1946.
21. ASCE. *Blast protection of buildings. ASCE SEI 59-11.* American Society of Civil Engineers, Reston, VA, 2011.
22. Gel'fandetal B, Voskoboinikov I, Khomik S. Recording the position of a blast-wave front in air. *Combust Explos Shock Waves* 2004;40(6):734–736.
23. US Department of Defense. Structures to resist the effects of accidental explosions. Report no. UFC 3-340-02, United Facilities Criteria, Washington, DC, 2008.
24. Krauthammer T, Assadi-Lamouki A, Shanaa HM. Analysis of impulsively loaded reinforced concrete structural elements—I. *Theory Comput Struct* 1993;48(5):851–860.
25. Krauthammer T, Assadi-Lamouki A, Shanaa HM. Analysis of impulsively loaded reinforced concrete structural elements—II. *Implementation Comput Struct* 1993;48(5):861–871.
26. McVay MK. Spall damage of concrete structures. No. WES/TR/SL-88-22, Army Engineer Waterways Experiment Station Vicksburg MS Structures LAB, Vicksburg, MS, 1988.
27. Zhang D, Yao S, Lu F, Chen X, Lin G, Wang W, Lin Y. Experimental study on scaling of RC beams under close-in blast loading. *Eng Fail Anal* 2013;33:497–504.
28. Aoude H, Dagenais FP, Burrell RP, Saatcioglu M. Behavior of ultra-high performance fiber reinforced concrete columns under blast loading. *Int J Impact Eng* 2015;80:185–202.
29. Zhang XH, Zhang CW, Duan ZD. Numerical simulation on shock waves generated by explosive mixture gas from large nuclear blast load generator based on equivalent-energy principles. *J Exp Shock Waves* 2014;34(1):80–86.
30. Zhang XH, Zhang CW, Duan ZD. Numerical simulation on impact responses and failure modes of steel frame structural columns subject to blast loads. *J Shenyang Jianzhu Univ* 2009;25(4):656–662.
31. Zhang XH, Zhang ZD, Duan ZD, Zhang CW. Analysis for dynamic response and failure process of reinforced concrete beam under blast load. *J Northeast Forest Univ* 2009;37(4):50–53.
32. Zhang CW, Xu H, Liu J, Li L, Zhang X, Liu C, Wu Z, Liu J, Li J. Recent advances in structural vibration control and blast resistance research in hit blast resistance and protective engineering laboratory. In *Proceeding of 1st Asia Pacific Young Researchers and Graduates Symposium*, Kunsan, Republic of Korea, 2009, pp.132–187.
33. Zhang CW, Wang W, Zhang X, Lu S. Large scale blast emulator based explosive gas loading methods for structures and recent advances in experimental studies. In *Proceedings of The 8th International Conference on Shock & Impact Loads on Structures*, Adelaide, Australia, 2009, pp. 769–779.
34. ABAQUS. Analysis user's manual version 6.10. Volume I to VI, ABAQUS, Inc. an Dassault Systémes, Providence, RI, 2010.
35. LS-DYNA 971. Livermore Software Technology Corporation. Livermore, CA, 2015.
36. AUTODYN. *Interactive non-linear dynamic analysis software, version 12, user's manual.* SAS IP, Inc., Canonsburg, PA, 2009.
37. Pham TM, Hao H. Influence of global stiffness and equivalent model on prediction of impact response of RC beams. *Int J Impact Eng* 2018;113:88–97.
38. CEN. Eurocode 1: Actions on structures, European Committee for Standardization, Brussels, Belgium, 2003.
39. Li QM, Reida SR. Local impact effects of hard missiles on concrete targets. *Int J Impact Eng* 2005;32:224–284.
40. Bangash MYH. *Impact and explosion: structural analysis and design.* CRC Press LLC, Boca Raton, FL, 1993.
41. AASHTO. *Guide specifications and commentary for vessel collision design of highway bridges*, 2nd edn. American Association of State Highway and Transportation Officials, Washington, DC, 2009.

42. CEN. Actions on structures. Part 1-7: General actions-accidental actions. European Committee for Standardization; 2006. BS EN 1991-1-1, Brussels, Belgium, 2002.
43. CMR. General code for design of railway bridges and culverts, TB10002.1–2005, China Ministry of Railways, China Railway Press, Beijing, China (in Chinese), 2005.
44. SA/SNZ. Structural design actions Part 1: Permanent, imposed and other actions Sydney, NSW 2001; Wellington 6020: AS/NZS 1170.1:2002, 2002.
45. AS 1170.1. AS/NZS 1170.1:1988: Structural design actions-permanent, imposed and other actions, Australia, 1989.
46. UK's Highways Agency. The design of highway bridges for vehicle collision loads. BD 60/04. Department for Transport, UK, 2004.
47. Buth CE, Brackin MS, Williams WF, Fry GT. Collision loads on bridge piers: phase 2. Report of guidelines for designing bridge piers and abutments for vehicle collisions (FHWA/TX-11/9-4973-2). College Station, TX, 2011.
48. Agrawal AK, Xu X, Chen Z. *Bridge vehicle impact assessment (C-07-10)*, University Transportation Research Center, NY, USA, 2011.
49. Meir-Dornberg KE. Ship collision safety zones, and loading assumptions for structures on inland waterways. *VDI-Berichte* 1983;496:1–9.
50. Minorsky VU. An analysis of ship collisions with reference to protection of nuclear power plants. *J Ship Res* 1959;3(1):1–4.
51. Woisin G. *The collision test of the GKSS. Jahrbuch Der Schiffbautechnischen Gesellschaft*, Volume 70, Berlin, Springer-Verlag, 1976, pp. 465–487.
52. Pedersen PT, et al. Ship impacts: bow collisions. *Int J Impact Eng* 1993;V13(2):163–187.
53. Consolazio GR, Cook RA, Lehr GB. Barge impact testing of the St. George Island Causeway Bridge-Phase I: Feasibility study (Research Report No. BC-354 RPWO–23). Engineering and Industrial Experiment Station, University of Florida, Gainesville, FL, 2002.
54. Consolazio GR, Cook RA, Biggs AE, Cowan DR. Barge impact testing of the St. George Island Causeway Bridge-Phase II: Design of infrastructure systems (Research Report No. BC-354 RPWO–56). Engineering and Industrial Experiment Station, University of Florida, Gainesville, FL, 2003.
55. Consolazio GR, Cook RA, McVay MC. Barge impact testing of the St. George Island Causeway Bridge-Phase III: Physical testing and data interpretation, structures (Research Report No. BC-354 RPWO–76). Engineering and Industrial Experiment Station, University of Florida, Gainesville, FL, 2006.
56. Patev RC, Barker BC, Koestler LV. *Prototype barge impact experiments, Allegheny Lock and Dam 2*. Pennsylvania. US Army Corps of Engineers, Pittsburgh, 2003.
57. AASHTO. *AASHTO LRFD bridge design specifications (customary U.S. Units)*, 6th edn. American Association of State Highway and Transportation Officials, Washington, DC, 2012.
58. Do TV, Pham TM, Hao H. Dynamic responses and failure modes of bridge columns under vehicle collision. *Eng Struct* 2018;156:243–259.
59. Do TV, Pham TM, Hao H. Impact force profile and failure classification of reinforced concrete bridge columns against vehicle impact. *Eng Struct* 2019;183:443–458.
60. El-Tawil S, Severino E, Fonseca P. Vehicle collision with bridge piers. *J Bridge Eng* 2005;10(3):345–353.
61. Sharma H, Hurlebaus S, Gardoni P. Performance-based response evaluation of reinforced concrete columns subject to vehicle impact. *Int J Impact Eng* 2012;43:52–62.
62. Abdelkarim OI, ElGawady MA. Performance of bridge piers under vehicle collision. *Eng Struct* 140:337–352.
63. Chen L, El-Tawil S, Xiao Y. Reduced models for simulating collisions between trucks and bridge piers. *J Bridge Eng* 2016;21(6):04016020.
64. Chen L, Xiao Y, Xiao G, Liu C, Agrawal AK. Test and numerical simulation of truck collision with anti-ram bollards. *Int J Impact Eng* 2015;75:30–39.

65. AASHTO. Guide specifications and commentary for vessel collision design of highway bridges. American Association of State Highway and Transportation Officials, Washington, DC, 2009.

66. European Committee for Standardization (CEN). *Eurocode 1– Actions on structures. Parts 1–7: General actions–accidental actions due to impact and explosions*, 3rd draft, CEN, Brussels, Belgium, 2002.

67. Goble G, Schulz J, Commander B. Lock and dam# 26 field test report for the army corps of engineers. Bridge Diagnostics, Boulder, CO, 1990.

68. Arroyo JR, Ebeling RM, Barker BC. Analysis of impact loads from full-scale, low-velocity, controlled barge impact experiments. December 1998 (no. ERDC/ITL-TR-03-3). Engineer Research and Development Center Vicksburg MS Information Technology Lab, Vicksburg, MS, 2003.

69. Fan W, Zhang Y, Liu B. Modal combination rule for shock spectrum analysis of bridge structures subjected to barge collisions. *J Eng Mech –ASCE* 2016;142(2):04015083.

70. Consolazio GR, Cowan DR. Nonlinear analysis of barge crush behavior and its relationship to impact resistant bridge design. *Comput Struct* 2003; 81: 547–557.

71. Consolazio GR, Cowan DR. Numerically efficient dynamic analysis of barge collisions with bridge piers. *J Struct Eng–ASCE* 2005;131(2131):1256–1266.

72. Fan W, Yuan W. Ship bow force-deformation curves for ship-impact demand of bridges considering effect of pile-cap depth. *Shock Vib* 2014; 2014: 1–19.

73. Consolazio GR, Davidson MT. Simplified dynamic barge collision analysis for bridge design. *Transp Res Rec* 2008;2050(1):13–25.

74. Fan W, Yuan W, Yang Z, Fan Q. Dynamic demand of bridge structure subjected to vessel impact using simplified interaction model. *J Bridge Eng* 2010;16(1):117–126.

75. Sha Y, Hao H. A simplified approach for predicting bridge pier responses subjected to barge impact loading. *Adv Struct Eng* 2014;17(1):11–23.

76. Yuan P, Harik IE, Davidson MT. Multi-barge flotilla impact forces on bridges. Report No. KTC-08-13/SPR261-03-2F. Kentucky Transportation Center College of Engineering, University of Kentucky, Lexington, KY, 2008.

77. Gholipour G, Zhang C, Li M. Effects of soil–pile interaction on the response of bridge pier to barge collision using energy distribution method. *Struct Infrastruct E* 2018;14(11):1520–1534.

78. Gholipour G, Zhang C, Mousavi AA. Analysis of girder bridge pier subjected to barge collision considering the superstructure interactions: the case study of a multiple-pier bridge system. *Struct Infrastruct E* 2019;15(3):392–412.

79. Gholipour G, Zhang C, Mousavi AA. Reliability analysis of girder bridge piers subjected to barge collisions. *Struct Infrastruct E* 2019;15(9):1200–1220.

80. Jiang H, Wang J, Chorzepa MG, Zhao J. Numerical investigation of progressive collapse of a multispan continuous bridge subjected to vessel collision. *J Bridge Eng* 2017;22(5):04017008.

Chapter 2

State-of-the-art review on concrete structures subjected to impact loads

2.1 Introduction

Owing to rapid increase in the construction of reinforced concrete (RC) structures all over the world, the need for recognizing the responses of such structures exposed to dynamic and extreme loading conditions such as impact and blast loads is a very significant topic of importance. RC structures may be subjected to lateral impact loads arising from the falling of heavy rocks and objects, collision of vehicles and vessels, or the high-velocity impacts of relatively lighter projectiles and rockets. Impact loading is a dynamic and extreme loading type, the duration of which may reach 1,000 times shorter than earthquakes. Impact loads can be characterized into three types based on their intensity and duration (t_d). These types are: (i) quasi-static loading in which the structure reaches its maximum response before ending of the impact duration; (ii) dynamic loading in which the structure reaches its maximum response almost at the same time with the ending of the impact duration; (iii) impulsive loading in which impact duration ends before reaching the structure its maximum response. Accordingly, structural components can demonstrate different behaviors under concentrated lateral impact loads including localized and overall responses, as shown in Figure 2.1. When a structural member is subjected to high-rate impact loading with a very short duration relative to the structure's natural period (T), the stress wave propagation and inertia resistance of the structure are predominant on the responses. Under this loading condition, it is more expected to observe localized failure in the structure. However, when this ratio (t_d/T) is large, the structural responses and failure modes are dependent on the stiffness of the structure, and the structure tends to fail in overall modes [1].

Another classification of impact loadings based on their dissipative mechanism was proposed by Eurocode [3]. This classification includes (i) hard impact in which the initial kinetic energy was dissipated by striking objects such as colliding of vessels and vehicles with deformable bows with concrete structures and (ii) soft impact in which a major part of the initial kinetic energy was dissipated by the impacted structure such as impacting the rocks and rigid objects on concrete structures. Two simplified approaches using single-degree-of-freedom (SDOF) and two-degree-of-freedom (2-DOF) models were recommended by Eurocode [3] to formulate and identify the overall responses of concrete structures under soft and hard impacts as shown in Figure 2.2. In the proposed SDOF model, a partial mass of the impacted member exposed to a distributed (non-concentrated) impact load is idealized with a mass point element connected to a discrete (i.e., spring) element representing the global stiffness of the member under an

DOI: 10.1201/9781003262343-2

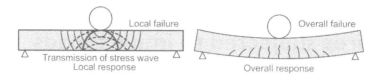

Figure 2.1 Impact response modes of RC beams [2].

equal concentrated dynamic load. However, the stiffness of both global and local structural responses is idealized connected to the relevant partial masses.

Although many simplified approaches exist in the current design codes to predict the responses of RC structures under impact loads, they are not able to obtain the brittle damage behaviors of concrete structures during high-rate and impulsive impact loads. Concrete structures might suffer localized failure modes and damages such as brittle spalling, scabbing, perforation, and punching shear failure [5, 6] under high-rate impact loads, or overall failure modes under rather low-rate impact loads as shown in Figure 2.3.

Generally, several design codes define the estimation of impact loads and the simplified responses of structures subjected to different types of lateral impact load especially those arising from vessels [7–9] and vehicles [8, 10–12] collisions using equivalent static and quasi-static analyses. However, the amplifications of dynamic effects such as inertia and strain rate have not been taken into account by these guidelines. Table 2.1 summarizes the impact load provisions considered by several design guidelines.

Studying the impact responses of structures under impact loads is possible through the three main approaches, which include simplified analytical methods, finite element (FE) numerical simulation, and experimental tests. In general, there exist some limitations on the use of simplified analytical approaches. As such, these techniques not only omit the structural dynamic behaviors such as inertia and strain rate effects of the materials, but they also are not able to capture the brittle failures and damage behaviors of structures under extreme loads. Besides, although the experimental tests give chances for accurately and reasonably evaluating the structural responses in the real world, conducting such tests needs notable professional equipment and economical resources. Compared to experimental approaches, FE numerical methods provide appropriate alternatives to conduct the test scenarios by reducing the time and costs along with obtaining accurate and reasonable results. There are many available commercial software codes such as LS-DYNA [16] and ABAQUS [17] to numerically

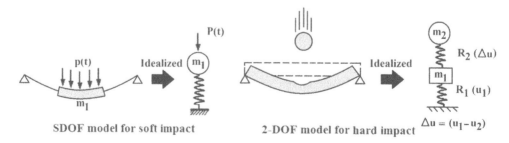

Figure 2.2 Simplified models recommended by Eurocode [3] for the design of structures under impact loads [4].

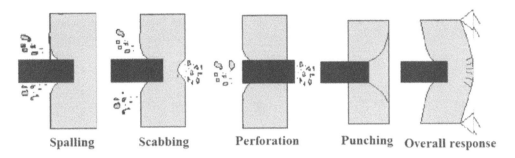

Spalling   Scabbing   Perforation   Punching   Overall response

Figure 2.3 Typical failure modes of concrete structures under different impact loads [5, 6].

simulate impact tests by adopting various contact algorithms between the striking and stuck components.

Impact loading tests can also be simulated in laboratory scales using different designed experimental facilities as follows:

- Drop weight impact facility: The impactor with a certain mass is vertically released from a specified height regarding the desired impact energy as shown in Figure 2.4a [2]. This facility is the most common experimental method used

Table 2.1 Summary of current guidelines considering impact loading

Guideline	Loading	Remarks and notes
AASHTO [7, 10]	Vessel collision	Equivalent static load based on deformation-force data and kinetic energy of head-on vessel collisions
	Vehicle collision	Equivalent static impact force full-scale crash tests of tractor-trailers/truck-barriers collisions (derived not directly from head-on impact tests). Consideration of 1,800-kN static force applied to the height of 1.35 m from the column base
JSCE [13]	Rock falls	Performance-based design structures under especially falling objects (e.g., rock falls) using the equivalent mean impact forces and absorbed energy
AS 1170.1 [14]	Vehicle collision	Equivalent static impact load based on the kinetic energy of vehicles with masses between the ranges 1,500 and 2,000 kg
CEN [3, 8]	Vehicle collision	Equivalent static force considering the effects of impact velocity, impact angle, mass distribution, deformation behavior, and damping characteristics of both impact and structure (ranges of maximum impact forces: 1,000 kN for truck and 500 kN for car impacts)
	Vessel collision	Equivalent static force based on the deformation energy of a vessel considering the influences of an impact angle
UK's Highways Agency [15]	Vehicle collision	Equivalent nominal loads applying horizontally to bridge piers based on experimental tests (between the ranges 250 and 1,000 kN)
CMR [9]	Vessel collision	Equivalent static load based on the kinetic energy of impacting ships (considering impact angle)

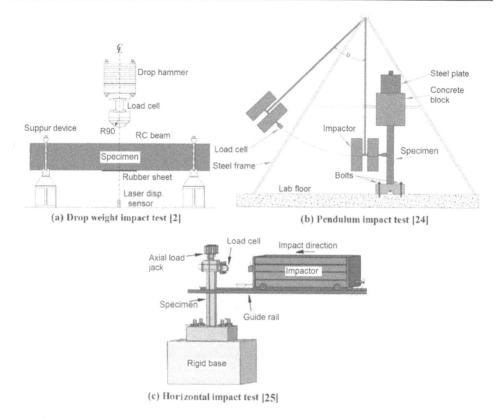

(a) Drop weight impact test [2]

(b) Pendulum impact test [24]

(c) Horizontal impact test [25]

Figure 2.4 Impact loading test facilities.

    to study the impact responses of concrete members placed horizontally such as beams [2, 18–20] and slabs [21, 22].
- Pendulum impact facility: The impactor can be released from different angles to generate different initial impact energy as shown in Figure 2.4b [23, 24].
- Horizontal impact facility: The impactor collides with the structure horizontally with a specified initial velocity as shown in Figure 2.4c [25, 26].

    This chapter aims to present a state-of-the-art review on the responses and failure behaviors of various types of concrete structures and structural members, including columns, bridge piers, beams, and slabs under different types of impact loads arising from the collision of rigid objects (i.e., soft impacts) or vehicles and vessels (i.e., hard impacts). The influences of different structural- and loading-related parameters on the impact resistance of concrete structures are reviewed. In addition, the theoretical background, current design guidelines, and existing approaches for analyzing structures under impact loadings are reviewed.

2.2 Structural columns subjected to impact loads

The vulnerability of relatively small-size RC columns commonly used in low- and medium-rise RC framed buildings subjected to lateral impact loads has been widely studied in the literature [26–31].

Liu et al. [29] experimentally and numerically carried out a series of low-velocity impact tests on the axially loaded circular RC columns. An improved FE method was proposed to overcome the drawbacks of existing conventional FE modeling approaches in the prediction of impact responses of RC structures. In this method, the impact loadings were applied to RC columns in which their concrete materials were modeled using a modified model providing proper confinement effects (by modifying soften behavior of the concrete), crack opening and closing (by modifying the concrete modulus), and bond-slip behaviors (by assigning discrete elements along with the longitudinal reinforcements). In addition, from a parametric study, significant influences of the reinforcement ratios on both overall and local failures, and positive effects of the axial load on the column impact resistance for small deformations were concluded. However, no sensitivity level was determined for the positive effects of axial load. To fill this gap existing in the previous studies, Gholipour et al. [28] carried out a parametric study on the impact responses, failure modes of square RC columns under different lateral impact loadings varying in terms of axial load ratio, impact velocity, and the height of impact location. It was found that the ranges between 0.3 and 0.5 for the axial load ratio caused a substantial increase in the impact resistance of the columns as shown in Figure 2.5. Besides, reducing the height of impact location led to the increase of the peak impact force (PIF) and changing of the column failure mode from a global flexural mode to local shear failures. In-line with these studies, the vulnerability of RC columns varying in terms of shear reinforcement ratio and axial load level subjected to different-energy drop-weight impact tests was experimentally investigated by Yilmaz et al. [31]. Reducing the values of peak and residual

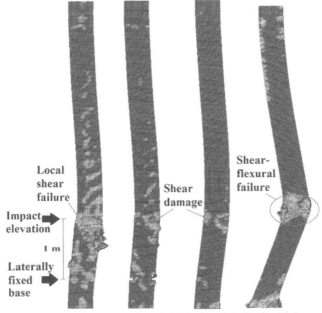

Ratio = 0.1   Ratio = 0.3   Ratio = 0.5   Ratio = 0.8

Figure 2.5 Failure behaviors of RC columns with different axial loads under middle-rate impact loading [28].

Right-side Frontal-side Left-side   Right-side Frontal-side Left-side   Right-side Frontal-side Left-side

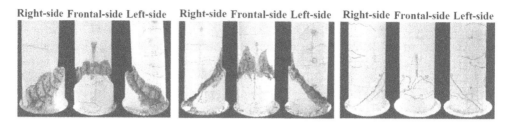

Figure 2.6 The governance of shear failure on the responses of RC columns under lateral impact loads [26].

displacements, and the magnitude of the columns' absorbed energy was obtained with the increase of the axial load level. Besides, although both shear and flexural capacities of the columns were increased by enhancing the axial load level until a balanced level identified using the moment-axial load interaction diagrams, the ductility and energy dissipation of RC columns were reduced. Beyond this level, the axial load had a negative influence on the resistance capacity of columns.

In assessing the shear mechanisms of RC columns, Demartino et al. [26] experimentally studied the impact-induced responses of shear-deficient RC columns with different hoop spacing, and boundary conditions subjected to different-rate lateral impact loadings with velocities from 2.25 to 4.5 m/s. The governance of diagonal shear failures originated from the column base to the impact point was mostly observed as shown in Figure 2.6. It was obtained that the initial impact phase was profoundly dependent on the inertial forces and characteristics of the contact surface. In addition, more severe damages were observed in columns with a fixed base. Similarly, from a series of dropping mass impact tests of shear-deficient axially loaded RC columns conducted by Remennikov and Kaewunruen [30], the occurrence of brittle shear failures around the impact zone was mainly observed as shown in Figure 2.7 due to the mobilization of inertial forces during the initial impact phase and reducing the bending moment at the mid-span.

Unlike the studies mentioned earlier, the predominance of flexural failure modes on the responses of square RC columns with different cross-sectional dimensions

Figure 2.7 Failure behavior of an RC column under mid-span impact load [30].

subjected to horizontal impact loads was obtained by Cai et al. [25] when the impact loads were applied to the top positions of columns (columns' head). In addition, the influences of the columns' slenderness ratio, impact weight, and velocity parameters were evaluated on the impact responses and damage patterns of the columns. The positive influences of the impacting weight on the average value of impact forces, the impact velocity on the impact durations, and the column cross-sectional dimension on the impact forces were concluded.

Despite many studies focused on the impact responses of RC columns, the vast majority of previous works have focused on the protective design, and investigating the performance of retrofitted concrete columns using high-strength composite materials such as fiber-reinforced polymer (FRP) [32–35], carbon FRP (CFRP) [23, 36–39], and ultrahigh-performance fiber-RC (UHPFRC) [40–43], or using steel jackets surrounding the core concrete such as concrete-filled steel tube (CFST) [44–54], concrete-filled double-skin steel tube (CFDST) [52–59] subjected to lateral impact loads. In some cases, the combinations of both approaches were adopted such as CFST-CFRP [39, 48, 60, 61] and CFST-FRP [62, 63] as a strengthening method. The positive effects of the aforementioned retrofitting techniques in the enhancement of impact resistance and mitigating the damage levels of concrete columns were mostly concluded in the previous works. However, the effectiveness of employing such approaches in enhancing the axial load carrying capacity of columns when they are subjected to lateral impact loads has not been investigated in the literature. From the evaluation of the impact responses of axially loaded CFDST columns, Aghdamy et al. [57] found that initial PIF is most sensitive to initial impact velocity. Moreover, the duration of impact load was extremely dependent on the impact location, initial impact velocity, axial load ratio (limited to less than 0.3), and the impactor-to-column mass ratio. It was concluded that the axial load until a ratio of 0.3 had positive influences on the magnitude of impact forces, the column flexural capacity, and negative effects on both the peak and residual values of the column displacements. Table 2.2 presents a summary of previous works that studied the influences of various parameters on the impact responses of RC columns.

2.3 Beams and slabs subjected to impact loads

Although applying lateral impact loads arising from the collision of vehicles or vessels with columns is more likely than other structural members, beams or slabs utilized in framed buildings or bridges constructed in mountainous areas may also be subjected to impact loads arising from the falling objects and rocks. Hence, many researchers have attempted in the literature investigating the impact forces, structural responses, and failure behaviors of RC beams analytically [2, 64–69], numerically [1, 70–81], and experimentally [18, 20, 71, 82–92] under low-rate [1, 2, 76, 80, 90, 92] and relatively high-rate [93–96] impact loads by focusing on the influence of the structural-related parameters such as beam inertia [75, 80, 81, 93, 94], longitudinal [2] and transverse [18, 91, 97] reinforcements, and loading-related parameters such as impact weight and velocity [90, 91]. A summary of previous studies investigating the effectiveness of various parameters on the impact responses of RC beams is presented in Table 2.3.

Fujikake et al. [2] experimentally and analytically studied the impact force and maximum mid-span deflection of RC beams varying in the ratios of longitudinal reinforcements under different drop weight impact loadings with low-to-medium ranges of the

Table 2.2 Summary of studies on the influence of various parameters on the impact resistance of column members

Study	Analysis	Parameter	Effectiveness
Liu et al. [29]	Experimental and numerical	• Reinforcement ratios • Axial load	• Significant effects on both overall and local failures • Positive effect on the impact resistance when the column deformations are small
Gholipour et al. [28]	Numerical	• Axial load ratio • Impact velocity • Height of impact location	• Positive on the column resistance for the ratios between 0.3 and 0.5 • Negative effect on the length of plastic hinges • Reducing the impact height changed the failure modes from global flexural mode to local shear failures
Yilmaz et al. [31]	Experimental	• Shear reinforcement spacing • Axial load level	• Increased the maximum and the residual displacement values, the energy absorption capacities, and shear cracks • Positive on the columns impact resistance for the levels less than a specific balanced level
Demartino et al. [26]	Experimental	• Hoop spacing • Boundary conditions • Contact surface • Column inertia	• Significant effects on the column shear capacity • More severe damage for fixed base columns • Significant effects on the initial impact phase
Cai et al. [25]	Experimental	• Height of impact location • Column cross-sectional dimension • Impact weight • Impact velocity	• Changing the column failure mode from shear to flexural failures with increasing the height of impact load • Positive on the magnitude of impact forces • Positive on the average values of impact forces • Positive on the impact durations
Aghdamy et al. [57]	Experimental and numerical	• Impact location • Impact velocity • Impactor-to-column mass ratio • Axial load ratio	• Significant effects on the impact duration • Significant positive on the impact forces, the column flexural capacity until a ratio of 0.3
Alam et al. [48]	Numerical	• Axial load	• Positive on the column impact resistance
Alam et al. [61]	Numerical	• Axial load	• Positive until 45% of the column capacity
Chen et al. [62]	Experimental	• Axial load	• Positive on the column impact resistance

Table 2.3 Summary of studies on the various parameters on the impact resistance of beam members obtained by various studies

Study	Analysis	Parameter	Effectiveness
Pham and Hao [1]	Numerical and analytical	• Beam's global stiffness by the beam span and reinforcements	• Significant on the impact responses and failure modes at later stages than initial impulse
		• Beam's local stiffness	• Significant on the beam responses in the first impact impulse
Fujikake et al. [2]	Experimental and analytical	• Drop height	• The beam failures tended to local failures with increasing the drop height
		• Amount of longitudinal reinforcement	• Lower amount caused overall flexural failures, and higher amounts caused both local and overall failures
Gholipour et al. [73]	Numerical	• Impact velocity	• The beams tended to local failures with increasing impact loading rate
Jin et al. [74]	Experimental	• Stirrup ratio	• Significant effects on the local damage of concrete; marginal effects on the impact force and the beam deflections
		• Impactor's mass	• Significant positive on the impact duration
		• Beam span length	• Significant negative on the peak impact force; significant positive on the impact duration
Guo et al. [75]	Numerical and analytical	• Relative mass of an impactor to beam	• Significant positive on the peak impact force until a ratio of 1.0
		• Impact velocity	• Significant positive on the peak impact force
Li et al. [77]	Numerical	• Impactor's mass • Impact velocity • Inclination angle of an impactor • Concrete strength	• Significant positive on the impact force and the beam displacement • Significant effects on the impact force; marginal effects on the beam displacements
Li et al. [79]	Numerical	• Impactor geometry	• Flathead impactor generates the highest peak impact force and shortest duration, and had marginal effects on the beam displacements; curve-head impactor caused more sever damages
		• The curvature radius of impactor's head	• Significant positive on the peak impact force; negative on the impact duration
		• Inclination angle of an impactor	• Significant effects on the peak impact force generated by flathead impactors

(Continued)

Table 2.3 Summary of studies on the various parameters on the impact resistance of beam members obtained by various studies (Continued)

Study	Analysis	Parameter	Effectiveness
Pham and Hao [80]	Numerical	• Plastic hinge • Boundary conditions	• Marginal on the impact force and duration; significant on the beam residual displacement and the damage levels
		• Concrete strength	• Significant effects on the beam failure modes; marginal effects on the impact force and the beam displacement
Pham and Hao [81]	Numerical	• Beam inertia • Plastic hinge position	• Significant effects on the beam impact behavior and demand
Isaac et al. [86]	Experimental	• Span/depth ratio • Impact velocity	• Significant on the force wave propagation in the beam
Yan et al. [88]	Experimental and numerical	• Impact velocity	• Changing the beam failure mode from flexural to shear mode with increasing impact velocity
Adhikary et al. [90]	Experimental and numerical	• Beam-to-impactor mass ratio	• Large mass impacting with low-velocity caused smaller peak impact forces and larger peak deflections
		• Longitudinal reinforcement ratio • Concrete strength • Boundary conditions	• Positive on the peak impact force and negative on the beam peak deflections • Fixed-end beams endured more peak impact forces than pinned-end beams
Zhao et al. [91]	Experimental and numerical	• Beam span	• Positive on the predomination of stress wave propagation on the impact response
		• Transverse reinforcement ratio • Impact velocity	• Positive on the shear resistance of beams • Positive on the occurrence of local shear failures
Cotsovos et al. [93]	Experimental and numerical	• Loading rate	• Extremely dependent on the beam inertial forces; negative effects on the length of plastic hinges
Ožbolt and Sharma [97]	Numerical	• Amount of shear reinforcement • Loading rate	• Significantly affects the crack pattern • Significant effects on the failure modes
Pham et al. [98]	Experimental and numerical	• Contact stiffness	• Significant effects on the peak impact, the beam demands force, and the impact duration; marginal effects on the impact impulse, energy, and the beam displacements

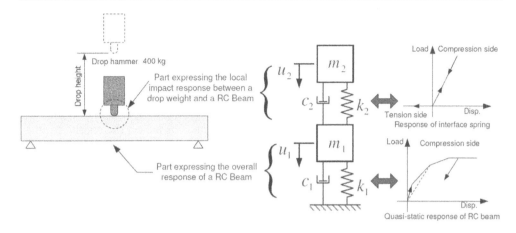

Figure 2.8 Simplified two-degree-of-freedom model of responses of RC beams under impact loads proposed by Fujikake et al. [2].

impact energy and velocity. To evaluate the flexural failure modes of RC beams, they sufficiently reinforced with transverse reinforcements against shear failure modes. It was found that well-reinforced RC beams against shear failures can undergo overall response mode under low-rate impact loads, and localized response mode along with compressive damages in the concrete cover with increasing impact velocity. In addition, a simplified 2-DOF model was proposed to calculate both localized and overall response phases based on load-displacement responses of RC beams under impact loads as shown in Figure 2.8.

Pham and Hao [1] numerically and analytically studied the effect of global stiffness of structure on the impact behavior of RC beams. For impact velocities more than 1.0 m/s, it was concluded that the initial impulse and PIF were not sensitive to the structure of global stiffness. While the global stiffness governed the following impact impulses during the free-vibration phase of the beam response, a delayed activation of the flexural stiffness in the following impact phases was declared as the main cause of such a conclusion. In addition, it was revealed that the secondary peaks of impact forces are profoundly related to the stress wave propagation in the impactor and the impacted beam. Also, the negligible effects of several structural parameters such as the ratio of longitudinal rebars, the beam span length on the initial impact impulse were concluded. Besides, Pham and Hao [98] found that the impact force and the responses of RC beams are very sensitive to the contact stiffness and conditions between the impactor and the beam. Thereafter, the significant influences of the impactor geometry and interlayer between the impactor and the beam were concluded from a numerical study of RC beams done by Li et al. [79] under drop-weight impact loads. Accordingly, the curvature radius of the impact had positive effect on the PIF, while it had negative effect on the duration of impact force. Also, more severe damages were observed on the beam under hemispherical and curved impactors rather than those exposed to a flathead impactor.

The influences of several structural-related parameters including the beam span length, cross-sectional area, shear span effective depth ratios, longitudinal and transverse ratios, and shear to bending resistance ratios on the impact responses RC beams under low-rate impact loadings were evaluated by Adhikary et al. [90]. It was found that although the increase of longitudinal reinforcement ratio enhanced the flexural resistance of the beams by reducing the beam deflections, the predominance of shear failures was obtained by observing diagonal cracks and shear plug damages.

Despite the previously reviewed studies investigating the behaviors of RC beams under low- and medium-rate impact loads, there exist many research works addressing the impact responses of RC beams under high-rate impact loads. Zhan et al. [96] experimentally investigated the failure behaviors of RC beams under high-rate impact loadings with impact velocities between 6.0 and 13.0 m/s. Based on the results from a parametric study, two empirical formulas quantitatively describing the relationships between the impact loading energy and the impact responses of beams such as deflection and flexural load-carrying capacity were developed.

Understanding the mechanism of stress wave propagation in RC structures plays the most significant key role in the determination of their shear capacity [93], demands [81], and failure behaviors [99, 100]. The effects of stress wave propagation in evaluating the impact responses of RC beams were considered by Cotsovos [93] to calculate the shear resistance, by Pham and Hao [81] to estimate the shear force diagram, and by Zhao et al. [99] to assess the shear failure and damages. Cotsovos et al. [93] concluded that the responses and resistance capacity of RC structures under high-rate impact loading are profoundly affected by the inertia forces and stress wave propagation at the initial response phase of the structure activated in a partial length of RC beam called "effective length" (L_{eff}) that led to different bending moment diagrams compared to those achieved under low-rate impact loadings as shown in Figure 2.9. Afterward, the load-carrying capacity of RC beams under high-rate impact loads was investigated by Cotsovos [94] using a proposed simplified method based on the concept of stress wave propagation and its travel time in RC structures. According to this method, stress waves generated under high-rate impact loads do not necessarily reach the beam supports during the initial phase of the response. Under such loading conditions, the appearance of negative moments on the upper surface of the beam with a distance of L_{eff} as shown in Figure 2.9 is very likely. It was concluded that the length of L_{eff} decreased with the increase of the velocity of impact loading.

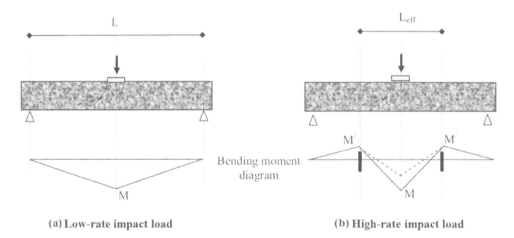

(a) Low-rate impact load   (b) High-rate impact load

Figure 2.9 Schematic of bending moment diagrams of the RC beam under (a) high-rate and (b) low-rate impact loads [93].

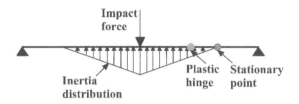

Figure 2.10 Estimation of plastic hinge location in RC beams under impact load by Pham and Hao [28].

In-line with the works carried out by Cotsovos et al. [93, 94], Pham and Hao [81] investigated the position of plastic hinges formed in RC beams, and the dynamic demand diagrams for the shear force and bending moment under impact loads by assuming the linear distribution of inertia force along the beam. By assuming the mobilization of inertial forces in a partial length of the beam (i.e., effective length), the formation of plastic hinges was expected inside the effective length between the stationary points where the beam's accelerations and inertia were zero as shown in Figure 2.10. Besides, the influences of the plastic hinge and boundary conditions on the behavior of RC beams under low-rate impact loads were assessed by Pham and Hao [80]. Consideration of the location of plastic hinges in the determination of the equivalent stiffness of RC beams was extremely recommended. Also, it was obtained that the boundary conditions had marginal effects on the peak value and the duration of the impact force.

As beams are basically designed to fail in flexural modes, recognizing their shear mechanisms is the topic of importance, which has been investigated by several previous studies experimentally [20, 86, 91], numerically [97, 99], and analytically [100, 101]. The shear failure mechanism of RC beams under drop-weight impact loading was experimentally and numerically studied by Zhao et al. [91] by using various parameters, including beam span, transverse reinforcement ratio, impact mass, and impact velocity. It was mainly concluded that the beams tended to fail in shear modes through the occurrence of localized shear plug and diagonal cracks around the impact zone shortly after the onset of impact loading with increasing the impact velocity. Based on the experimental observations, shear failure modes were categorized into three types as shown in Figure 2.11, including (i) diagonal cracks and shear plug failure around the impact zone under high-rate impact loading (Type I), (ii) inclined flexural-shear cracks and damages propagated from the supports to the impact point under low-rate impact loading (Type II), (iii) a combined failure mode (Type I + II).

The inability of existing simplified SDOF methods in assessing the shear failures of beams under impact loads motivated Yi et al. [100] to propose a simplified approach predicting the shear resistance and evaluating the occurrence probability of shear failures in an effective length of RC beams under impact loads considering the effects of stress wave propagation. As the proposed method was only based on shear capacity and demand of beams without considering shear deformation, quantification of the shear damage was not possible. Hence, Zhao et al. [99] improved their previous approach proposed in [91] to a simplified three-degree-of-freedom model as shown in Figure 2.12 by considering the beam deformations in both shear and flexural response modes. The positive influences of the beam span length on the duration of the impact forces were obtained compared to its negative effects on the mean of impact forces.

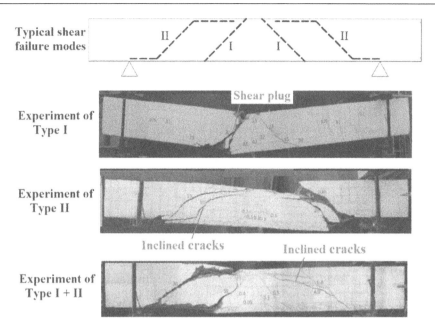

Figure 2.11 Typical shear failure modes in RC beams under different-rate impact loadings [91].

From an experimental investigation by Saatci and Vecchio [18] on the shear behaviors of RC beams varying in the ratio of shear reinforcements under high-rate impact loading with an impact velocity of 8 m/s, it was observed that all specimens with different shear resistances suffered severe shear cracks and shear plug failures. Moreover, by measuring the velocity of impact force wave propagating from the impact point to supports, it was revealed that the velocity of force wave could be significantly smaller than that of longitudinal and shear wave velocity with totally different inherent. From another experimental study done by Kishi et al. [20], failure behaviors of 27 simply

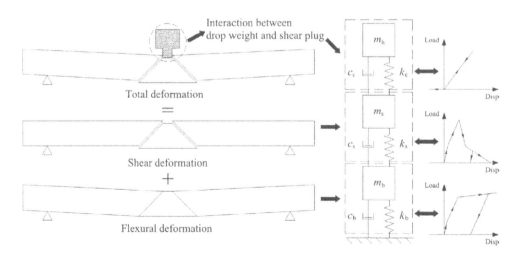

Figure 2.12 The simplified three-degree-of-freedom model proposed by Zhao et al. [99].

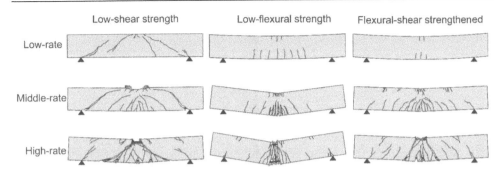

Figure 2.13 Typical failure modes of RC beams varying in shear and flexural strengths under different-rate impact loads [70].

supported RC beams without shear rebar were investigated under drop-weight impact loading. The occurrence probability of flexural failure modes in the beams without shear rebar was concluded under low-rate impact loads. However, the beams tended to fail in shear modes by increasing the impact velocity. Figure 2.13 shows the typical failure modes including global and localized shear and flexural failures in RC beams with simple supports varying in shear and flexural strengths under different-rate impact loads based on the information and observations from the previous works [2, 20, 86, 90, 93]. It is observed that RC beams with low-shear and low-flexural strengths under low- and middle-rate impact loads [20, 86, 90] suffer overall shear and flexural failures. Moreover, beams with sufficient flexural-shear strength endure minor flexural cracks under low-rate impact and global flexural-shear damages under middle-rate impacts [2]. When RC beams are subjected to high-rate impact loads, compared to the formation of a plastic hinge in the beam with low flexural strength, the occurrence of localized shear plugs around the impact zone is observed in the beams with sufficient flexural strengths [20, 86].

With the purpose of protective design of RC structures against impact loads, several research works can be found in the literature investigating the performance of RC beams retrofitted and reinforced by high-strength composite materials such as FRP [102, 103], steel fiber RC (SFRC) [104], CFRP laminates [105–107], UHPFRC [92], recycled aggregate concrete (RAC) [89], glass FRP (GFRP) rebars [108], engineered cementitious composites (ECC) containing polyvinyl alcohol (PVA) fibers [109], high-strength steel wire mesh and high-performance mortar (HSSWM-HPM) [87], coconut fiber RC (CFRC) beams strengthened with flax FRP (FFRP) [110], or strengthened by steel jackets [46, 110]. Also, there exist some limited numbers of research works studying the performance of precast concrete beams under impact loads [77, 88].

Slabs are also one of the most common structural members that are commonly used in connection with supporting beams. Many studies existing in the literature evaluating the dynamic responses, and failure behaviors of concrete slabs and plates under low- [111–117], moderate- [118], and high-rate [119] impact loads analytically [119–121], numerically [118, 120, 122–124], experimentally [22, 113, 116, 123, 125]. Generally, concrete slabs underwent two typical failure modes that include globally distributed crack patterns [114, 116] under low-rate impact loads, and localized failure and punching shear failure [22] under high-rate and projectile [126–133] impact loads, as shown in Figure 2.14a and b. Like those retrofitting approaches used to protect concrete

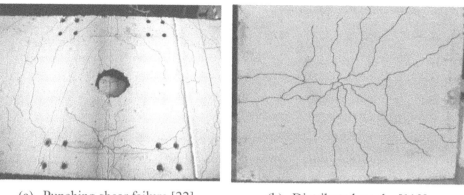

(a) Punching shear failure [22]   (b) Distributed cracks [113]

Figure 2.14 Typical failure modes of concrete slabs under different impact loads: (a) punching shear failure [22] and (b) distributed cracks [113].

columns and beams against impact loads, several research works focused on investigating the performance of concrete slabs strengthened with composite materials such as steel fibers [111], FRP [117], CFRP [134], ultrahigh performance concrete (UHPC) [112], UHPFRC [113], slurry-infiltrated fibrous concrete (SIFCON) [117], hybrid bamboo fiber (HBF) [135], coconut fiber [117, 136], and reinforced by FRP-bars [137].

The main objectives of utilizing these protective design techniques are to mitigate the brittle damages and enhancing both shear and flexural resistance of concrete structures in harsh environments or under extreme loadings. Therefore, special recommendations should be considered in utilizing these methods concerning the applications of the structures. For instance, the use of excessive high-strength materials in the tensile surface of RC beams may lead to the occurrence of early shear failures in beams before flexural modes.

2.4 Bridge piers subjected to impact loads

Columns are mainly axial load-carrying structural members that are commonly used in large-scale civil structures and infrastructures such as high-rise buildings, highway bridges, and subways. Impact loads arising from the collision of vessels or vehicles with RC columns used in bridges, low-rise buildings, and isolated traffic structures can be taken place during accidental events or intentional terrorist attacks. There exist many research works in the literature investigating the responses of RC bridge piers subjected to lateral impact loads from the collisions of vehicles [8, 10–12, 138–160], vessels [7–9, 28, 161–174], shipping objects [175], and falling rocks [176, 177]. Due to the significant discrepancies between the force-deformation behaviors of striking vehicles and vessels, and also different structural characteristics of impacted bridge piers (e.g., pier size and dimensions, substructure, and boundary conditions), it is expected to capture relatively different dynamic responses under vehicle collisions compared to those from vessel collisions. Hence, it is focused in this section on the review of existing analysis approaches and previous studies on the impact responses of RC bridge piers under vehicle and vessel collisions.

2.4.1 Bridge piers subjected to vehicle collisions

From the review study on the failure causes of 114 bridges by Harik et al. [178] during a 38-year period from 1951 to 1988 in the USA, it was found that 15% of these failures occurred due to truck collisions. In addition, based on a report by Wardhana and Hadipriono [179], about 3% of 503 bridges in the USA were failed due to vehicle collision during an 11-year period from 1989 to 2000 in the USA. Two examples of bridge pier and superstructure failures are illustrated in Figure 2.15.

The influences of various structural- and loading-related parameters on the vehicle collision force and the structural responses of the impacted pier have been widely evaluated in the literature. A parametric study was done by Zhou et al. [150] on the impact responses of RC bridge piers varying in terms of several parameters including impact velocity, impact mass, and the strengths of the pier concrete and steel reinforcements. Compared to the marginal influences of the concrete strength on the magnitude and duration of impact force, and the pier global deformations, the strength of steel reinforcements had substantial effects on the pier deformations. Also, three performance levels including the local damage, flexural-shear failure, and cross-sectional fracture with shear failure of the pier were investigated using the FE simulations. Furthermore, a damage index based on the ratio of impact force to the shear capacity of bridge piers was proposed by Zhou and Li [151] using FE numerical simulations to describe different damage levels of bridge piers subjected to vehicle collisions.

Abdelkarim and ElGawady [143] carried out an extensive parametric study of RC bridge piers numerically by the collision of different vehicles. The positive influences of reinforcement ratio, column cross-sectional dimensions, axial load level, material strain rate effects, and impact velocity and weight of vehicles were found on the peak value of impact force. However, it was not affected by the variability of concrete strength, pier boundary condition, and the depth of soil surrounding the pier base.

Compared to several proposed simplified models of the vehicle-pier collision system proposed by Al-Thairy and Wang [181], Milner et al. [182], and Vrouwenvelder [183], Chen et al. [146] proposed a more sophisticated system using a reduced coupled mass-spring-damper (CMSD) model. From the evolution of different parameters, it was revealed that except for the marginal effects of vehicle weight, other parameters, including impact velocity, pier geometry, and the material properties of the pier, had

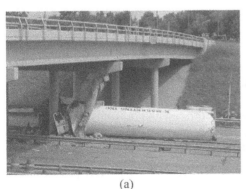

(a)	(b)

Figure 2.15 (a) Failure of bridge column due to truck collision [180]; (b) collapse of bridge superstructure after colliding a tractor-trailer [138].

significant influences on the impact force results. Afterward, a simplified CMSD model was proposed by Chen et al. [147] to assess the validation of a proposed spectrum-based design method in the prediction of impact responses of piers. Besides, an equivalent frame model (scaled model) of a large-size truck was designed by Chen et al. [148] to use in collision with RC bridge piers experimentally. The influences of cross-sectional shape and dimensions of RC piers, striking truck cargo weight, impact velocity, impact position, and vehicle type were numerically evaluated by Chen et al. [154] on the impact force results during heavy-truck collisions with RC bridge piers. It was obtained that the first peak of impact force was more sensitive to the impact velocity, while the following peaks were more sensitive to the impact weight. Moreover, the shape and diameter of the impacted pier had marginal influences on the impact force results. Yi et al. [156] concluded that RC columns with circular cross-sectional shapes suffered larger displacements and more severe damage levels under truck collision rather than those with square shapes. Moreover, from a sensitivity analysis of column impact resistance to the axial load ratio and the concrete strength, no influence trend of these parameters was obtained.

Fan et al. [40] numerically studied the impact responses and the performance of UHPFRC-strengthened bridge piers compared to those with normal concrete in the presence of superstructure load subjected to vehicle collisions. The influence of superstructure axial load and the top boundary conditions on the pier impact responses were evaluated by developing different simplified pier models in the forms of single columns under equivalent axial load, and the pier-bent model (with multicolumns) in the presence of the superstructure equivalent mass. The importance of the stress initialization analysis phase to reach an equilibrium state of the bridge under the bridge self-weight before the transient impact loading phase was revealed. More reasonable results were captured from the pier-bent model rather than those of single piers. Moreover, a parametric study in assessing the influences of different parameters, including reinforcement ratios, UHPFRC strength, the thickness of UHPFRC jacket, and the impact velocity, was carried out. The thickness of UHPFRC was realized as the most effective factor in the impact resistance of piers rather than others when the impact velocity was rather low.

Despite the previous studies focused on the impact force and global deformation of piers, the damage mechanisms and failure behaviors (including both local and global failures) of RC bridge piers were evaluated by several research works. Different damage states and failure mechanisms of RC bridge piers under a truck collision (with a weight of 66 kN, a velocity of 31.3 m/s, and an impact angle of 20°) were numerically evaluated by Agrawal et al. [144] as shown in Figure 2.16. According to this classification, the pier suffered spalling damages in the concrete cover immediately after the onset of collision. Afterward, with progressing the damages to the concrete core, the pier endured the rebar severance, breakage at the impact level, concrete erosion at the footing, and the formation of plastic hinges at both top and bottom end of the pier, respectively.

Auyeung and Alipour [145] numerically evaluated the failure behaviors of RC bridge piers by varying vehicle mass, velocity, pier diameter, and transverse reinforcement. Figure 2.17a–c illustrates different examples of the failure modes of bridge columns such as pure flexure, combined shear-flexure, and pure shear, respectively. While the pier diameter governed the global failure modes, the levels of local failure modes were extremely sensitive to transverse reinforcements. Thereafter, Auyeung et al. [152]

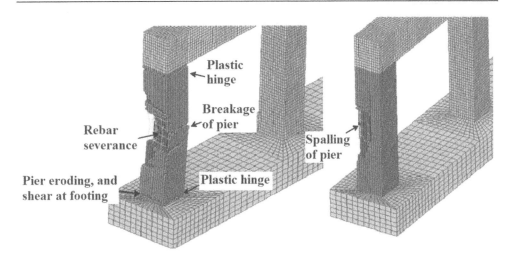

Figure 2.16 Different damage behaviors of RC piers under vehicle collision [144].

proposed a damage index based on the structural characteristics of the bridge pier and the kinetic energy of colliding vehicles to investigate different performance levels of the piers' responses under the vehicle collisions. From a parametric study, the vehicle impact velocity was recognized as the most effective loading parameter on the impact and the pier shear forces.

As the transferring time of shear forces due to the application of impact loads arising from vehicle collisions is very short, recognizing the local shear failure mechanisms and the effectiveness of some key dynamic factors such as the pier inertia on the impact responses is very necessary from the design point of view. However, omitting the damage states that are profoundly dependent on the severity of impact loadings and the dynamic shear capacity of RC piers is a very notable gap of the current design codes. Hence, several attempts have been carried out by the previous works to present

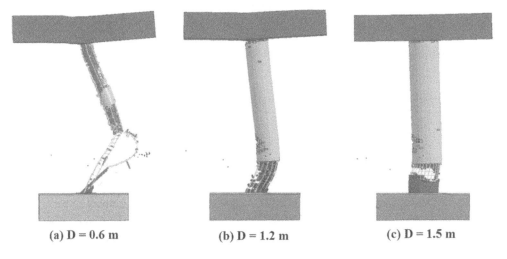

(a) D = 0.6 m   (b) D = 1.2 m   (c) D = 1.5 m

Figure 2.17 Failures of bridge piers with different diameters (D) under vehicle collision [145].

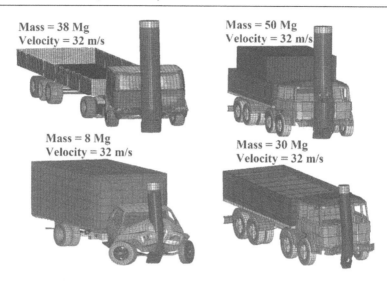

Figure 2.18 FE simulation of different vehicle collisions with RC columns with shear failures [48].

some efficient design frameworks considering the shear failure mechanisms of RC piers under vehicles.

As an improvement in the existing methodologies, Sharma et al. [157] proposed an approach to estimate the shear force capacities of RC bridge piers subjected to different vehicle collisions for different damage states and performance levels as shown in Figure 2.18. Table 2.4 presents the damage states in the corresponding performance levels of RC columns under vehicle collision. It was found that the dynamic shear force capacities estimated by the proposed method were more than those calculated by ACI-318 [184]. Compared to the assumption by Tsang and Lam [149] in which the time required to full contact was larger than the duration of shear wave velocity, while it was revealed by Sharma et al. [157] that these durations were almost similar. Afterward, the fragility of RC columns was assessed by Sharma et al. [139] using a proposed probabilistic method based on the shear capacity of columns. Shear capacities of RC columns were modeled based on the performance levels to use in a probabilistic assessment by Sharma et al. [159] in terms of different loading and structural uncertainties.

Table 2.4 Performance levels of RC columns subjected to vehicle collision [157]

Damage level	Damage description	Performance level	Performance level description
DLI	Insignificant damage	←PLI	Fully operational with no damage
DL2	Minor spalling of concrete, yielding of longitudinal steel	←PL2	Operational structure with damage
DL3	Significant cracking of concrete, spiral and longitudinal bar exposed, buckling of bars	←PL3	Total collapse of the structure
DL4	Loss of axial load capacity, longitudinal bar fracture		

Do et al. [24] investigated the impact responses and failure behaviors of RC bridge columns subjected to vehicle collision using FE simulations in LS-DYNA. While the initial PIF was profoundly affected by the engine, the following peaks were more sensitive to the total mass of the vehicle (mostly contains cargo weight). Also, it was revealed that the mass of the engine has a key role in the determination of the pier failure modes and the value of the impact force. Also, the influence of the pier axial load on the PIFs and the failure behaviors were evaluated through different pier-superstructure interaction models. In a mutual action, the substantial positive influences of impact forces on the axial force of the pier were concluded. From a series of the FE simulations of vehicle-pier collisions, various failure behaviors, including flexural failure, shear failure, and punching shear damage, were observed for the piers that successfully represented the numerical models of those observed in real impact events given in Ref. [185], as shown in Figure 2.19a–c. Furthermore, two catastrophic flexural and shear failures of bridge columns at the mid-height, leading to the collapse of bridge piers, are illustrated in Figure 2.20a and b, respectively [185].

Similar to the conclusion of the study done by Do et al. [24], the dependency of the highest PIF on the truck's engine block was also concluded by Cao et al. [140] from a series of FE simulations of a heavy-truck collision with bridge piers. However, the catastrophic pier failure was observed during the secondary impact of the truck-trailer. Moreover, a simplified impact loading function (i.e., the impact loading profile) as given in Equation (2.1) based on the FE simulation data of colliding truck weight, velocity, and impacted pier cross-sectional dimensions was proposed by Cao et al. [141] to use in the design of bridge piers against the collision of heavy trucks. Then, the performance levels of different bridge piers under the proposed pulse model were evaluated by Cao et al. [142]. The proposed impact pulse model by Cao et al. [141] and the application heights of the pulse model are shown in Figure 2.21a and b. This model includes three pulses generated due to the impacts of the bumper (Pulse^1), engine (Pulse^2), and trailer (Pulse^3).

$$F_i = f\left\{\alpha_r(V)^{\beta_r}(W)^{\gamma_r}\left(\frac{b}{900}\right)^{\varepsilon_r}\right\} \qquad (2.1)$$

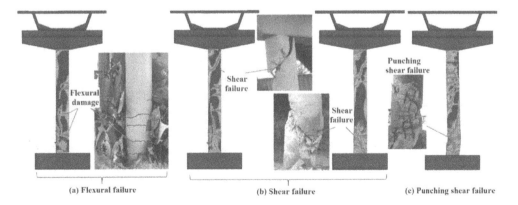

(a) Flexural failure   (b) Shear failure   (c) Punching shear failure

Figure 2.19 Different failure behaviors of RC bridge piers under vehicle collision obtained FE simulations done by Do et al. [24] compared to those observed in real events [185].

(a) Flexural failure   (b) Shear failure

Figure 2.20 Typical failures occurred in bridge columns at the mid-height under vehicle collisions [185].

where F_i denotes the peak force. α_r, β_r, γ_r, and ε_r are regression parameters. V and W are the truck impact velocity (km/h) and weight (kN). b denotes the pier width (mm).

Afterward, Do et al. [159] numerically studied the profile of impact forces from pier-vehicle collisions with respect to various structural- and loading-related parameters. Compared to the impact loading model proposed by Cao et al. [141], a simplified impact force model was proposed considering the pier shear capacity as illustrated in Figure 2.22a and b. According to this model, the first impact phase (Phase-1) is dependent on the length between the bumper and the engine, the truck velocity, the column width. During this phase, the *PIF* (i.e., F_1) can be calculated using the engine's mass and the impact velocity as given in Equation (2.2). The pier endures a punching

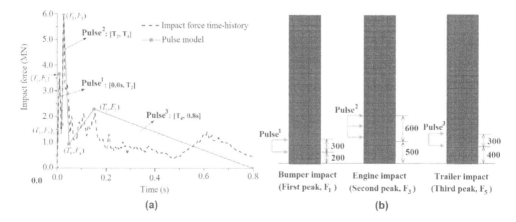

Figure 2.21 (a) The impact pulse model proposed by Cao et al. [141]; (b) application heights of the pulses.

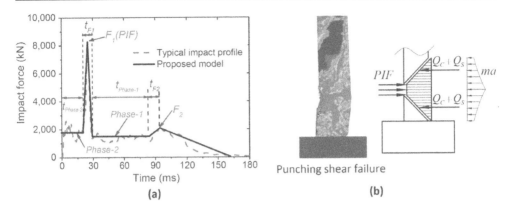

Figure 2.22 (a) Simplified model of the impact force, (b) the mechanism of punching shear failure [159].

shear failure when *PIF* reaches the maximum dynamic shear capacity (Q_{dyn}^{max}) as calculated by Equation (2.3). The second phase (Phase-2) was taken equal to 1,290 kN.

$$F_1(\text{kN}) = 969.3\sqrt{0.5 m_e V^2} - 7{,}345.9 \leq Q_{dyn}^{max} \ (16.7 \text{ m/s} < V < 40 \text{ m/s}) \qquad (2.2)$$

where m_e is the mass of the engine (ton), and V is the truck impact velocity.

$$Q_{dyn}^{max} = 2(DIF_c \times Q_c + DIF_s \times Q_s) + \Sigma ma \qquad (2.3)$$

where DIF_c and DIF_s are the dynamic increase factors of the concrete and steel material strength in the diagonal section, respectively. Q_c and Q_s are the contribution of the concrete and the steel reinforcement to resist the shear force, respectively. *m* and *a* are the mass and acceleration of the shear plug, respectively.

From the vast majority of pier-vehicle collision studies as summarized in Table 2.5, the significant influences of the impacting vehicle characteristics such as vehicle type (including the variability of the bow configuration), impact velocity, and vehicle weight (including the mass of both engine and cargo portions) were concluded. However, the design collision force provided by AASHTO specifications does not consider the effects of dynamic vehicle-pier interactions, impact velocity, vehicle weight, and characteristics. From an FE numerical study of truck collisions with different bridge piers by El-Tawil et al. [138], it was concluded that dynamic PIFs could be larger than those predicted by AASSHTO-LRFD [10] and the design approach proposed by this design code could be unconservative.

Compared to the extensive investigation on the effects of various structural parameters, the influences of the axial load parameter have not been comprehensively explored. Besides, most of the previous studies concluded the positive influences of this parameter on the impact resistance of bridge piers when it was in its service levels. Therefore, more attempts are needed to explore the sensitivity levels of the impact responses of bridge piers to the axial load parameter in future works.

According to the conclusions of several works investigating precast concrete segmental bridge columns (PCSBC), such types of columns can provide more ductility and suffer fewer damages under lateral impact loads compared to conventional RC

Table 2.5 Summary of the influences of various parameters on the impact responses of RC bridge piers to vehicle collisions

Study	Analysis	Parameter	Effectiveness
Zhou et al. [150]	Numerical	• Impact velocity • Impact mass	• Substantial positive on the peak impact force
		• Concrete strength	• Marginal negative on the pier global deformations
		• Reinforcements strength	• Substantial negative on the pier global deformations
Abdelkarim and ElGawady [143]	Numerical	• Concrete strength	• No effect on the peak impact force
		• Reinforcement ratio	• Positive on the peak impact force
		• Column cross-sectional dimensions • Axial load • Impact velocity • Impact weight • Material strain rate	• Positive on the peak impact force
		• Pier boundary condition • Depth of soil surrounding pier substructure	• No effect on the peak impact force
Chen et al. [146]	Analytical	• Impact velocity • Pier cross-sectional dimensions • Pier material properties	• Substantial positive on the peak impact force
Chen et al. [148]	Experimental	• Impact velocity • Impact mass	• Significant positive on the peak impact force
Chen et al. [154]	Numerical	• Pier cross-sectional shape and dimensions	• Marginal effects on the impact force
		• Impact velocity	• Significant effects on the first peak impact force
		• Cargo weight	• Significant effect on the following peaks of impact force
		• Impact position • Vehicle type	• Significant positive on the peak impact force
Yi et al. [156]		• Cross-sectional shape	• More severe damages and displacements for round-shape piers than square-shape ones
		• Axial load • Concrete strength	• No influence on impact resistance
Fan et al. [40]		• Axial load	• Positive in the service level for small deformations; negative for large deformations
		• Reinforcement ratios • UHPFRC strength	• Marginal positive on the pier resistance
		• UHPFRC thickness	• Substantial positive on the pier lateral impact resistance and axial load capacity for low-velocity impacts
		• Impact velocity	• Significant positive on the pier damage level

(Continued)

Table 2.5 Summary of the influences of various parameters on the impact responses of RC bridge piers to vehicle collisions (Continued)

Study	Analysis	Parameter	Effectiveness
Auyeung and Alipour [145]	Numerical	• Impact mass • Impact velocity • Pier diameter	• Substantial positive on the peak impact force • Substantial negative on the pier global failure modes (i.e., governed the global failures)
		• Transverse reinforcement	• Substantial negative on the pier local failure modes (i.e., governed the local failures)
Auyeung et al. [152]	Numerical	• Impact velocity	• Significant positive on the impact and the pier shear forces
Do et al. [24]	Numerical	• Vehicle engine's mass	• Significant positive on the peak impact force, shear forces, and moment
		• Vehicle velocity	• Significant positive on the peak impact force
Cao et al. [140]	Numerical	• Vehicle engine's mass	• Significant positive on the highest peak impact force
		• Vehicle trailer's mass	• Significant positive on the following peaks of impact force, and damage levels

bridge piers [155, 160, 186–190]. The performance of precast segmental columns under vehicle collisions was numerically assessed by Do et al. [160] varying in prestressing level, the number of segments, concrete strength, and vehicle velocity. It was found that the number of segments and initial prestressing level had marginal influences on the impact force, while they significantly affected the residual displacements and the damage behaviors. Moreover, it was revealed that the impact velocity has not always absolute positive influences on the impact force of segmental columns. In line with this study, impact behaviors of two different bridge piers, including RC pier and precast modular pier, were numerically evaluated by Chung et al. [155] subjected to impact loading functions derived from a series of vehicle-pier collision simulations. Larger displacement and stresses were obtained in precast pier than those of RC piers under the same peak dynamic loading. Afterward, Do et al. [187] carried out a comparative study between the impact performance of PCSBC and monolithic bridge columns. The better flexural and shear performances were obtained for PCSBC due to the existence of shear slippage and joint rocking between concrete segments. Also, PCSBC columns suffered localized shear and compression damages limited to the impacted segments compared to the global flexural and shear damages observed in the monolithic columns. With the intent of reducing the stresses and relative displacements between the segments of the column under lateral impact loads, Zhang et al. [186] utilized a new shear design with smoothed curvature.

2.4.2 Bridge piers subjected to vessel collisions

Bridge piers spanning across the navigable coastal channels are potentially at the risk of vessel collisions that can cause severe damages or even collapse of such structures. Based on a report given by AASHTO [7], during a period from 1960 to 2002, vessel

| (a) Queen Isabella Causeway Bridge [188] | (b) I-40 Bridge [189] | (c) Sunshine Skyway Bridge [190] |

Figure 2.23 The collapse of several bridges in the United States due to vessel collisions.

collisions caused the collapse of 31 bridges and 342 fatalities. Figure 2.23 shows the collapse of several bridges due to vessel collisions. Several design codes for bridges [7–9] include the vessel collision loads based on static analysis methods considering the type of striking vessel and waterways. However, the provisions provided by these guidelines are not able to accurately estimate the vessel collision loads due to omitting dynamic characteristics, and the geometry of both striking vessels and stuck bridge structures.

As a prominent and famous experimental study, Consolazio et al. [191] conducted a series of full-scale barge collision tests on the old piers of St. George Island Bridge. Many open-access technical reports were published based on the information from these experimental tests capturing novel insights into various aspects of the vessel-bridge collision event [192–199]. However, conducting full-scale experimental tests of vessel collisions with bridges requires special strategic plans and notable financial resources. Hence, analytical and numerical approaches using high fidelity computer software can be considered a proper alternative to estimate the responses of both vessels and impacted structures, accurately and reasonably.

Generally, the dynamic analysis of vessel-bridge pier collision can be categorized into three techniques as follows:

- High-resolution FE technique consists of a multiple-degree-of-freedom (MDOF) model of vessel versus an MDOF model of bridge pier as shown in Figure 2.24a.
- Medium-resolution technique consists of an SDOF model of vessel versus the MDOF model of a bridge pier as shown in Figure 2.24b.
- The low-resolution technique consists of an SDOF model of a vessel versus an SDOF model of a bridge pier as shown in Figure 2.24c.

In the first technique, the dynamic impact loads, the nonlinear dynamic responses, and failure behaviors of both bridge structure and impacting vessel such as the crush

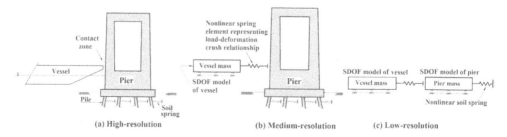

| (a) High-resolution | (b) Medium-resolution | (c) Low-resolution |

Figure 2.24 Different dynamic analysis techniques of vessel-pier collision [200].

deformations are quantified considering detailed collision mechanics and nonlinear interactions between the high-resolution FE models of impacting vessel and bridge pier. However, adopting such techniques will not be appropriate for design applications because they require significant computational costs and resources. In the second technique (i.e., medium-resolution), the stiffness (load-deformation) of the vessel bow is simplified using a nonlinear discrete element in the vessel SDOF system. Although this technique is not able to obtain the vessel crush behaviors, it can be properly used for design intents due to considering the high-resolution model and all structural characteristics of the impacted pier. Besides, low-resolution techniques representing the simplified models of vessel-pier collisions are particularly utilized to predict the dynamic impact loads as the preliminary stage of a design process.

There exist various analytical approaches in the literature simplifying the models of the striking vessel and impacted bridge. Consolazio and Cowan [201] proposed a coupled vessel impact analysis (CVIA) method in which an SDOF vessel model collides with an MDOF pier model (i.e., medium-resolution technique). In this technique, various loading and structural-related parameters of vessel-pier collision system, including vessel bow stiffness, vessel mass, impact velocity, impact angle, pier stiffness, pier mass, pier geometry, and soil conditions, were taken into account to capture time history results of impact force, bridge displacements, and internal structural forces generated in bridge pier. The stiffness of the vessel bow modeled with a nonlinear spring represents the load-deformation crush relationships captured from the high-resolution analysis [202]. Thereafter, the proposed coupled CVIA used to analyze the dynamic responses of an equivalent one-pier, two-span (OPTS) simplified bridge model proposed by Consolazio and Davidson [203] in which the effects and characteristics of adjacent piers and spans were considered using a series of equivalent translational and rotational springs attached to a lumped mass of adjusted piers and spans as shown in Figure 2.25. The capability of this model was concluded to successfully predict the pier impact responses considering the dynamic amplification

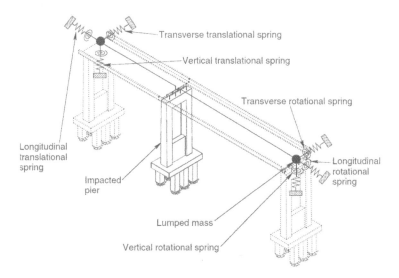

Figure 2.25 Analytical model for bridge-ship collisions based on the proposed macro-element by Consolazio and Davidson [203].

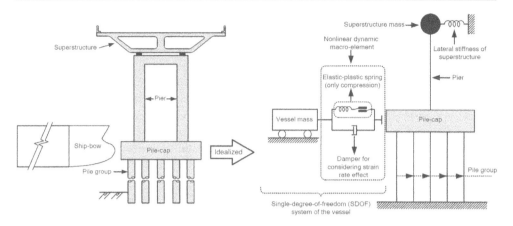

Figure 2.26 Macro-element model of bridge pier using spring and lumped mass elements [204].

characteristics along with a significant reduction in the analysis time. Fan et al. [204] proposed a nonlinear dynamic macro-element model of bridge pier to quantify the pier demand subjected to ship collisions as shown in Figure 2.26. In this model, the crush behavior of ship bow considering with the strain rate effects was modeled using the combination of an elastic-plastic spring and a dashpot element attached in parallel. The calculated results from the proposed method were agreed well with those from the high-resolution analysis. Moreover, it was revealed that the design impact forces predicted by the current design codes can be underestimated due to neglecting the dynamic amplification factors such as the material strain rate effects.

A vast majority of vessel-bridge collision studies focused on investigating the force-deformation results of vessels considering the shape and size of the impacted pier under low- to medium-rate [201, 202, 205–208], and high-rate impacts [209, 210]. Besides, many attempts were carried out to evaluate the structural responses of bridge piers using simplified analytical methods [169, 172, 203, 211, 212], and FE high-resolution techniques [28, 162, 166, 174, 213]. Also, several approaches were proposed to analyze the impact loads and bridge dynamic demands using shock spectrum analysis [211, 214–216], structural reliability analysis [163, 170, 217–221], and equivalent static analysis [222, 223] methods. Table 2.6 presents a summary of vessel-pier collisions with regard to the influences of various parameters.

In Table 2.6, it can be found that the impact forces and responses are affected not only by loading-related parameters, including vessel mass, velocity, weight, and bow configurations (stiffness-related), but also by structural-related parameters, including the pier inertia, axial load ratio (superstructure inertia-related), geometry, and soil-structure interaction behavior. However, the force-deformation relationships provided by the current design codes such as AASHTO have not taken into account the key dynamic factors of both vessel and impact pier such as the strain rate effects of materials, geometry, shape, and size parameters. Hence, these deficiencies existing in the guidelines may lead to unconservative and inaccurate impact loads and structural responses. Besides, although the effectiveness of the pier superstructure inertia has been considered by several vessel-pier collision studies, except very limited works in the literature [28], the sensitivity of the impact resistances of bridge piers to axial load parameter has not been rigorously investigated.

Table 2.6 Summary of the influences of various parameters on the impact responses of RC bridge piers to vessel collision

Study	Analysis	Parameter	Effectiveness
Consolazio and Cowan [201]	Numerical	• Barge mass	• Marginal positive influence on the pier demand and the peak impact force; significant positive effects on the impact duration
		• Impact angle	• Marginal influence on the impact force; significant negative effects on the pier deflection
		• Pier stiffness • Pier mass	• Marginal positive on the impact force and the pier resistance
		• Pier geometry	• Marginal positive on the impact force; significant effects on the pier deflection (greater impact resistance by circular-shaped columns than those of square-shaped columns)
Consolazio et al. [202]	Numerical	• Pier geometry	• Higher impact forces for round-faced columns than those for flat-faced columns
Consolazio and Cowan [205]	Numerical	• Pier width/diameter	• Marginal influence on the impact force
		• Pier geometry	• Significant effects on the impact forces (higher forces for Flat-faced columns than round-faced columns)
Yuan and Harik [206]	Numerical and analytical	• Pier geometry • Pier width/diameter	• Significant positive on the peak impact force of rectangular piers; marginal positive on the peak impact force of circular piers
Getter and Consolazio [207]	Numerical	• Impact angle	• Significant negative on the bow force-deformation relationship for impact on wide piers; similar effectiveness for impact angles of 5° or more, when piers have less width
Fan and Yuan [208]	Numerical	• Pile-cap depth	• Important role in quantifying impact demand of bridge piers
Kantrales et al. [209]	Experimental and analytical	• Pier geometry	• Larger impact forces were obtained for flat-faced piers than those for round-faced piers
Luperi and Pinto [210]	Numerical	• Pier width/diameter • Pier geometry	• More significant positive on the bow force-deformation relationships for flat-faced piers than round-faced-piers
Wang and Morgenthal [169]	Numerical and analytical	• Barge mass	• Significant positive on the peak impact force until 25% loaded barge; ascendingly positive on the pier deflection; marginal on the pier moment
		• Impact velocity	• Significant positive on the peak impact force and the pier deflection
		• Height of impact location	• Marginal on the peak impact force; significant positive on the pier deflection
		• Column height	• Marginal on the peak impact force and the pier deflection
		• Diameter of longitudinal reinforcement	• Marginal on the peak impact force; significant negative on the pier deflection

(Continued)

Table 2.6 Summary of the influences of various parameters on the impact responses of RC bridge piers to vessel collision (Continued)

Study	Analysis	Parameter	Effectiveness
Fan et al. [172]	Numerical and analytical	• Strain rate of steel used for vessel bow	• Significant positive on the impact forces; significant negative on the impact duration
Yuan et al. [212]	Numerical and analytical	• Pier stiffness	• Significant effect on the peak impact forces; marginal effect on the mean impact force
		• Number of barges in a flotilla	• Marginal effect on the mean impact force; positive on the impact duration
		• Pier geometry	• Larger impact forces produced by square columns than those by a circular column
		• Pier width	• Significant positive on the impact forces
		• Kinetic impact energy	• Significant effect on the impact duration
Gholipour et al. [28]	Numerical	• Axial load ratio	• Marginal positive on the peak impact force; negative on the length of plastic hinges; positive on the column resistance until a ratio of 0.5
		• Impact velocity	• Significant positive on the peak impact forces; negative on the length of plastic hinges
Gholipour et al. [162]	Numerical	• Superstructure to pier mass ratio	• Significant positive on the peak impact forces
		• Height of impact location	• Significant positive on the peak impact forces and the pier demand for ratios more than 1.0
Sha and Hao [166]	Numerical	• Pier concrete nonlinearity	• Significant effect on the impact force and barge crush depth
		• Impact velocity	• Significant positive on the peak impact forces and durations
		• Barge mass	• Marginal effect on the impact force for high-velocity impacts
Getter et al. [222]	Numerical and analytical	• Superstructure inertia	• Significant effect on the pier demand
Gholipour et al. [161]		• Soil-pier interaction	• Significant effect on the relatively light piers
Sha and Hao [164]	Experimental and numerical	• Impact velocity	• Significant effect on the peak impact forces
		• Vessel mass	• Significant effect on the impact durations
		• Pier height	• Marginal influences on the impact force
		• Superstructures mass	
		• Height of impact location	
Zhang et al. [168]	Numerical	• Initial kinetic energy of collision	• Significant positive on the pier demand and the soil deformations
		• Soil damping surrounding the piles	• Negative effect on the pier demand

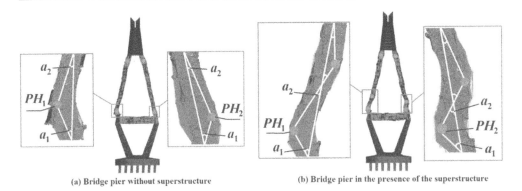

(a) Bridge pier without superstructure   (b) Bridge pier in the presence of the superstructure

Figure 2.27 Failure behaviors of the bridge pier with and without the superstructure subjected to ship collision [28].

The significant effects of the different boundary conditions of impacted bridges at the top affected by the inertia of overlaying superstructures [162, 173] and at the bottom surrounded by soil layer [161, 168, 171, 223–226] were explored in the literature. The substantial influences of bridge superstructure on the impact responses and failure behaviors of the impacted pier especially on the location of plastic hinge formed in the pier columns were concluded by Gholipour et al. [28]. Figure 2.27 illustrates the influence of the bridge superstructure on the failure modes of the impacted pier. It is observed that the pier indicates different characteristics for the formed plastic hinges (PH_i), including their locations and the relative curvatures (a_i). Based on a study done by Davidson et al. [173], the amplification dynamic effects of bridge superstructures during a vessel collision event can be categorized into: (i) inertial resistance of superstructure amplification that is mobilized shortly after the onset of impact loading and causes maximum shear forces in bridge pier and (ii) superstructure momentum driven-sway amplification due to increasing the velocity at the pier top that leads to producing maximum bending moments in the pier.

Despite many previous studies considering the effects of dynamic characteristics of both vessel and pier mentioned earlier, the material nonlinearity and structural damages have been taken into account in predicting impact responses and failure behaviors of the impacted piers [28, 163, 166, 174, 227]. With growing FE computer codes in recent years and the feasibility of modeling the nonlinear behaviors of concrete materials considering strain rate effects, several studies were focused on investigating the damage and failure behaviors [23, 28, 164, 166, 169], and progressive collapse [174, 224, 228] bridges under vessel impact loads.

The nonlinear dynamic responses and progressive damage process of a cable-stayed concrete bridge pier were numerically and analytically investigated by Gholipour et al. [227] under ship collision considering the nonlinearity of concrete and steel materials. Moreover, the strain rate effects of concrete and steel materials were formulated to use in a proposed 2-DOF simplified system. It was found that the proposed simplified method could accurately estimate the impact force and pier displacement response compared to those from FE simulations. Moreover, among different damage indices proposed to describe the damage states of the impacted pier, a damage index based on pier deflection was captured more efficiently than other approaches. Besides, the progressive damage process of the impacted pier from the appearance of minor tensile and flexural cracks, developing shear damages, cross-sectional fracture,

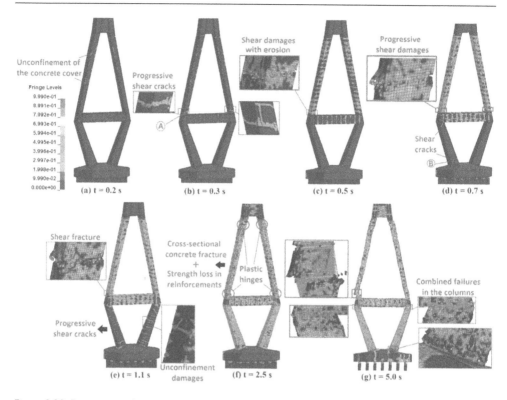

Figure 2.28 Progressive damage process of the bridge pier under ship collision [227].

and consequently the formation of plastic hinges in the pier columns were observed, respectively, as shown in Figure 2.28.

Sha and Hao [164] experimentally studied the impact responses of scaled models of fixed-base (i.e., neglecting soil-structure interactions at the bottom boundary condition) circular bridge piers subjected to pendulum impact loading in the presence of the equivalent mass of the superstructure. Consideration of fixed base boundary conditions led to underestimate pier responses. Besides, the influences of several parameters on the impact forces and the pier responses were evaluated through a parametric study based on the FE simulations of vessel-pier collisions. Compared to the notable sensitivity of the PIFs to impact velocity, impact durations were more affected by vessel mass parameter. Besides, structural parameters, including pier height, superstructures mass, and the height of impact location, had marginal influences on the impact force. Furthermore, the performance of RC bridge piers subjected to ship collision was evaluated by Wan et al. [167] using experimental tests on scaled models, and FE simulations of ship-pier collision considering nonlinear material models. The significant influences of the material nonlinearity on the impact results and the pier responses were concluded in high-energy collision scenarios.

2.5 Chapter review

This chapter presents a state-of-the-art review of responses and behaviors of different concrete structures subjected to lateral impact loads. First, the basic theories of impact loadings and the response mechanisms of concrete structures to such extreme loads were introduced. Then, the specifications existing in the current design codes and guidelines regarding impact loads and their limitations were discussed. In addition, the dynamic responses and failure behaviors of concrete structures including bridge piers subjected to vehicle and vessel collisions, or isolated structural members, such as concrete columns (commonly used in low- and medium-rise buildings), beams, and slabs under lateral impact loads of rigid objects analyzed using simplified analytical methods, FE simulations, and experimental methods, were comprehensively reviewed. Moreover, the influences of various loading- and structural-related parameters on the impact responses of concrete structures were studied.

It was revealed that the impact loads predicted by the current guidelines may be unconservative due to omitting the amplification dynamic effects such as inertia and strain rate effects. Also, despite many attempts in recognizing the failure modes and especially shear failures of RC bridge piers under vehicle collisions, there is no recommendation in the current design codes considering the damage states of the impacted structure in the prediction of equivalent impact loads.

From the review on the influences of various parameters on the impact responses of RC columns and piers under lateral impact loads arising from vessel or vehicle collisions, it was obtained that the effectiveness of the axial load parameter has not been rigorously explored. However, most of the previous research works have studied the impact capacities of axially loaded structures when they were exposed to their service levels of axial loads.

By reviewing vessel-pier collision studies, the vast majority of these research works concluded the significant influences of the dynamic parameters of both striking vessels and struck structures such as the pier inertia, axial load ratio (superstructure inertia-related), geometry, and soil-structure interaction behavior on the impact responses of bridge piers. Hence, the inability of the current design codes in estimating accurate and reasonable impact loads and responses under vessel collisions was found due to neglecting the effects of these parameters.

References

1. Pham TM, Hao H. Influence of global stiffness and equivalent model on prediction of impact response of RC beams. *Int J Impact Eng* 2018;113:88–97.
2. Fujikake K, Li B, Soeun S. Impact response of reinforced concrete beam and its analytical evaluation. *J Struct Eng* 2009;135(8):938–950.
3. CEN. Eurocode 1: actions on structures. European Committee for Standardization, Brussels, Belgium, 2003.
4. Saatci S. Behaviour and modelling of reinforced concrete structures subjected to impact loads. PhD Dissertation, University of Toronto, Canada, 2007.
5. Li QM, Reida SR. Local impact effects of hard missiles on concrete targets. *Int J Impact Eng* 2005;32:224–284.
6. Bangash MYH. *Impact and explosion: structural analysis and design.* CRC Press LLC, Boca Raton, FL, 1993.
7. AASHTO. *Guide specifications and commentary for vessel collision design of highway bridges*, 2nd edn. American Association of State Highway and Transportation Officials, Washington, DC, 2009.
8. CEN. Actions on structures. Part 1–7: general actions-accidental actions. European Committee for Standardization; 2006. BS EN 1991-1-1, Brussels, Belgium, 2002.
9. CMR. General code for design of railway bridges and culverts. In *TB10002.1-2005, China Ministry of Railways.* China Railway Press, Beijing, 2005 (in Chinese).

10. AASHTO-LRFD. *LRFD Bridge Design Specifications*. American Association of State Highway and Transportation Officials, Washington, DC, 2012.
11. Abdelkarim OI, ElGawady MA. Performance of hollow-core FRP–concrete–steel bridge columns subjected to vehicle collision. *Eng Struct* 2016;123:517–531.
12. Abdelkarim OI, ElGawady MA. Performance of bridge piers under vehicle collision. *Eng Struct* 2017;140:337–352.
13. JSCE. *Subcommittee of impact problems of JSCE: practical methods for impact test and analysis*. Edited by Kishi N. Japan Society of Civil Engineers, Tokyo, Japan, 2004.
14. AS 1170.1. AS/NZS 1170.1:1988: structural design actions-permanent, imposed and other actions, Australia, 1989.
15. UK's Highways Agency. The design of highway bridges for vehicle collision loads. BD 60/04. Department for Transport, UK, 2004.
16. LS-DYNA 971. Livermore Software Technology Corporation, Livermore, CA, 2017.
17. ABAQUS. Analysis user's manual version 6.10, volume I to VI, ABAQUS, Inc. an Dassault Systémes, Providence, RI, 2010.
18. Saatci S, Vecchio FJ. Effects of shear mechanisms on impact behavior of reinforced concrete beams. *ACI Struct J* 2009;106(1):78–86.
19. Bhatti AQ, Kishi N, Mikami H, Ando T. Elasto-plastic impact response analysis of shear-failure-type RC beams with shear rebars. *Mater Des* 2009;30(3):502–510.
20. Kishi N, Mikami H, Matsuoka KG, Ando T. Impact behavior of shear-failure-type RC beams without shear rebar. *Int J Impact Eng* 2002;27(9):955–968.
21. Xiao Y, Li B, Fujikake K. Behavior of reinforced concrete slabs under low velocity impact. *ACI Struct J* 2017;114(3):643–658.
22. Zineddin M, Krauthammer T. Dynamic response and behavior of reinforced concrete slabs under impact loading. *Int J Impact Eng* 2007;34(9):1517–1534.
23. Sha YY, Hao H. Laboratory tests and numerical simulations of CFRP strengthened RC pier subjected to barge impact load. *Int J Struct Stab Dy* 2015;15(02):1450037.
24. Do TV, Pham TM, Hao H. Dynamic responses and failure modes of bridge columns under vehicle collision. *Eng Struct* 2018;156:243–259.
25. Cai J, Ye JB, Chen QJ, Liu X, Wang YQ. Dynamic behaviour of axially-loaded RC columns under horizontal impact loading. *Eng Struct* 2018;168:684–697.
26. Demartino C, Wu JG, Xiao Y. Response of shear-deficient reinforced circular RC columns under lateral impact loading. *Int J Impact Eng* 2017;109:196–213.
27. Thilakarathna HMI, Thambiratnam DP, Dhanasekar M, Perera N. Numerical simulation of axially loaded concrete columns under transverse impact and vulnerability assessment. *Int J Impact Eng* 2010;37:1100–1112.
28. Gholipour G, Zhang C, Mousavi AA. Effects of axial load on nonlinear response of RC columns subjected to lateral impact load: ship-pier collision. *Eng Fail Anal* 2018;91:397–418.
29. Liu B, Fan W, Guo W, Chen B, Liu R. Experimental investigation and improved FE modeling of axially-loaded circular RC columns under lateral impact loading. *Eng Struct* 2017;152:619–642.
30. Remennikov A, Kaewunruen S. *Impact resistance of reinforced concrete columns: experimental studies and design considerations*, Faculty of Engineering, University of Wollongong, Wollongong, Australia, 2006.
31. Yilmaz T, Kiraç N, Anil Ö. Experimental investigation of axially loaded reinforced concrete square column subjected to lateral low-velocity impact loading. *Struct Concr* 2019;20(4):1358–1378.
32. Parvin A, Brighton D. FRP composites strengthening of concrete columns under various loading conditions. *Polymers* 2014;6(4):1040–1056.
33. Harris B, Beaumont PWR, Moncunill de Ferran E. Strength and fracture toughness of carbon fibre polyester composites. *J Mater Sci* 1971;6(3):238–251.

34. Hayes SV, Adams DF. Rate sensitive tensile impact properties of fully and partially loaded unidirectional composites. *J Test Eval* 1982;10(2):61–68.
35. Hancox NL. Izod impact testing of carbon-fibre-reinforced plastics. *Composites* 1971;2(1):41–45.
36. Kimura H, Itabashi M, Kawata K. Mechanical characterization of unidirectional CFRP thin strip and CFRP cables under quasi-static and dynamic tension. *Adv Compos Mater* 2001;10(2/3):177–187.
37. Cantwell WJ, Smith K. The static and dynamic response of CFRP-strengthened concrete structures. *J Mater Sci Lett* 1999;18(4):309–310.
38. Yan X, Yali S. Impact behaviors of CFT and CFRP confined CFT stub columns. *J Compos Constr* 2012;16(6):662–670.
39. Alam MI, Fawzia S, Zhao XL, Remennikov AM. Experimental study on FRP-strengthened steel tubular members under lateral impact. *J Compos Constr* 2017;21(5):04017022.
40. Fan W, Xu X, Zhang Z, Shao X. Performance and sensitivity analysis of UHPFRC-strengthened bridge columns subjected to vehicle collisions. *Eng Struct* 2018;173:251–268.
41. Fan W, Guo W, Sun Y, Chen B, Shao X. Experimental and numerical investigations of a novel steel-UHPFRC composite fender for bridge protection in vessel collisions. *Ocean Eng* 2018;165:1–21.
42. Yoo DY, Banthia N. Mechanical and structural behaviors of ultra-high-performance fiber-reinforced concrete subjected to impact and blast. *Constr Build Mater* 2017;149:416–431.
43. Fan W, Shen D, Yang T, Shao X. Experimental and numerical study on low-velocity lateral impact behaviors of RC, UHPFRC and UHPFRC-strengthened columns. *Eng Struct* 2019;191:509–525.
44. Wang R, Han LH, Hou CC. Behavior of concrete filled steel tubular (CFST) members under lateral impact: experiment and FEA model. *J Constr Steel Res* 2013;80:188–201.
45. Huo JS, Zheng Q, Chen BS, Xiao Y. Tests on impact behaviour of micro-concrete-filled steel tubes at elevated temperatures up to 400°C. *Mater Struct* 2009;42(10):1325–1334.
46. Han LH, Hou CC, Zhao XL, Rasmussen KJ. Behaviour of high-strength concrete filled steel tubes under transverse impact loading. *J Constr Steel Res* 2014;92:25–39.
47. Bambach MR. Design of hollow and concrete filled steel and stainless steel tubular columns for transverse impact loads. *Thin Wall Struct* 2011;49(10):1251–1260.
48. Alam MI, Fawzia S, Zhao XL. Numerical investigation of CFRP strengthened full scale CFST columns subjected to vehicular impact. *Eng Struct* 2016;126:292–310.
49. Zhang X, Chen Y, Wan J, Wang K, He K, Chen X, Wei J, Jiang G. Tests on residual ultimate bearing capacity of square CFST columns after impact. *J Construct Steel Res* 2018;147:27–42.
50. Zhu AZ, Xu W, Gao K, Ge HB, Zhu JH. Lateral impact response of rectangular hollow and partially concrete-filled steel tubular columns. *Thin Wall Struct* 2018;130:114–131.
51. Aghdamy S, Thambiratnam DP, Dhanasekar M, Saiedi S. Computer analysis of impact behavior of concrete filled steel tube columns. *Adv Eng Softw* 2015;89:52–63.
52. Bambach MR, Jama H, Zhao XL, Grzebieta RH. Hollow and concrete filled steel hollow sections under transverse impact loads. *Eng Struct* 2008;30(10):2859–2870.
53. Qu H, Li G, Chen S, Sun J, Sozen MA. Analysis of circular concrete-filled steel tube specimen under lateral impact. *Adv Struct Eng* 2011;14(5):941–951.
54. Shan JH, Chen R, Zhang WX, Xiao Y, Yi WJ, Lu FY. Behavior of concrete filled tubes and confined concrete filled tubes under high speed impact. *Adv Struct Eng* 2007;10(2):209–218.
55. Wang R, Han LH, Zhao XL, Rasmussen KJ. Experimental behavior of concrete filled double steel tubular (CFDST) members under low velocity drop weight impact. *Thin Wall Struct* 2015;97:279–295.
56. Wang R, Han LH, Zhao XL, Rasmussen KJ. Analytical behavior of concrete filled double steel tubular (CFDST) members under lateral impact. *Thin Wall Struct* 2016;101:129–140.

57. Aghdamy S, Thambiratnam DP, Dhanasekar M, Saiedi S. Effects of load-related parameters on the response of concrete-filled double-skin steel tube columns subjected to lateral impact. *J Constr Steel Res* 2017;138:642–662.
58. Aghdamy S, Thambiratnam DP, Dhanasekar M, Saiedi S. Effects of structure-related parameters on the response of concrete-filled double-skin steel tube columns to lateral impact. *Thin Wall Struct* 2016;108:351–368.
59. Du G, Andjelic A, Li Z, Lei Z, Bie X. Residual axial bearing capacity of concrete-filled circular steel tubular columns (CFCSTCs) after transverse impact. *Appl Sci* 2018;8(5):793.
60. Shakir AS, Guan ZW, Jones SW. Lateral impact response of the concrete filled steel tube columns with and without CFRP strengthening. *Eng Struct* 2016;116:148–162.
61. Alam MI, Fawzia S, Liu X. Effect of bond length on the behaviour of CFRP strengthened concrete-filled steel tubes under transverse impact. *Compos Struct* 2015;132:898–914.
62. Chen C, Zhao Y, Li J. Experimental investigation on the impact performance of concrete-filled FRP steel tubes. *J Eng Mech* 2014;141(2):04014112.
63. Wang R, Han LH, Tao Z. Behavior of FRP–concrete–steel double skin tubular members under lateral impact: experimental study. *Thin Wall Struct* 2015;95:363–373.
64. Soleimani SM, Sayyar Roudsari S. Analytical study of reinforced concrete beams tested under quasi-static and impact loadings. *Appl Sci* 2019;9(14):2838.
65. Fujikake K, Senga T, Ueda N, Ohno T, Katagiri M. Study on impact response of reactive powder concrete beam and its analytical model. *J Adv Concr Technol* 2006;4(1):99–108.
66. Pham TM, Hao H. Prediction of the impact force on reinforced concrete beams from a drop weight. *Adv Struct Eng* 2016;19(11):1710–1722.
67. Bischoff PH, Perry SH, Eibl J. Contact force calculations with a simple spring-mass model for hard impact: a case study using polystyrene aggregate concrete. *Int J Impact Eng* 1990;9(3):317–325.
68. Zhao W, Qian J, Jia P. Peak response prediction for RC beams under impact loading. *Shock Vib* 2019;2019:1–12.
69. Kishi N, Mikami H. Empirical formulas for designing reinforced concrete beams under impact loading. *ACI Struct J* 2012;109(4):509–519.
70. Bhatti AQ, Kishi N. Impact response of RC rock-shed girder with sand cushion under falling load. *Nucl Eng Des* 2010;240(10):2626–2632.
71. Tachibana S, Masuya H, Nakamura S. Performance based design of reinforced concrete beams under impact. *Nat Hazard Earth Sys* 2010;10(6):1069–1078.
72. Zhang C, Gholipour G, Mousavi AA. Nonlinear dynamic behavior of simply-supported RC beams subjected to combined impact-blast loading. *Eng Struct* 2019;181:124–142.
73. Gholipour G, Zhang C, Mousavi AA. Loading rate effects on the responses of simply supported RC beams subjected to the combination of impact and blast loads. *Eng Struct* 2019;201:109837.
74. Jin L, Xu J, Zhang R, Du X. Numerical study on the impact performances of reinforced concrete beams: a mesoscopic simulation method. *Eng Fail Anal* 2017;80:141–163.
75. Guo J, Cai J, Chen W. Inertial effect on RC beam subjected to impact loads. *Int J Struct Stab Dy* 2017;17(04):1750053.
76. Tantrapongsaton W, Hansapinyo C, Wongmatar P, Chaisomphob T. Flexural reinforced concrete members with minimum reinforcement under low-velocity impact load. *Int J Geomate* 2018;14(46):129–136.
77. Li H, Chen W, Hao H. Dynamic response of precast concrete beam with wet connection subjected to impact loads. *Eng Struct* 2019;191:247–263.
78. Kong X, Fang Q, Chen L, Wu H. Nonlocal formulation of the modified K&C model to resolve mesh-size dependency of concrete structures subjected to intense dynamic loadings. *Int J Impact Eng* 2018;122:318–332.
79. Li H, Chen W, Hao H. Influence of drop weight geometry and interlayer on impact behavior of RC beams. *Int J Impact Eng* 2019;131:222–237.

80. Pham TM, Hao H. Effect of the plastic hinge and boundary conditions on the impact behavior of reinforced concrete beams. *Int J Impact Eng* 2017;102:74–85.
81. Pham TM, Hao H. Plastic hinges and inertia forces in RC beams under impact loads. *Int J Impact Eng* 2017;103:1–11.
82. Kishi N, Nakano O, Matsuoka KG, Anto T. Experimental study on ultimate strength of flexural-failure-type RC beams under impact loading. In Transactions of 16th International Conference on Structural Mechanics in Reactor Technology, International Association for Structural Mechanics in Reactor Technology, Raleigh, NC, 2001.
83. Li Y, Wang X, Guo X. Experimental study on anti-impact properties of a partially prestressed concrete beam. *Explos Shock Waves* 2006;26:256.
84. Ishikawa N, Enrin H, Katsuki S, Ohta T. Dynamic behavior of prestressed concrete beams under rapid speed loading. *Struct Under Shock Impact V* 1998;35:717–726.
85. Ishikawa N, Katsuki S, Takemoto K. Incremental impact test and simulation of prestressed concrete beam. *Struct Mater* 2002;11:489–498.
86. Isaac P, Darby A, Ibell T, Evernden M. Experimental investigation into the force propagation velocity due to hard impacts on reinforced concrete members. *Intl J Impact Eng* 2017;100:131–138.
87. Liao W, Li M, Zhang W, Tian Z. Experimental studies and numerical simulation of behavior of RC beams retrofitted with HSSWM-HPM under impact loading. *Eng Struct* 2017;149:131–146.
88. Yan Q, Sun B, Liu X, Wu J. The effect of assembling location on the performance of precast concrete beam under impact load. *Adv Struct Eng* 2018;21(8):1211–1222.
89. Guo J, Cai J, Chen Q, Liu X, Wang Y, Zuo Z. Dynamic behaviour and energy dissipation of reinforced recycled aggregate concrete beams under impact. *Constr Build Mater* 2019;214:143–157.
90. Adhikary SD, Li B, Fujikake K. Low velocity impact response of reinforced concrete beams: experimental and numerical investigation. *Int J Protect Struct* 2015;6(1):81–111.
91. Zhao DB, Yi WJ, Kunnath SK. Shear mechanisms in reinforced concrete beams under impact loading. *J Struct Eng* 2017;143(9):04017089.
92. Guo W, Fan W, Shao X, Shen D, Chen B. Constitutive model of ultra-high-performance fiber-reinforced concrete for low-velocity impact simulations. *Compos Struct* 2018;185:307–326.
93. Cotsovos DM, Stathopoulos ND, Zeris CA. Behavior of RC beams subjected to high rates of concentrated loading. *J Struct Eng* 2008;134(12):1839–1851.
94. Cotsovos DM. A simplified approach for assessing the load-carrying capacity of reinforced concrete beams under concentrated load applied at high rates. *Int J Impact Eng* 2010;37(8):907–917.
95. Cotsovos DM, Pavlović MN. Numerical investigation of concrete subjected to compressive impact loading. Part 1: A fundamental explanation for the apparent strength gain at high loading rates. *Comput Struct* 2008;86(1–2):145–163.
96. Zhan T, Wang Z, Ning J. Failure behaviors of reinforced concrete beams subjected to high impact loading. *Eng Fail Anal* 2015;56:233–243.
97. Ožbolt J, Sharma A. Numerical simulation of reinforced concrete beams with different shear reinforcements under dynamic impact loads. *Int J impact Eng* 2011;38(12):940–950.
98. Pham TM, Hao Y, Hao H. Sensitivity of impact behaviour of RC beams to contact stiffness. *Int J Impact Eng* 2018;112:155–164.
99. Zhao DB, Yi WJ, Kunnath SK. Numerical simulation and shear resistance of reinforced concrete beams under impact. *Eng Struct* 2018;166:387–401.
100. Yi WJ, Zhao DB, Kunnath SK. Simplified approach for assessing shear resistance of reinforced concrete beams under impact loads. *ACI Struct J* 2016;113(4):747–756.

101. Fan W, Liu B, Huang X, Sun Y. Efficient modeling of flexural and shear behaviors in reinforced concrete beams and columns subjected to low-velocity impact loading. *Eng Struct* 2019;195:22–50.

102. Pham TM, Hao H. Behavior of fiber-reinforced polymer-strengthened reinforced concrete beams under static and impact loads. *Int J Protect Struct* 2017;8(1):3–24.

103. Pham TM, Hao H. Impact behavior of FRP-strengthened RC beams without stirrups. *J Compos Constr* 2016;20(4):04016011.

104. Hao Y, Hao H, Chen G. Experimental investigation of the behaviour of spiral steel fibre reinforced concrete beams subjected to drop-weight impact loads. *Mater Struct* 2016;49(1–2):353–370.

105. Jerome DM. *Dynamic response of concrete beams externally reinforced with carbon fiber reinforced plastic*. University of Florida, Gainesville, FL, 1996.

106. Wang WW, Dai JG, Harries KA. Performance evaluation of RC beams strengthened with an externally bonded FRP system under simulated vehicle loads. *J Bridge Eng* 2011;18(1):76–82.

107. Erki MA, Meier U. Impact loading of concrete beams externally strengthened with CFRP laminates. *J Compos Constr* 1999;3(3):117–124.

108. Goldston MW, Remennikov A, Saleh Z, Sheikh MN. Experimental investigations on the behavior of GFRP bar reinforced HSC and UHSC beams under static and impact loading. *Structure* 2019;22:109–123.

109. Anil Ö, Durucan C, Erdem RT, Yorgancilar MA. Experimental and numerical investigation of reinforced concrete beams with variable material properties under impact loading. *Constr Build Mater* 2016;125:94–104.

110. Wang W, Chouw N. Behaviour of CFRC beams strengthened by FFRP laminates under static and impact loadings. *Constr Build Mater* 2017;155:956–964.

111. Hrynyk TD, Vecchio FJ. Behavior of steel fiber-reinforced concrete slabs under impact load. *ACI Struct J* 2014;111(5):1213.

112. Verma M, Prem PR, Rajasankar J, Bharatkumar BH. On low-energy impact response of ultra-high performance concrete (UHPC) panels. *Mater Des* 2016;92:853–865.

113. Anil Ö, Kantar E, Yilmaz MC. Low velocity impact behavior of RC slabs with different support types. *Constr Build Mater* 2015;93:1078–1088.

114. Goswami A, Adhikary SD, Li B. Predicting the punching shear failure of concrete slabs under low velocity impact loading. *Eng Struct* 2019;184:37–51.

115. Yılmaz T, Kıraç N, Anil Ö, Erdem RT, Sezer C. Low-velocity impact behaviour of two way RC slab strengthening with CFRP strips. *Constr Build Mater* 2018;186:1046–1063.

116. Othman H, Marzouk H. An experimental investigation on the effect of steel reinforcement on impact response of reinforced concrete plates. *Int J Impact Eng* 2016;88:12–21.

117. Wang W, Chouw N. Experimental and theoretical studies of flax FRP strengthened coconut fibre reinforced concrete slabs under impact loadings. *Constr Build Mater* 2018;171:546–557.

118. Thai DK, Kim SE. Numerical simulation of pre-stressed concrete slab subjected to moderate velocity impact loading. *Eng Fail Anal* 2017;79:820–835.

119. Thai DK, Kim SE, Bui TQ. Modified empirical formulas for predicting the thickness of RC panels under impact loading. *Constr Build Mater* 2018;169:261–275.

120. Micallef K, Sagaseta J, Ruiz MF, Muttoni A. Assessing punching shear failure in reinforced concrete flat slabs subjected to localised impact loading. *Int J impact Eng* 2014;71:17–33.

121. Guo Q, Zhao W. Displacement response analysis of steel-concrete composite panels subjected to impact loadings. *Int J Impact Eng* 2019;131:272–281.

122. Ožbolt J, Ruta D, İrhan B. Impact analysis of thermally pre-damaged reinforced concrete slabs: verification of the 3D FE model. *Int J Impact Eng* 2019;133:103343.

123. Iqbal MA, Kumar V, Mittal AK. Experimental and numerical studies on the drop impact resistance of prestressed concrete plates. *Int J Impact Eng* 2019;123:98–117.
124. Kumar V, Iqbal MA, Mittal AK. Study of induced prestress on deformation and energy absorption characteristics of concrete slabs under drop impact loading. *Constr Build Mater* 2018;188:656–675.
125. Tahmasebinia F, Remennikov A. Simulation of the reinforced concrete slabs under impact loading. In Australian Structural Engineering Conference, Melbourne, Australia, 2008.
126. Kong X, Fang Q, Li QM, Wu H, Crawford JE. Modified K&C model for cratering and scabbing of concrete slabs under projectile impact. *Int J Impact Eng* 2017;108:217–228.
127. Rajput A, Iqbal MA. Ballistic performance of plain, reinforced and pre-stressed concrete slabs under normal impact by an ogival-nosed projectile. *Int J impact Eng* 2017;110:57–71.
128. Iqbal MA, Rajput A, Gupta NK. Performance of prestressed concrete targets against projectile impact. *Int J impact Eng* 2017;110:15–25.
129. Almusallam TH, Siddiqui NA, Iqbal RA, Abbas H. Response of hybrid-fiber reinforced concrete slabs to hard projectile impact. *Int J Impact Eng* 2013;58:17–30.
130. Pavlovic A, Fragassa C, Disic A. Comparative numerical and experimental study of projectile impact on reinforced concrete. *Compos Part B Eng* 2017;108:122–130.
131. Xu X, Ma T, Ning J. Failure analytical model of reinforced concrete slab under impact loading. *Constr Build Mater* 2019;223:679–691.
132. Almusallam TH, Abadel AA, Al-Salloum YA, Siddiqui NA, Abbas H. Effectiveness of hybrid-fibers in improving the impact resistance of RC slabs. *Int J Impact Eng* 2015;81:61–73.
133. Vossoughi F, Ostertag CP, Monteiro PJ, Johnson GC. Resistance of concrete protected by fabric to projectile impact. *Cement Concrete Res* 2007;37(1):96–106.
134. Mousa MA, Uddin N. Response of CFRP/AAC sandwich structures under low velocity impact. *ACI Mater J* 2014;111(1):99–109.
135. Wang XD, Zhang C, Huang Z, Chen GW. Impact experimental research on hybrid bamboo fiber and steel fiber reinforced concrete. *Appl Mech Mater* 2013;357:1049–1052.
136. Wang W, Chouw N. The behaviour of coconut fibre reinforced concrete (CFRC) under impact loading. *Constr Build Mater* 2017;134:452–461.
137. Sadraie H, Khaloo A, Soltani H. Dynamic performance of concrete slabs reinforced with steel and GFRP bars under impact loading. *Eng Struct* 2019;191:62–81.
138. El-Tawil S, Severino E, Fonseca P. Vehicle collision with bridge piers. *J Bridge Eng* 2005;10(3):345–353.
139. Sharma H, Gardoni P, Hurlebaus S. Probabilistic demand model and performance-based fragility estimates for RC column subject to vehicle collision. *Eng Struct* 2014;74:86–95.
140. Cao R, El-Tawil S, Agrawal AK, Xu X, Wong W. Heavy truck collision with bridge piers: computational simulation study. *J Bridge Eng* 2019;24(6):04019052.
141. Cao R, El-Tawil S, Agrawal AK, Xu X, Wong W. Behavior and design of bridge piers subjected to heavy truck collision. *J Bridge Eng* 2019;24(7):04019057.
142. Cao R, El-Tawil S, Agrawal AK, Xu X, Wong W. Performance-based design framework for bridge piers subjected to truck collision. *J Bridge Eng* 2019;24(7):04019064.
143. Abdelkarim OI, ElGawady MA. Design of short reinforced concrete bridge columns under vehicle collision. *Transp Res Rec* 2016;2592(1):27–37.
144. Agrawal AK, Liu GY, Alampalli S. Effects of truck impacts on bridge piers. *Adv Mater Res* 2013;639:13–25.
145. AuYeung S, Alipour A. Evaluation of AASHTO suggested design values for reinforced concrete bridge piers under vehicle collisions. *Transp Res Rec* 2016;2592(1):1–8.
146. Chen L, El-Tawil S, Xiao Y. Reduced models for simulating collisions between trucks and bridge piers. *J Bridge Eng* 2016;21(6):04016020.

147. Chen L, El-Tawil S, Xiao Y. Response spectrum-based method for calculating the reaction force of piers subjected to truck collisions. *Eng Struct* 2017;150:852–863.
148. Chen L, Xiao Y, El-Tawil S. Impact tests of model RC columns by an equivalent truck frame. *J Struct Eng* 2016;142(5):04016002.
149. Tsang HH, Lam NT. Collapse of reinforced concrete column by vehicle impact. *Comput Aided Civ Inf* 2008;23(6):427–436.
150. Zhou D, Li R, Wang J, Guo C. Study on impact behavior and impact force of bridge pier subjected to vehicle collision. *Shock Vib* 2017;2017:1–12.
151. Zhou D, Li R. Damage assessment of bridge piers subjected to vehicle collision. *Adv Struct Eng* 2018;21(15):2270–2281.
152. Auyeung S, Alipour A, Saini D. Performance-based design of bridge piers under vehicle collision. *Eng Struct* 2019;191:752–765.
153. Cai C, He Q, Zhu S, Zhai W, Wang M. Dynamic interaction of suspension-type monorail vehicle and bridge: numerical simulation and experiment. *Mech Syst Signal Pr* 2019;118:388–407.
154. Chen L, Wu H, Fang Q, Zhang T. Numerical analysis of collision between a tractor-trailer and bridge pier. *Int J Protect Struct* 2018;9(4):484–503.
155. Chung CH, Lee J, Gil JH. Structural performance evaluation of a precast prefabricated bridge column under vehicle impact loading. *Struct Infrastruct E* 2014;10(6):777–791.
156. Yi NH, Choi JH, Kim SJ, Kim JHJ. Collision capacity evaluation of RC columns by impact simulation and probabilistic evaluation. *J Adv Concr Technol* 2015;13(2):67–81.
157. Sharma H, Hurlebaus S, Gardoni P. Performance-based response evaluation of reinforced concrete columns subject to vehicle impact. *Int J Impact Eng* 2012;43:52–62.
158. Sharma H, Gardoni P, Hurlebaus S. Performance-based probabilistic capacity models and fragility estimates for RC columns subject to vehicle collision. *Comput Aided Civ Inf* 2015;30(7):555–569.
159. Do TV, Pham TM, Hao H. Impact force profile and failure classification of reinforced concrete bridge columns against vehicle impact. *Eng Struct* 2019;183:443–458.
160. Do TV, Pham TM, Hao H. Numerical investigation of the behavior of precast concrete segmental columns subjected to vehicle collision. *Eng Struct* 2018;156:375–393.
161. Gholipour G, Zhang C, Li M. Effects of soil–pile interaction on the response of bridge pier to barge collision using energy distribution method. *Struct Infrastruct E* 2018;14(11):1520–1534.
162. Gholipour G, Zhang C, Mousavi AA. Analysis of girder bridge pier subjected to barge collision considering the superstructure interactions: the case study of a multiple-pier bridge system. *Struct Infrastruct E* 2019;15(3):392–412.
163. Gholipour G, Zhang C, Mousavi AA. Reliability analysis of girder bridge piers subjected to barge collisions. *Struct Infrastruct E* 2019;15(9):1200–1220.
164. Sha Y, Hao H. Laboratory tests and numerical simulations of barge impact on circular reinforced concrete piers. *Eng Struct* 2013;46:593–605.
165. Song Y, Wang J. Development of the impact force time-history for determining the responses of bridges subjected to ship collisions. *Ocean Eng* 2019;187:106182.
166. Sha Y, Hao H. Nonlinear finite element analysis of barge collision with a single bridge pier. *Eng Struct* 2012;41:63–76.
167. Wan Y, Zhu L, Fang H, Liu W, Mao Y. Experimental testing and numerical simulations of ship impact on axially loaded reinforced concrete piers. *Int J Impact Eng* 2019 125:246–262.
168. Zhang J, Li X, Jing Y, Han W. Bridge structure dynamic analysis under vessel impact loading considering soil-pile interaction and linear soil stiffness approximation. *Adv Civil Eng* 2019;2019:1–11.
169. Wang W, Morgenthal G. Dynamic analyses of square RC pier column subjected to barge impact using efficient models. *Eng Struct* 2017;151:20–32.

170. Wang W, Morgenthal G. Reliability analyses of RC bridge piers subjected to barge impact using efficient models. *Eng Struct* 2018;166:485–495.

171. Fan W, Yuan WC. Numerical simulation and analytical modeling of pile-supported structures subjected to ship collisions including soil-structure interaction. *Ocean Eng* 2014;91:11–27.

172. Fan W, Yuan W, Yang Z, Fan Q. Dynamic demand of bridge structure subjected to vessel impact using simplified interaction model. *J Bridge Eng* 2010;16(1):117–126.

173. Davidson MT, Consolazio GR, Getter DJ. Dynamic amplification of pier column internal forces due to barge–bridge collision. *Transp Res Rec* 2010;2172(1):11–22.

174. Jiang H, Wang J, Chorzepa MG, Zhao J. Numerical investigation of progressive collapse of a multispan continuous bridge subjected to vessel collision. *J Bridge Eng* 2017;22(5):04017008.

175. Madurapperuma MA, Wijeyewickrema AC. Response of reinforced concrete columns impacted by tsunami dispersed 20 and 40 shipping containers. *Eng Struct* 2013;56:1631–1644.

176. He S, Yan S, Deng Y, Liu W. Impact protection of bridge piers against rockfall. *B Eng Geol Environ* 2019;78(4):2671–2680.

177. Lu Y, Zhang L. Analysis of failure of a bridge foundation under rock impact. *Acta Geotech* 2012;7:57–68.

178. Harik IE, Shaaban AM, Gesund H, Valli GYS, Wang ST. United States Bridge failures, 1951-1988. *J Perform Constr Fac* 1990;4(4):272–277.

179. Wardhana K, Hadipriono FC. Analysis of recent bridge failures in the United States. *J Perform Constr Fac* 2003;17(3):144–150.

180. Staples AM. Pier protection. In LRFD Bridge Design Workshop, Gainesville, FL, 2007.

181. Al-Thairy H, Wang YC. An assessment of the current Eurocode 1 design methods for building structure steel columns under vehicle impact. *J Constr Steel Res* 2013;88:164–171.

182. Milner R, Grzebieta R, Zou R. Theoretical study of a motor vehicle-pole impact. In *Proceeding of Road Safety Research, Policing and Education Conference*, Monash University, Melbourne, VIC, Australia, 2001.

183. Vrouwenvelder T. Stochastic modelling of extreme action events in structural engineering. *Probab Eng Mech* 2000;15(1):109–117.

184. ACI 318-14. *Building code requirements for structural concrete and commentary*. American Concrete Institute, Farmington Hills, MI, 2014.

185. Buth CE, Williams WF, Brackin MS, Lord D, Geedipally SR, Abu-Odeh AY. Analysis of large truck collisions with bridge piers: phase 1. Report of Guidelines for Designing Bridge Piers and Abutments for Vehicle Collisions. Texas Transportation Institution, College Station, TX, 2010.

186. Zhang X, Hao H, Li C. Experimental investigation of the response of precast segmental columns subjected to impact loading. *Int J Impact Eng* 2016;95:105–124.

187. Do TV, Pham TM, Hao H. Impact response and capacity of precast concrete segmental versus monolithic bridge columns. *J Bridge Eng* 2019;24(6):04019050.

188. Wikipedia. Queen Isabella Causeway bridge disaster. <https://en.wikipedia.org/wiki/Queen_Isabella_Causeway>, 2001.

189. Wikipedia. I-40 bridge disaster. <http://en.wikipedia.org/wiki/I-40_bridge_disaster>, 2002.

190. Stpetecatalyst. <https://stpetecatalyst.com/remembering-the-skyway-bridge-disaster-39-years-later/>, 2019.

191. Consolazio GR, Cowan DR, Biggs A, Cook RA, Ansley M, Bollmann HT. Full-scale experimental measurement of barge impact loads on bridge piers. *Transp Res Rec* 2005;1936(1):80–93.

192. Consolazio GR, Cook RA, Lehr GB. Barge impact testing of the St. George Island causeway bridge. Phase I: feasibility study. Research Report No. BC-354 RPWO-23, Engineering and Industrial Experiment Station, University of Florida, Gainesville, FL, 2002.

193. Consolazio GR, Cook RA, Biggs AE, Cowan DR. Barge impact testing of the St. George Island causeway bridge. Phase II: design of instrumentation systems. Research Report No. BC-354 RPWO-56, Engineering and Industrial Experiment Station, University of Florida, Gainesville, FL, 2003.

194. Consolazio GR, Cook RA, McVay MC, Cowan D, Biggs A, Bui L. Barge impact testing of the St. George Island Causeway Bridge, Phase III: physical testing and data interpretation. Research Report No. BC-354 RPWO-76, Engineering and Industrial Experiment Station, University of Florida, Gainesville, FL, 2006.

195. McVay MC, Wasman SJ, Bullock PJ. Barge impact testing of St. George Island Causeway Bridge geotechnical investigation. Research Report No. BD-545 RPWO-05, Engineering and Industrial Experiment Station, University of Florida, Gainesville, FL, 2005.

196. Consolazio GR, McVay MC, Cowan DR, Davidson MT, Getter DJ. Development of improved bridge design provisions for barge impact loading. Research Report No. BD-545 RPWO-29. Dept of Civil and Coastal Engineering, University of Florida, Gainesville, FL, 2008.

197. Consolazio GR, Getter DJ, Davidson MT. A static analysis method for barge-impact design of bridges with consideration of dynamic amplification. Research Report No. BD-545 RPWO-85. Dept of Civil and Coastal Engineering, University of Florida, Gainesville, FL, 2009.

198. Consolazio GR, Davidson MT, Getter DJ. Vessel crushing and structural collapse relationships for bridge design. Research Report No. 2010/72908/74039. Dept of Civil and Coastal Engineering, University of Florida, Gainesville, FL, 2010.

199. Consolazio GR, Getter DJ, Kantrales GC. Validation and implementation of bridge design specifications for barge impact loading. Research Report No. BDK75-977-31. Dept of Civil and Coastal Engineering, University of Florida, Gainesville, FL, 2014.

200. Hendrix JL. Dynamic analysis techniques for quantifying bridge pier response to barge impact loads. Master of Engineering Thesis, University of Florida, Gainesville, FL, 2003.

201. Consolazio GR, Cowan DR. Numerically efficient dynamic analysis of barge collisions with bridge piers. *J Struct Eng* 2005;131(8):1256–1266.

202. Consolazio GR, Davidson MT, Cowan DR. Barge bow force-deformation relationships for barge–bridge collision analysis. *Transp Res Rec* 2009;2131:3–14.

203. Consolazio GR, Davidson MT. Simplified dynamic analysis of barge collision for bridge design. *Transp Res Rec* 2008;2050(1):13–25.

204. Fan W, Yuan WC, Zhou M. A nonlinear dynamic macro-element for demand assessment of bridge substructures subjected to ship collision. *J Zhejiang Univ Sci A* 2011;12(11):826–836.

205. Consolazio GR, Cowan DR. Nonlinear analysis of barge crush behavior and its relationship to impact resistant bridge design. *Comput Struct* 2003;81:547–557.

206. Yuan P, Harik IE. Equivalent barge and flotilla impact forces on bridge piers. *J Bridge Eng* 2010;15(5):523–532.

207. Getter D, Consolazio G. Relationships of barge bow force-deformation for bridge design: probabilistic consideration of oblique impact scenarios. *Transp Res Rec* 2011;2251(1):3–15.

208. Fan W, Yuan W. Ship bow force-deformation curves for ship-impact demand of bridges considering effect of pile-cap depth. *Shock Vib* 2014;2014:1–19.

209. Kantrales GC, Consolazio GR, Wagner D, Fallaha S. Experimental and analytical study of high-level barge deformation for barge–bridge collision design. *J Bridge Eng* 2016;21(2):04015039.

210. Luperi FJ, Pinto F. Structural behavior of barges in high energy collisions against bridge piers. *J Bridge Eng* 2016;21(2):04015049.

211. Sha Y, Hao H. A simplified approach for predicting bridge pier responses subjected to barge impact loading. *Adv Struct Eng* 2014;17(1):11–23.

212. Yuan P, Harik IE, Davidson MT. Multi-barge flotilla impact forces on bridges. Report No. KTC-08-13/SPR261-03-2F. Kentucky Transportation Center College of Engineering, University of Kentucky, Lexington, KY, 2008.
213. Consolazio GR, Hendrix JL, McVay MC, Williams ME, Bollmann HT. Prediction of pier response to barge impacts with design-oriented dynamic finite element analysis. *Transp Res Rec* 2004;1868(1):177–189.
214. Fan W, Liu Y, Liu B, Guo W. Dynamic ship-impact load on bridge structures emphasizing shock spectrum approximation. *J Bridge Eng* 2015;21(10):04016057.
215. Fan W, Zhang Y, Liu B. Modal combination rule for shock spectrum analysis of bridge structures subjected to barge collisions. *J Eng Mech* 2016;142(2):04015083.
216. Fan W, Yuan WC. Shock spectrum analysis method for dynamic demand of bridge structures subjected to barge collisions. *Comput Struct* 2012;90:1–12.
217. Davidson MT, Consolazio GR, Getter DJ, Shah FD. Probability of collapse expression for bridges subject to barge collision. *J Bridge Eng* 2013;18(4):287–296.
218. Cheng J. Reliability analysis of the Sutong Bridge Tower under ship impact loading. *Struct Infrastruct E* 2014;10(10):1320–1329.
219. Shao JH, Zhao RD, Geng B. Probabilistic analysis of bridge collapse during ship collisions based on reliability theory. *J Highway Transp Res Dev* 2015;9(1):55–62.
220. Koh HM, Lim JH, Kim H, Yi J, Park W, Song J. Reliability-based structural design framework against accidental loads–ship collision. *Struct Infrastruct E* 2017;13(1):171–180.
221. Kameshwar S, Padgett JE. Response and fragility assessment of bridge columns subjected to barge-bridge collision and scour. *Eng Struct* 2018;168:308–319.
222. Getter DJ, Consolazio GR, Davidson MT. Equivalent static analysis method for barge impact-resistant bridge design. *J Bridge Eng* 2011;16(6):718–727.
223. Bui LH. Static versus dynamic structural response of bridge piers to barge collision loads. PhD Thesis, University of Florida, Gainesville, FL, 2005.
224. McVay MC, Wasman SJ, Consolazio GR, Bullock PJ, Cowan DG, Bollmann HT. Dynamic soil–structure interaction of bridge substructure subject to vessel impact. *J Bridge Eng* 2009;14(1):7–16.
225. Aziz HY, Yong HY, Mauls BH. Dynamic response of bridge-ship collision considering pile-soil interaction. *Civil Eng J* 2017;3(10):965–971.
226. Wang W, Morgenthal G. Parametric studies of pile-supported protective structures subjected to barge impact using simplified models. *Marine Struct* 2019;63:138–152.
227. Gholipour G, Zhang C, Mousavi AA. Nonlinear numerical analysis and progressive damage assessment of a cable-stayed bridge pier subjected to ship collision. *Mar Struct* 2020;69:102662.
228. Tian L, Huang F. Numerical simulation for progressive collapse of continuous girder bridge subjected to ship impact. *Trans Tianjin Univ* 2014;20(4):250–256.

Chapter 3

State-of-the-art review on concrete structures subjected to blast loads

3.1 Introduction

The need for recognizing the response mechanisms of concrete structures and exploring protective solutions from the design viewpoint has been raised in recent years with the increasing risk of applied blast loads to concrete structures during accidental events or intentional terrorist attacks. As has been reported in the literature, many people lost their lives during explosions and terrorist attacks on strategic buildings [1] or bridges [2, 3]. This section introduces the theoretical background of blast load, existing provisions in the prediction of blast loads in the current design codes, available techniques to analyze the target structures under explosions, and typical damage mode and failure behaviors of various concrete structures.

3.1.1 Theoretical background

Blast loading is an extreme loading type that would generate very high strain rates in the ranges from 10^2 s^{-1} to 10^4 s^{-1}. The ranges of strain rate produced by different types of loading are illustrated in Figure 3.1.

Figure 3.2 shows a typical plot of blast wave pressure-time history arising from a free-air burst, which includes two impulse phases on a target structure. According to this plot, the shock wave arrives at the shock front surface of a target structure at t_a. Then, the blast pressure wave propagates in the structure as a positive impulse phase during $t_a + t_o$ and with a peak overpressure of P_{so}. Afterward, the blast wave is reflected from the boundaries and tensile zone of the structure during a negative impulse phase until $t_a + t_o + t_o^-$ which results in suction forces [5]. The characteristics of this plot are profoundly dependent on blast-loading parameters such as shape and weight of explosive, standoff distance and the height of explosion from the ground, and also on structural parameters such as the geometry, weight, and size of the target.

The most common criterion to classify blast loadings based on Hopkinson-Cranz law [6] is known as the scaled distance (Z) of detonation from the target defining a relationship between equivalent weight (W_{TNT}) and standoff distance (R) of explosive charge as follows:

$$Z = \frac{R}{W_{TNT}^{1/3}} \qquad (3.1)$$

where R is the standoff distance, and W_{TNT} is the equivalent weight of charge.

DOI: 10.1201/9781003262343-3

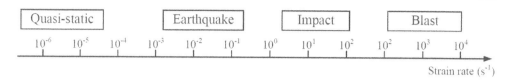

Figure 3.1 The ranges of strain rates produced by different types of loading [4].

Based on the scaled distance parameter, blast loads can be categorized into near-field and far-field detonations as shown in Figure 3.3. Under a far-field detonation, the reflected blast pressure is applied to the target structure as a uniformly distributed loading. However, the shape of distributed blast loading tends to be more concentrated around explosion effective area with decreasing the scaled distance for near-field detonations. There exist different definitions in classifying blast loads on scaled distance. According to the American Society of Civil Engineers (ASCE SEI 59-11) [7], blast loads with scaled distances less than 1.2 m/kg$^{1/3}$ are identified as close-in detonations. There is another criterion defined by Gel'fandetal et al. [8] based on the dimension of the charge r_0. Based on this criterion, blast loads with standoff distance R between the ranges 0 and $20r_0$ ($0 < R < 20r_0$) are recognized as close-in detonations. In addition, UFC 3-340-02 [9] considered a scaled distance equal to 0.4 m/kg$^{1/3}$ as the sensitive level for scaling blast loads. As such, blast loads with a scaled distance less than this value were considered as close-in explosions.

Blast loading types can be also classified based on the inactions and relationships between blast loading pressure and its impulse response on target. These diagrams present vital information about blast loading duration versus its amplitude, which can be very helpful in assessing the blast responses of structures [10, 11] and considering recommendations on protective design [7, 12–24]. Figure 3.4 shows a typical pressure-impulse diagram with two asymptotes for each of pressure and impulse parameters, which are divided into three regions including impulsive, dynamic, and quasi-static. According to this classification, blast loads with a very short duration relative to the natural period of the structure are categorized into the impulsive region in which the structure responses are sensitive only to the associated impulse [15].

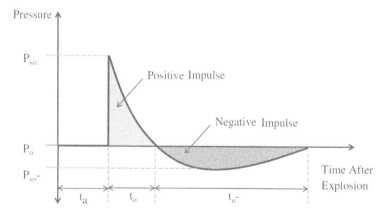

Figure 3.2 A typical plot of free-field blast pressure-time history [5].

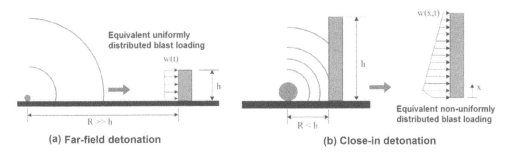

(a) Far-field detonation   (b) Close-in detonation

Figure 3.3 Reflected pressures on the target structures regarding the scaled distance of blast loading [4].

Besides, blast loads with a duration almost equal to the natural period are categorized into a dynamic region in which structural responses are sensitive to both pressure and impulse factors. The quasi-static region represents a blast loading type with a longer duration relative to the natural period of the structure in which the structural responses are only dependent on the blast pressure.

3.1.2 Analysis of structures subjected to blast loads

Methods for analysis of the structural responses to blast loads can be categorized into three main approaches including simplified analytical methods, numerical simulations, and experimental tests. The effects of blast loads acting on different structures have been formulated in the previous works [24, 25] by simplifying both reflected pressure and target systems. Figure 3.5a–e shows examples of simplified analytical models under idealized blast pressures using an equivalent single-degree-of-freedom (SDOF) and multi-degree-of-freedom (MDOF) systems. To determine the resistance and functions and load-transformation factors to use in simplified SDOF and MDOF systems, most of the analytical studies considered idealized continuous systems, and lumped inelasticity models as shown in Figure 3.5d and e.

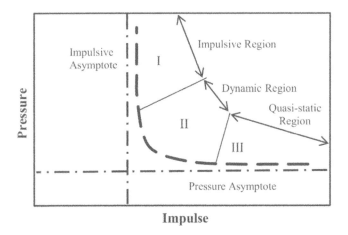

Figure 3.4 A typical pressure-impulse diagram [15].

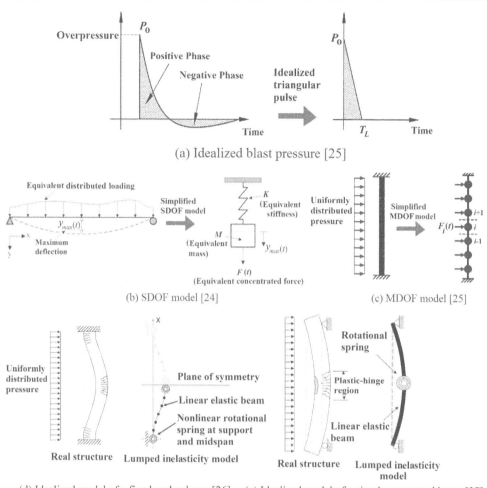

(a) Idealized blast pressure [25]

(b) SDOF model [24]

(c) MDOF model [25]

(d) Idealized model of a fixed-end column [26]   (e) Idealized model of a simply-supported beam [27]

Figure 3.5 Examples of simplified models adopted by different analytical studies. (a) Idealized blast pressure [25], (b) SDOF model [24], (c) MDOF model [25], (d) idealized model of a fixed-end column [26], and (e) idealized model of a simply-supported beam [27].

Although the accuracy of analytical methods in predicting global responses of concrete structures under blast loads was obtained in the previous studies [24–27], such approaches are not able to capture localized damages of structures such as concrete spalling. Therefore, the need for finite element (FE) numerical and experimental investigations of RC structures was realized.

To model blast loads using FE numerical methods, several software codes such as ABAQUS [28], LS-DYNA [29], AUTODYN [30] have been developed recently. In LS-DYNA, there exist two methods to simulate blast loads on structures: load-blast-enhanced (LBE) and multi-material Arbitrary Lagrangian-Eulerian (MM-ALE) [31]. In the first method, the blast pressure is directly applied to the shock-front surface of the structure. However, in the second method, the blast pressure is carried out by using an FE medium from the explosive to the structure [32, 33].

To simulate blast loads on structures, there exist the following approaches utilizing different facilities::

- Blast loads by air burst using explosive charges (Figure 3.6a): In this method, the structure is exposed to an explosive charge, such as trinitrotoluene (TNT), which can be different in chemical compounds, geometry, weight, and size.
- Blast loads by shock tubes (Figure 3.6b): Using shock tube facilities, the shockwaves are generated by a compression chamber. Then, they were released toward the target by using a variable-length deriver. The firing of the shockwave is controlled by using a differential pressure diaphragm in a spool section. Finally, the shockwaves with desired pressure travel toward a rigid end frame using an expansion section that has been attached to the shock-front surface of the target structure.
- Blast loads by the University of California, San Diego (UCSD) simulator (Figure 3.6c): The UCSD developed an explosive loading facility in which distributed blast load is simulated using hydraulic-based actuators and applied

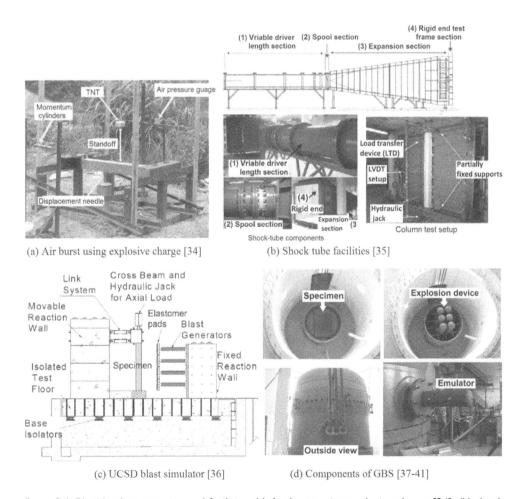

(a) Air burst using explosive charge [34]   (b) Shock tube facilities [35]

(c) UCSD blast simulator [36]   (d) Components of GBS [37–41]

Figure 3.6 Blast loading experimental facilities. (a) Air burst using explosive charge [34], (b) shock tube facilities [35], (c) UCSD blast simulator [36], and (d) components of GBS [37–41].

as the distributed impact loads to the structure through the elastomer pads attached to actuators' head.

- Blast loads by Gas Blast Simulator (GBS) (Figure 3.6d): This device was developed by the Anti-Explosion and Protective Engineering Ministry Key Laboratory at Harbin Institute of Technology, China, in which specimen is placed inside a multifunctional system subjected to a series of gas blast shock waves.

3.1.3 Typical damage modes of RC structures subjected to blast loads

Damage levels of RC structures resulted from blast loads can be classified based on the detonation scaled distance factor. Based on this classification, concrete structures under contact or very near-field explosions can undergo different severity levels of localized spalling at the back surface (i.e., tensile zone) of the structure due to suction forces and tensile stresses when the blast waves are reflected from these boundaries as shown in Figure 3.7.

The concept of concrete spallation in RC structures under explosions proposed by McVay [43]. Accordingly, the intensity of spall damages in concrete structures due to the reflected stress waves from the tensile zone of the structure can be classified as given in Table 3.1. Also, under near-field detonations, the combinations of localized spall damages with global deformations of RC structures are likely. Besides, when RC structures are subjected to uniformly distributed blast loads arising from far-field detonations, observing global flexural and tensile damages due to ductile response modes is likely. A classification on the flexural failure modes of RC columns based on the displacement-ductility ratio presented by McVay is given in Table 3.2 [43].

There exist many studies in the literature in investigating the damage states and failure behaviors of RC structural members such as beams [34, 44–51], columns [52–56], and slabs [57–61] under blast loads. The blast responses of large-scale structures and infrastructures such as framed buildings [62, 63] and bridges [64–73] have been widely reported in the literature. Damage levels of RC structures resulted from blast loads can be classified based on the detonation-scaled distance factor as given in Table 3.3. According to this classification, concrete structures under contact or very near-field explosions would undergo different severe localized spalling, the combinations of localized spall damages with global deformations under near-field detonations, or global flexural damages due to ductile response modes under far-field detonations.

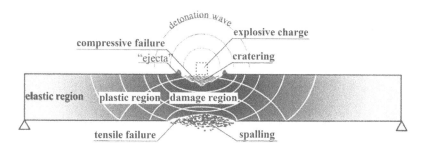

Figure 3.7 Localized failures in an RC structure under contact detonation [42].

Table 3.1 Spalling damage classifications [43]

Damage state	Damage description	Scheme of damage
No damage	From no change in the condition of the wall to a few barely visible cracks.	
Threshold spall	From a few cracks and a hollow sound to a large bulge in the concrete with a few small pieces on the floor.	
Medium spall	From a very shallow spall to a third of the wall thickness.	
Severe spall	From just over one-third the wall thickness to almost breach.	
Breach	From a small hole which barely lets light through to a large hole.	

Table 3.2 Flexural failure modes based on the displacement-ductility ratio [43]

Failure mode	Damage description	Scheme of damage
Light flexure	From no permanent displacement but a few flexural cracks to a ductility ratio of 3.	
Medium flexure	From a ductility ratio of 3–10.	
Severe flexure	From a ductility ratio of 10 to almost breach.	

Note: The ductility ratio represents the ratio of the column maximum mid-span displacement to the first yield displacement in the load-displacement curve.

3.1.4 Design codes for structures against blast loads

Despite many guidelines for the design of buildings against blast loads such as the US Department of the Army [5, 16, 17], US Department of Defense [9, 17, 18], US General Services Administration (GSA) [18], FEMA [12, 13], and American Society of Civil Engineers (ASCE) [7, 20, 21], there exist very limited design codes presenting necessary recommendations for blast resistance of bridge components. However, some design information based on the simplified approaches has been given by the National Cooperative Highway Research Program (NCHRP) [22, 23] in which the failure behaviors of bridges under explosions have not been included. Also, the state-of-the-art guidance on security planning, blast phenomenology, blast response mechanism of bridge structural members, material performance, and protective design recommendations was presented by Federal Highway Administration (FHWA) [3, 24].

Table 3.3 Typical failure behaviors of different RC members under different blast loads

Failure mode	Damage classifications based on scaled distance		
	Severe spallation (under contact and very close-in blasts)	*Combination of localized and global failure (under close-in blasts)*	*Global flexural damages (under far-field blasts)*
Column	[22, 64]	[56, 22]	[71]
Beam	[50]	[45]	[51]
Slab	[59]	[60]	[61]

65 cm

Table 3.4 Summary of the current guidelines for blast loading on structures

Guideline	Remarks and notes
TM-5-855-1 [5]	Provides design and analysis procedures for the protective structures exposed to the effects of conventional weapons and for use in designing hardened facilities
TM-5-1300 [16]	Provides design approaches for structures to resist the effects of blast waves and fragments by considering blast load parameters and structural response modes
UFC 3-340-02 [9]	Prediction of idealized close-in and far-field blast loads by using shock and gas considering dynamic increase factors (DIFs), which provide both flexural and shear failure-based design approaches
UFC 4-010-01 [18]	Provides appropriate, implementable, and enforceable measures to establish a level of protection against terrorist attacks for all depart of defense and military buildings
FEMA 427 [12]	Provides an extensive qualitative design approach to mitigate the effects of terrorist attacks by explosions and considering chemical, biological, and radiological attacks
FEMA 428 [13]	Predicts the expected overpressure on buildings by using explosive weight and stand-off distance in both horizontal and vertical distances arising from various vehicles' explosions
ASCE 1997 [20]	Provides a structural design guideline for blast resistance of petrochemical facilities
ASCE/SEI 59-11 [7]	Considers dynamic increasing factors for structures for only far-range blast loads using single-degree-of-freedom (SDOF) analysis, which provide flexural failure-based design approaches
ASCE (7-10) [21]	Provides the concepts and analysis methods of progressive collapse of integrated and redundant structural systems under explosions
NCHRP 12-72 [23]	Provides effective methods, structural design, and retrofit guidelines to mitigate the risk of terrorist attacks against critical bridges

Table 3.4 lists a summary of existing recommendations in several design guidelines for blast design of structures mainly based on SDOF analyses.

3.2 Structural columns subjected to blast loads

Axially loaded RC columns commonly used in low- and medium-rise buildings are always at risk of blast loads and bomb explosions in public places near the buildings. Due to the low redundancy of column members compared to other structural members such as beams and slabs investigating the responses, residual capacities, and vulnerability of such components against explosions is very important to mitigate the human casualties and damages. Hence, the vast majority of blast studies have been devoted to investigating the blast responses and of RC columns analytically [25, 53, 74–76], numerically [49, 55, 77–83], and experimentally [56, 64, 84–87].

Astarlioglu et al. [53] proposed an advanced SDOF model considering both flexural and shear behaviors of RC columns subjected to combined axial and transverse blast loadings. The accuracy of this method was examined by comparing the results between those from the SDOF model and FE simulations. By investigating the responses of the RC columns under combined axial-quasi-static loadings compared to those under combined axial-blast loadings, it was found that the axial load had significant negative effects on the resistance of RC columns under axial-blast loading even in its small ranges. El-Dakhakhni et al. [25] studied the blast responses and pressure-impulse diagrams of RC columns by developing a MDOF model under idealized triangular pulses considering high-strain rate effects of the materials. Besides, the effects of the axial load were taken into account by adopting an additional moment in the model. Structural damage levels of idealized columns were assessed by developing pressure-impulse curves for different levels of axial load. Yu et al. [74] studied different failure modes of non-dimensional analytical models of RC columns including flexural failure, shear failure, and combined flexural-shear failure through pressure-impulse diagrams under idealized triangular pulses. Substantial influences of the tangent modulus of direct shear-slip relation on both shear and combined flexural-shear failure modes were concluded. However, it had marginal effects on the pure flexural failure modes. Fallah and Louca [75] used an SDOF model by considering plastic softening and hardening behaviors of RC components to categorize the responses of the structure into elastic, elastic-plastic hardening, elastic-plastic softening, rigid-plastic-hardening, and rigid-plastic-softening by using pressure-impulse diagrams. The negative influences of hardening and softening indices were obtained on pressure-impulse diagrams.

Due to the deficiencies of simplified analytical methods in describing the damage levels of concrete structures based on their deformation responses, a damage index based on the residual load-carrying capacity of RC columns under blast loads was defined by Shi et al. [55]. Also, a simplified method was proposed to generate the pressure-impulse (*P-I*) diagrams based on the numerical simulations. Based on FE simulations, different failure modes of typical RC columns under three different types of blast load were classified by Mutalib and Hao [77] based on pressure-impulse diagrams as shown in Figure 3.8a and b. In Figure 3.8, when the column is subjected to an impulsive blast loading, the occurrence of shear failure is likely. However, the column under a quasi-static blast loading fails in flexural mode. Besides, columns suffered a combination of shear and flexural failure modes under dynamic blast load having the duration between those of the aforementioned loading types. The influences of several structural parameters including the column dimension, concrete strength, longitudinal, and transverse reinforcement ratio on the pressure-impulse diagrams were evaluated

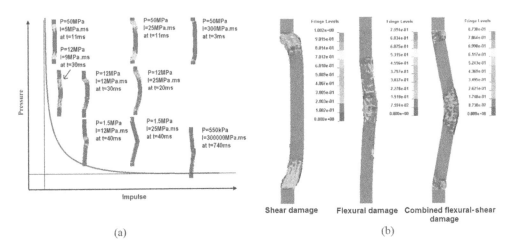

Figure 3.8 Damage modes of RC columns under different blast loads (a) damage classifications using pressure-impulse diagram [74] and (b) typical failure modes of RC columns [55].

by Mutalib and Hao [77] through a parametric study. Despite the positive effects of the column depth, concrete strength, and reinforcement ratio on the column resistance, increasing the column height caused the increase of the damage level in the column under identical blast loading. In addition, a marginal positive influence of the column width on the blast resistance of columns was concluded compared to the significant effects of the column depth, which substantially affects the cross-section moment of inertia. Although the increase of the column width caused the enhancement of the cross-section resistance, the larger front-face area of the column was exposed to uniformly distributed blast loading. Mutalib and Hao [77] numerically assessed the damage levels of fiber-reinforced polymer (FRP) strengthened RC columns under different types of blast loadings categorized by pressure-impulse diagrams and the influences structural factors on pressure-impulse diagrams through a parametric study by utilizing a similar analysis approach employed by Shi et al. [55].

The influences of structural- and loading-related parameters have been evaluated by many research works and parametric studies in the literature. Bao and Li [53] numerically evaluated the residual axial resistance of RC columns under a close-in blast load varying in terms of various parameters including the ratio of transverse and longitudinal reinforcements, column height, and the axial load ratio (ALR). The positive effects of the ALR on the column resistance were concluded when the column deformations were in small ranges. However, after a critical value, enhancing the ALR led to an increase of the column lateral displacements and consequently decreasing the residual resistance. In addition, the residual axial capacity of the column increased with decreasing the column height and increasing the ratio of both transverse and longitudinal reinforcements. Kyei and Braimah [79] numerically investigated the effects of transverse reinforcement spacing and axial loading on the behaviors of typical RC columns commonly used in buildings under close-in and rather far-standoff blast loads by using LBE method in LS-DYNA. The significant influences of these parameters on the failure behaviors and the responses of the columns were concluded when the standoff distance of the explosive was relatively close. As such, decreasing the spaces of transverse reinforcement and ALR on the columns in the service ranges (i.e., the gravity loads arising from the upper stories of buildings) increased the column lateral resistance against close-in blast loads. However, high ranges of ALR indicated negative influences on the column lateral resistance. The spall damage levels of RC columns by varying the dimensions and reinforcement configurations under close-in blast loads were numerically investigated by Li and Hao [54]. Substantial influences of column depth and the reinforcements in the mitigation of the column spalling were captured. However, the marginal influence of the column height governing the flexural stiffness of the column on the concrete spallation level was concluded. By assessing the circular RC columns numerically in LS-DYNA by Liu et al. [88] under close-in blast loads applied to the columns at different heights, it was revealed that although the columns suffered different damage forms, global flexural deformations and failure modes were predominant on the column responses. These flexural deformations were amplified by increasing the mass of the explosive. The effects of transverse reinforcement, ALR, and the scaled distances of the explosion in close ranges for less than 1.0 $m/kg^{1/3}$ (i.e., 0.22, 0.54, and 0.86 $m/kg^{1/3}$) on the failure modes of RC columns experimentally were evaluated by Siba [86]. The positive influences of transverse reinforcement on the resistance of the columns were revealed. Yuan et al. [64] numerically and experimentally studied RC columns with different sizes and shapes subjected to contact blast loads.

The localized severe damages and spallation around the explosive were observed in comparison with insignificant global responses. Moreover, it was declared that square-shaped columns suffer more severe damages than circular-shaped columns. Also, it was concluded that the responses of columns could be significantly different under contact blast compared to those under even very close-in explosions. The effectiveness of various parameters (mainly structural-related parameters) on the blast resistance of RC columns has been summarized in Table 3.5. Compared to mainly positive influences of structural parameters on the blast resistance of columns, the axial load parameter presents different trends of effectiveness from the conclusions of various studies. As such, no specific sensitivity levels have been reported in the literature indicating the effectiveness of this parameter on the blast responses of RC columns. Hence, more

Table 3.5 Influence of various parameters on the blast response of RC columns

Study	Analysis	Parameter	Effectiveness on the blast resistance
Kyei and Braimah [79]	Numerical	• Transverse reinforcement spacing • Axial load ratio	• Negative on lateral resistance • Negative on lateral resistance (within the service levels)
Bao and Li [52]	Numerical	• Reinforcement ratio • Column height • Axial load ratio	• Positive on axial resistance • Negative on axial resistance • Positive on lateral resistance in small deformations; negative on lateral resistance after a critical value
Astarlioglu et al. [53] Li and Hao [54]	Analytical Numerical	• Axial load • Column depth • Column height • Reinforcement ratio	• Negative • Substantial positive • Marginal positive • Substantial positive
Cui et al. [80]	Numerical	• Cross-sectional dimensions • Stirrup ratio	• Positive • Negative
Wu et al. [85]	Numerical	• Column depth • Column height • Reinforcement ratio • Explosive location (height from the base) • Axial load ratio	• Positive • No influence • Positive • Changing the column failure mode from the flexural mode to shear failure at the base support by reducing the height of the charge • Positive between the ratios of 0.2 and 0.4 (i.e., in the service levels)
Siba [86]	Experimental	• Transverse reinforcement	• Positive
Liu et al. [88]	Numerical	• Explosive location (height from the base) • Mass of explosive	• Different failure modes • Positive on the flexural deformations
Yan [89]	Numerical	• Concrete strength • Reinforcement ratio	• Positive • Positive

attempts are required for future works to fill this gap of knowledge. Among the various parameters, it is found that although considering the axial load between its service levels, both axial load and the column height parameters do not present any certain trend of effectiveness on the blast resistance of RC columns.

A parametric study on the residual capacity of axially-loaded RC columns under close-in blast loads was carried out by Wu et al. [85] varying in terms of the structural-related parameters including the column depth, height, and the ratio of transverse and longitudinal reinforcements, and also in terms of loading-related conditions including of explosive location, and the ALR based on FE numerical simulations. Increasing the axial, flexural, and shear resistance of RC columns was revealed in proportion to the increase of the column depth, and the ratio of longitudinal and transverse reinforcements. However, the residual capacity and damage state of the column under blast load were not affected by the column height. Moreover, the enhancement of residual axial capacity and shear resistance of the column was concluded by increasing the ALR from 0.2 to 0.4 in the ranges of service axial loads. Also, it was observed that the column failure mode changed from a flexural mode to brittle shear failure mode as illustrated in Figure 3.9 and the residual capacity of the column was reduced by decreasing the height of explosive location.

Similar to the damage index proposed by Shi et al. [55], Yan [89] as utilized a damage index (*DI*) defined based the column residual axial load carrying capacity as follows:

$$DI = 1 - \frac{P_r - P_L}{P_N - P_L} \tag{3.2}$$

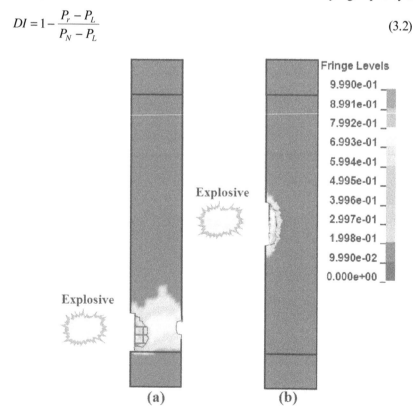

Figure 3.9 Failure modes of the RC column under blast loads varying in the height of explosive location [85].

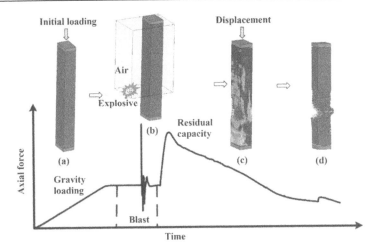

Figure 3.10 Loading steps to capture the residual axial load capacity of RC columns [89].

where P_N denotes the column bearing capacity before the damage, P_L represents the initial axial stress, and P_r is the column residual bearing capacity after damage.

The residual axial load capacity under blast loading was captured by using four loading steps proposed as follow:

- First, the axial load is applied to the column top end to reach the desired service level (arising from gravity loads of upper structures) during a stress initialization phase before the onset of blast loading (Figure 3.10a).
- The blast load is laterally applied to the column by using the ALE method in LS-DYAN (Figure 3.10b).
- The dynamic responses of the column are calculated until reaching a stable state (Figure 3.10c).
- A displacement-based axial load is gradually applied to the top end of the column to capture the residual capacity of the column (Figure 3.10d).

Accordingly, the damage levels of axially-loaded subway columns were assessed utilizing the proposed damage index under different close-in blast loads by varying the concrete strength and the ratio of longitudinal and transverse reinforcements. The bearing capacity of the column was enhanced as the ratio of reinforcements and the concrete strength increased. Moreover, the damage levels of the column under explosion based on the damage index values were classified as given in Table 3.6.

Cui et al. [80] proposed a damage index based on the ratio of column residual deflection relative to the column depth to investigate the failure behaviors of RC columns under close-in by using FE simulations. By evaluating the influences of different parameters on the damage levels of columns, it was found that unlike positive effects of the column cross-sectional dimensions on the blast resistance of RC columns, stirrup spacing and the thickness of concrete cover had negative influences. Also, mainly predominant local damages as the spalling of the concrete at the back surface of the column under close-in blast loads was concluded.

Table 3.6 Damage states of RC columns described by the proposed damage index [89]

Damage index	Damage	Remarks
0–0.2	Minor damage	No visible cracks and localized concrete damage, reinforcement intact, and no large lateral deformations and full residual axial capacity
0.2–0.5	Moderate damage	Minor-to-moderate flexural and diagonal shear cracks, but localized concrete damage, reinforcement intact, and no large lateral deformations, and almost full residual axial capacity
0.5–0.8	Terrible damage	Severe concrete cracking and spallation with a direct shear phenomenon, main reinforcement intact, ruptured stirrups, large lateral deformations, and low residual axial capacity
0.8–1.0	Collapse	Column severely damaged and collapsed (blown off or total compression failure) and zero residual axial capacity

For the protective design of concrete columns, numerous research works attempted to find optimal protection and strengthening solutions of concrete columns and mitigating the blast shock waves. To this end, some studies focused on designing composite columns to protect the core concrete against localized concrete spallation such as concrete-filled steel tubes (CFSTs) [90–92], concrete-filled double skin steel tube (CFDST) [14, 93–95], and concrete-steel composite (CSC) columns [78, 96, 97] under blast loads to enhance the global and residual resistance of columns. Wang et al. [91] found that the use of steel tubes in circular CFST columns had more positive influences on the axial resistance of the columns than their lateral resistance under blast loads. Zhang et al. [92] revealed that the increase of the axial load on CFST columns significantly reduced the severity of localized damage and shear failure under close-range explosions. A critical ALR of 0.16 was identified by Zhang et al. [14] as the sensitivity threshold of the columns' lateral resistance. As such, the columns experienced notable deflections when exposed to axial loads ratios beyond this value. From an experimental study on the performance of CFDST column under close-in blast loads by Li et al. [95], it was obtained that although the surrounding steel tube prevented the spall damage in the core concrete due to its confinement function, the core concrete absorbed the main part of explosion energy.

As another alternative to enhance the strength of concrete structures under explosions, many studies assessed the performance of various retrofitting composite materials such as FRP [98], carbon FRP (CFRP) [99], glass FRP (GFRP) [27], ultra-high performance concrete (UHPC) [100, 101], ultra-high-performance fiber-reinforced concrete (UHPFRC) [35, 102, 103], and steel fiber-reinforced concrete (SFRC) [26]. By utilizing longitudinal GFRP sheets in the retrofitting of RC columns by Jacques et al. [27], the peak displacement of columns was significantly decreased. Besides, employing additional transverse sheets provided more confinement for longitudinal sheets and reinforcements and prevented concrete cover spalling. Aoude et al. [35] concluded that the increase of fiber content until a range of 4% in the UHPFRC column significantly enhanced the blast resistance of columns. In addition, the strain rate effects of UHPFRC materials on the blast responses of retrofitted columns were efficiently formulated by using a simplified analysis with the SDOF model. Astarlioglu and Krauthammer [103] found that the UHPFRC decreased by about 30% of the average displacements of columns and increased the blast resistance of the column more than four times. From an experimental study on the blast responses of the SFRC column carried out by Burrell et al. [26], the inability of SFRC in the prevention of compression rebar buckling was concluded despite significant enhancement in the blast resistance of columns.

Moreover, the combination of the aforementioned approaches may be used as a protection solution such as UHPC-CFDST [93] and FRP-concrete-steel double-skin tubular columns (DSTCs) [104] against blast loads. Examples of damage mitigations observed in retrofitted columns subjected to blast loads under blast loads through adopting different strengthening methods are presented in Table 3.7. Zhang et al. [14] concluded that the use of UHPC instead of normal strength concrete as the filler in CFDST substantially increased the lateral resistance of columns. From a series of

Table 3.7 Influence of different retrofitting methods on the damage mitigation of RC columns under blast loads

Study	Damage mode of non-retrofitted columns	Damage mode of retrofitted columns
Crawford [108]		 Steel-jacketed CFRP-wrapped
Mutalib and Hao [77]	 Shear failure Flexural failure	 Longitudinal strips FRP-wrapped FRP and strips Longitudinal strips FRP-wrapped FRP and strips Against shear failure Against flexural failure

(Continued)

Table 3.7 Influence of different retrofitting methods on the damage mitigation of RC columns under blast loads (Continued)

Study	Damage mode of non-retrofitted columns	Damage mode of retrofitted columns
Burrell et al. [26]	Series-1 Series-2	Series-1 Series-2 — SFRC-strengthened
Aoude et al. [35]	Series-1 Series-2	Series-1 Series-2 — UHPFRC-strengthened

experimental tests on the blast resistance of UHPC-CFDST columns done by Zhang et al. [93], the predominance of global flexural response was obtained by using UHPC instead of normal strength concrete, which prevented the crushing of core concrete and buckling of steel jacket. Wang et al. [104] numerically evaluated the blast responses of a novel DSTC composed of an inner steel tube and an outer FRP tube. Compared to the marginal influence of the thickness of the outer FRP tube on blast resistance of columns, significant positive effects of the thickness of the inner steel tube were concluded. In some cases, FRP-based composites are utilized as the encasing and jacketing tubes to protect the core concrete of columns such as concrete-filled FRP tubes (CFFTs) [105–108]. Rodriguez-Nikl et al. [106] found that the use of FRP tubes changed the failure mode of columns from brittle shear to ductile flexure and significantly enhanced the load and displacement capacities. Besides, Qasrawi et al. [107] revealed the maximum allowable displacement rages from 0.211 to 0.261 m as the threshold level for energy absorbing capacity of FRP-jacketed columns. Table 3.8 summarizes the influences of various loading- and structural-related parameters on the blast resistance of retrofitted columns studied by previous works.

Table 3.8 Influence of various parameters on the blast resistance of retrofitted columns

Study and retrofitting approach	Analysis	Parameter	Effectiveness on the blast resistance
Mutalib and Hao [77] (FRP composite)	Numerical	• Concrete strength	• Significant positive on shear resistance
		• Column height	• Negative on flexural resistance
		• Column width	• Marginal positive
		• FRP thickness	• Significant positive
		• FRP strength	• Positive on shear and flexural resistances
Zhang et al. [90] (CFST)	Numerical	• Axial load	• Slight positive on flexural resistance
Wang et al. [91] (CFST)	Experimental	• Explosive charge weight	• Significant negative on lateral resistance
		• Thickness of steel tube	• Significant positive on axial resistance
		• Cross-section geometry	• More confinement effects of circular columns than those of square columns
Zhang et al. [92] (CFST)	Numerical	• Scaled distance	• Significant negative
		• Height of explosive charge	• More severe damage with decreasing the height of the charge
		• Explosive mass	• Negative
		• Column width	• Positive on both flexural and shear resistances
		• Steel ratio	• Positive on the confinement, flexural, and shear resistances
		• Axial load ratio	• Significant positive (especially on reducing the levels of localized and shear failures)
Zhang et al. [14] (CFDST)	Experimental	• Axial load ratio	• Marginal negative until a ratio of 0.16; Significant negative beyond 0.16 up to 0.5
		• Hollow section ratio	• Significant negative beyond a ratio of 0.5
		• Concrete strength	• Significant positive by reducing residual deflection
Zhang et al. [93] (UHPC-CFDST)	Experimental	• Explosive charge weight	• Significant positive on mid-span deflection
		• Axial load	• Slight negative on peak deflection
Wu et al. [96] (CSC)	Numerical	• Column depth	• Substantial positive
		• Column height	• No influence
		• Explosive location	• Column failure tends to a shear mode with decreasing the height of the location
		• Axial load	• Positive on shear resistance when it is increased from 0.2 to 0.4
		• Steel ratio	• Significant positive on residual axial capacity
Li et al. [95] (CFDST)	Experimental and Numerical	• Explosive charge weight	• Significant negative (positive on the severity of both localized and global failures)
Jacques et al. [27] (GFRP sheets)	Experimental	• GFRP longitudinal sheets	• Significant negative on peak displacement
		• GFRP transverse sheets	• Significant positive on confinement and spallation

(Continued)

Table 3.8 Influence of various parameters on the blast resistance of retrofitted columns (Continued)

Study and retrofitting approach	Analysis	Parameter	Effectiveness on the blast resistance
Li and Wu [101] (UHPC)	Numerical	• UHPC strength	• Positive on both shear and flexural resistance
		• Cross-section size	• Significant positive on flexural resistance
		• Reinforcement ratio	• Negative (positive on blast load effects)
		• Column height	
Aoude et al. [35] (UHPFRC)	Experimental	• Fiber content	• Significant positive until a range of 4%
Burrell et al. [26] (SFRC)	Experimental and Analytical	• Fiber type	• No influence from adopting high-strength fibers instead of normal fibers
Wang et al. [104] (FRP-concrete-steel DSTC)	Numerical	• Concrete strength	• Slight positive on tensile resistance
		• Scaled distance	• Significant negative (positive on peak and residual displacements)
		• FRP tube thickness	• Slight positive on peak and residual displacements; No influence on flexural resistance
		• Steel tube thickness	• Significant positive
		• Hollowness ratio	• Significant negative (positive on peak and residual displacements)
		• Axial load	• Significant negative beyond a ratio of 0.5
Qasrawi et al. [105] (CFFT)	Experimental	• Reinforcement ratio	• Significant negative on residual displacements
		• GFRP tube	• Significant positive
Rodriguez-Nikl et al. [106]	Experimental	• FRP tube	• Changed the column failure mode from brittle shear to ductile flexure

3.3 Beams and slabs subjected to blast loads

The damage behaviors and blast responses of RC slabs used in bridge superstructures and floors of buildings have been widely studied analytically [109–112], numerically [54, 113–117], and experimentally [57–61, 116, 118–120].

Low and Hao [110] studied the uncertainties of structural parameters of RC slabs under the uncertainties of blast loads using reliability analyses of an equivalent SDOF system. Slight influences of Young's modulus of elasticity of concrete and no influence of the length of slab on the resistance and failure behaviors of RC slabs under blast loads were concluded, respectively. Afterward, Low and Hao [111] used a coupled SDOF to model the flexural and direct shear responses of one-way RC slabs considering strain rate effects of materials.

Most of the numerical and experimental studies on the failure behaviors of RC slabs under explosions concluded the governing of localized spalling and punching failures under contact and close-in blast loads as shown in Figure 3.11a and b, respectively. Due to the high redundancy of slabs compared to columns and beams, such localized failures may not lead to the collapse of the whole structure. This fact was also demonstrated by the numerical simulation study by Tang and Hao [66, 67].

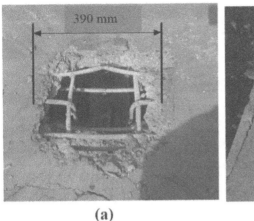

(a)   (b)

Figure 3.11 Localized spall damages in RC slabs subjected to (a) contact explosion [121] and (b) close-in blast load [59].

In attempts for rehabilitation and strengthening of concrete slabs against localized spall damages, several studies utilized some retrofitting materials such as fiber-reinforced concrete (FRC) [59], polyvinyl alcoholic fiber-reinforced concrete (PVAFRC), polyethylene fiber-reinforced concrete (PEFRC), and polypropylene fiber-reinforced concrete (PPFRC) [122], FRP [61, 123], CFRP [60], high-performance geopolymer concrete (HPGC) [124, 125], UHPC [120, 121], UHPFRC [126, 127] to enhance the blast performance of concrete slabs. Besides, the combination of high strength concrete (HSC) with high strength low alloy vanadium reinforcement (HSLA-VR) named (HSC-HSLA-VR) was utilized by Thiagarajan et al. [58] to enhance the blast resistance of slabs. Meng et al. [124] studied the performances of a novel composite material named high-performance alkali-activated geopolymer concrete (HPGC) under close-in blast loads. The HPGC substantially enhanced the blast resistance of slabs by increasing uniform strain distribution against detonation. Generally, the use of retrofitting materials such as fiber polymer enhances the resistance of concrete slabs especially against localized spallation and leads to global failure modes at supports. Table 3.9 lists examples of damage migration by adopting various composites in concrete slabs. In addition, the influences of various loading and structural parameters on the blast responses of concrete slabs have been summarized in Table 3.10.

Unlike the high redundancy of slabs, beams attached to slabs would play a more determinant role in the failure behaviors and the progressive collapse of a whole structure due to their vital role in transferring the distributed blast loading from slabs to columns. The blast responses and damage states of RC beams have been explored by many research studies analytically [128], numerically [38, 129–131], and experimentally [36, 37, 42–44, 97].

Simplified analytical models are unable to estimate the brittle damages of concrete structures under blast loads such as spallation. Besides, the use of accurate FE simulation methods with acceptable fine meshes under explosions consumes notable computational time and memory space in computers. To overcome such drawbacks of common simplified and FE simulation approaches in the estimation of the blast responses of structures, Li and Hao [47] proposed a two-step numerical method in which the responses of a structure were separately calculated during the loading and

Table 3.9 Influence of different retrofitting composite materials on the damage mitigation of slabs under blast loads

Study	Damage mode of non-retrofitted RC slabs	Damage mode of retrofitted slabs
Thiagarajan et al. [58]	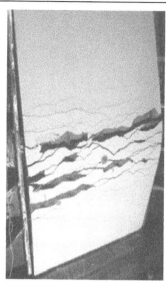	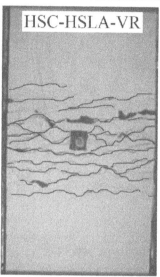
Foglar and Kovar [59]		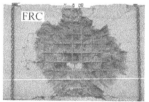
Wu et al. [60]		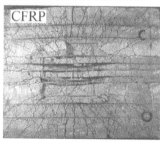
Zhou et al. [113]		

(Continued)

Table 3.9 Influence of different retrofitting composite materials on the damage mitigation of slabs under blast loads (Continued)

Study	Damage mode of non-retrofitted RC slabs	Damage mode of retrofitted slabs		
Ohtsu et al. [122]		PPFRC	PVAFRC	PEFRC
Meng et al. [124]		HPGC Crack		
Li et al. [121]	390 mm	UHPC 350 mm		
Foglar et al. [127]		UHPFRC		

free vibration steps. Moreover, the failure modes of the beams were studied varying in terms of the pressure of blast loading and the ratio of the loading duration (t_d) to the natural period time of the structure (T) (i.e., t_d/T). It was observed that the deflection mode of the beam changed from flexural elastic mode to combined flexural-shear mode with increasing of blast loading pressure. However, the domination of flexural failure mode on the beam responses increased with increasing of t_d/T under blast loads with the same pressures. Afterward, the influence of shear damages on the accuracy of the proposed method during the blast loading phase was numerically examined by Li and Hao [48]. Chen et al. [129] concluded that prestressed RC beams achieve higher blast resistances when the flexural modes are predominated on their responses.

Table 3.10 Influence of various parameters on the blast response of RC slabs

Study	Analysis	Parameter	Effectiveness
Foglar and Kovar [59]	Experimental	• Fibers	• Caused the change of failure modes from brittle failures to ductile modes
		• Concrete compressive strength	• Significant positive on the blast resistance
Dragos and Wu [109]	Numerical and Analytical	• Slab depth	• Significant effects on the pressure-impulse diagrams and shear failure modes
		• Span length	• No effects on the pressure-impulse diagrams and direct shear failure modes
		• Support conditions	• Simple supports provided larger values for pressure-impulse diagrams than those of fixed supports
Low and Hao [110]	Numerical and Analytical	• Young's modulus of elasticity of concrete	• Significant effects on the failure behaviors
		• Yield strength of rebars	• Marginal effects on the blast resistance
		• Strength of concrete	• Marginal effects on the blast resistance
		• Span length	• No influences on the failure modes
		• Slab thickness	• Positive on flexural resistance
		• Reinforcement ratio	• Positive on flexural resistance
Low and Hao [111]	Numerical and Analytical	• Span length	• The failure mode of slab tends to direct shear mode with decreasing the span length
Xu and Lu [114]	Numerical	• Boundary conditions	• No influence on concrete spalling for small standoff; Significant on global failure for large standoff
Jia et al. [115]	Numerical	• Explosive weight	• Significant positive on the damage level
		• Explosive position	• Damage level increases with moving from the center toward the boundaries
		• Boundary conditions	• Negative effects of simple supports
		• Length of negative reinforcement	• Positive effects on the blast resistance when the length is equal to the actual span
Tai et al. [116]	Numerical	• Reinforcement ratio	• The damage location tends to move from the center to the supports with increasing the ratio
		• Standoff distance	• Significant negative on the slab damage level
Wang et al. [118]	Experimental	• Explosive charge weight	• Changing the damage mode from flexural to punching failure with increasing charge weight
		• Slab size	• Larger slabs suffered more damages
Wang et al. [119]	Experimental and Numerical	• Amount of explosive charge	• Changing the failure modes from overall flexural failures to local punching modes with the increase of explosive charge
Li et al. [120]	Experimental and Numerical	• Compressive strength and steel fiber of UHPC	• Significantly reduce the spalling and punching damages
Mao et al. [126]	Experimental and Numerical	• Steel fiber	• Both have similar effects on blast resistances for far-field detonation; Steel bar has a significant positive effect on blast resistance of UHPFRC under close-in blast

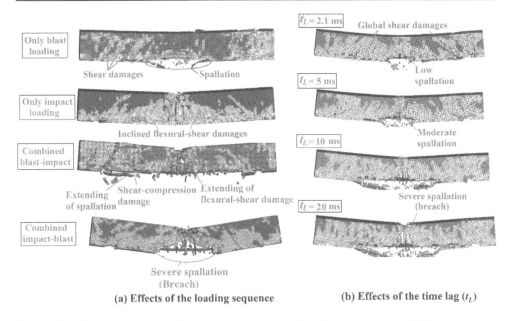

(a) Effects of the loading sequence   (b) Effects of the time lag (t_L)

Figure 3.12 Influences of the loading sequence and t_L on the failure behaviors of RC beams under combined actions of impact and blast loads [130].

In addition, it was found that although the increase of concrete strength and prestress level increases the flexural resistance of beams, it may enhance the occurrence possibility of shear damages around the beam supports.

From the investigation of RC beams subjected to the combination of impact and blast loads by Zhang et al. [130], different failure behaviors and more severe damage states were captured rather than those obtained under sole loadings. Owing to substantial discrepancies between the loading mechanisms of impact and blast, it was found that the accumulative impulse effects of the combined loads and the failure modes of RC beams were profoundly sensitive to the key loading parameters including the loading sequence, and the time lag (t_L) between the onsets of the applied loads as shown in Figure 3.12a and b, respectively. In order to quantify the damage levels of RC beams under combined loadings, a novel damage index was proposed based on the residual flexural capacity of RC beams as given in Equation (3.3) obtained through a three-step loading procedure in which the rotation-based flexural moments were applied to the ends of impact/blast-induced RC beam as shown in Figure 3.13. It was found that the beams suffered more severe spallation when the impact loadings were applied before blast loadings, and the time lag between the onsets of applied loads was increased. Afterward, Gholipour et al. [131] investigated the vulnerability of RC beams under the combined actions of impact and blast loads varying in terms of the impact loading rate (V), the loading sequence, and the time lag. It was observed that compared to the increase of the severity of the global shear failures with increasing of the impact velocity when subjected to blast-impact loadings (in which blast loading is applied before impact loading), the enhancement of impact velocity led to the increase of the severity of spallation in the beams subjected to impact-blast loadings (in which impact loading is applied before blast loading) as shown in Figure 3.14. Moreover, it was concluded that although RC beams endured more severe damages with increasing

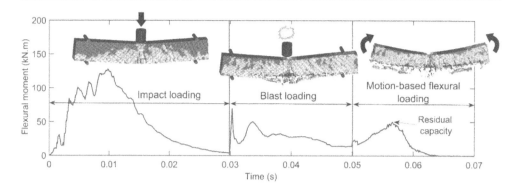

Figure 3.13 Three-step loading procedure to obtain the residual flexural capacity of RC beams [130].

time lag, shear failure modes were predominant when the sequential blast loadings were initiated simultaneously with the occurrence time of the initial peak impact force as shown in Figure 3.15. Furthermore, to quantitatively evaluation of the beam damage states, two novel damage indices based on the residual flexural moment, and shear force were proposed as given in Equations (3.3) and (3.4), respectively. Two three-step procedures based on applying repetitive impact loadings on the impact/blast-induced RC beams were also proposed to obtain the residual flexural and shear force capacities for impact-blast and blast-impact loading scenarios as shown in Figure 3.16a and b, respectively.

$$DI_{(M)} = 1 - \left(\frac{M_r}{M_N} \right) \tag{3.3}$$

$$DI_{(Q)} = 1 - \left(\frac{Q_r}{Q_N} \right) \tag{3.4}$$

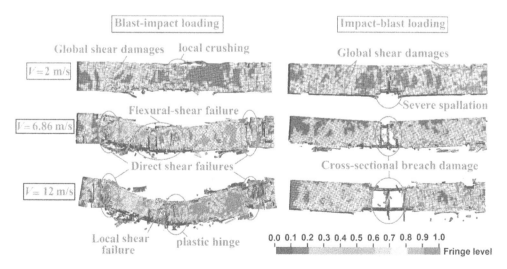

Figure 3.14 Influences of the loading sequence on the failure behaviors of RC beams under combined loadings varying in the impact velocity [131].

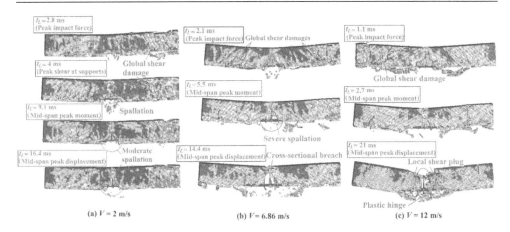

(a) $V = 2$ m/s   (b) $V = 6.86$ m/s   (c) $V = 12$ m/s

Figure 3.15 Influences of the time lag (t_L) on the failure behaviors of RC beams under combined loadings varying in the impact velocity [131].

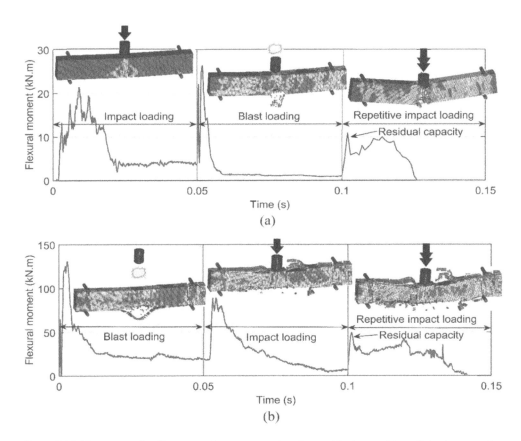

Figure 3.16 Three-step loading procedures to acquire the residual capacities of the beam under combined loadings with the onset priority for (a) impact loading and (b) explosion loading [131].

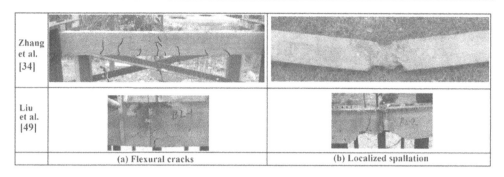

Figure 3.17 Different failure modes of RC beams under blast loads with (a) large standoff and (b) small standoff.

where M_r and Q_r denote the residual flexural and shear capacities, respectively. M_N and Q_N are the nominal flexural and shear capacities of the RC beam by considering the dynamic increase factors due to the strain rate effects of the materials.

Despite the high expensive costs of explosion field tests, many researchers experimentally investigated the blast responses of RC beams. Zhang et al. [34] conducted an experimental study on the damage levels of fix-ended RC beams under close-in blast loading. Also, the relationships between the thickness and the deflections generated in the mid-span of the beams and empirically quantified. It was concluded that the beam failure mode changed from minor flexural damages to localized spallation with decreasing of explosion scaled distance as shown in Figure 3.17a and b. Similarly, from an experimental study of RC beams under different blast loads varying in scaled distances and explosive mass by Liu et al. [49], the flexural failure mode of the beams tended to localized spall damages through the decrease of the scaled distance and increasing of the explosive mass as shown in Figure 3.17a and b. Yao et al. [44] studied the damage behaviors of fix-ended RC beams by varying the ratio and spacing of transverse reinforcements under close-in blast loads experimentally and numerically. Increasing the damage level of beams was observed with decreasing the reinforcement ratio and with the increase of stirrup spacing. Also, the concrete spalling and damage mechanism of RC beams under close-in explosions were numerically studied by Yan et al. [46]. Table 3.11 summarized the influences of various structural and loading parameters of the blast responses of RC beams.

In the use of high-strength concrete materials to enhance the flexural and shear strength of RC beams under blast loads by several research works in the literature [45, 132–135], it was revealed that beams with high reinforcement ratio (i.e., having high flexural strength) were suspect to fail in shear failure modes.

3.4 Bridge piers subjected to blast loads

Bridges are always at the risk of accidental events or intentional terrorist activities arising from the collision of vessels [136–140] and vehicles [141], or below-deck [142–147] and above-deck [66, 67, 73, 148, 149] explosions due to their open and accessibility and lower redundancy against the whole collapse compared to buildings. Despite many studies investigating the blast responses of RC columns commonly used in buildings that have smaller sizes rather relative to those used in bridges, there exist

Table 3.11 Influence of various parameters on the blast response of RC beams

Study	Analysis	Parameter	Effectiveness
Zhang et al. [34]	Experimental	• Scaled distance • Explosive mass	• Negative effects on the spallation area • Changing the damage mode from overall flexural to spallation with increasing the explosive mass
		• Beam size	• Positive effects on the local damage level
Yao et al. [44]	Experimental and Numerical	• Stirrup ratio • Explosive mass • Scaled distance	• Significant negative on the spalling level • Significant positive on the damage level • Significant negative on the deflection thickness ratio
Magnusson et al. [45]	Experimental	• Longitudinal reinforcement ratio • Steel fiber	• High ration led to shear failures • Positive effects on the shear and ductile resistances
Liu et al. [49]	Experimental and Analytical	• Scaled distance • Charge mass	• Changing the damage modes from minor flexural cracks to spallation • Positive on the concrete crushing on the top surface
Nagata et al. [50]	Numerical	• Scaled distance	• The severity of local failures significantly increased by reducing the scaled distance
Li et al. [51]	Experimental	• Concrete strength • Shear reinforcement ratio • Longitudinal reinforcement ratio	• Marginal positive on the blast resistance • Significant positive on preventing shear failure • Significant positive on increasing blast resistance
Chen et al. [129]	Numerical	• Prestressing level • Concrete strength	• Significant effects on delaying the flexural damages; Positive on increasing the probability of diagonal shear failure • Positive effects on the blast resistance
Rao et al. [132]	Experimental and Numerical	• Initiation way of the explosion • Scaled distance • Longitudinal reinforcement ratio	• The beam suffers more severe damages and flexural failures when subjected to double-end detonation • Decreasing both led to changing the failure modes from flexural modes to flexure-shear failures

a large number of researches in the literature focused on the analysis of the behaviors of bridge piers and exploring rehabilitation and strengthening methods against to explosions. As reported in the literature [24], below-deck detonations could potentially result in more catastrophic damages and failure of the whole bridge compared to those localized damages in the bridge superstructures during the above-deck detonation as shown in Figure 3.18. Hence, the investigations of blast responses of bridge substructures (i.e., bridge piers) and exploring more efficient protective solutions and strengthening approaches were substantially considered.

The influences of several key structural parameters on the design of bridge piers such as cross-sectional shape, length-to-depth ratio, transverse reinforcement, and splice location under different scaled distances of blast loads were evaluated by Williamson et al. [70, 71]. More blast resistances were provided by circular columns against blast loads relative to square and rectangular columns with the equivalent cross-sectional areas by reducing about 37% of blast impulse for close-in explosions. The reasons for such observations were: (i) larger reflected overpressure applied to

| Above-deck detonation | Below-deck detonation |

Figure 3.18 Different failure behaviors of RC bridges subjected to different-location blast loads [24].

the flat surfaces of square columns, (ii) smaller angle of incidence (α), and (iii) shorter standoff distance (R) of the explosive from the front surface of square columns compared to that of circular columns as shown in Figure 3.19. Accordingly, blast waves would be speared around the surface of circular columns easier than square columns. Similarly, Williams and Williamson [65] observed that square-shaped columns suffered more severe spalling in the side-cover concrete rather than those with circle cross-sections shown in Figure 3.20.

Compared to square columns under larger blast loads, Williamson et al. [71] observed more shear resistances at the base of circular columns as shown in Figure 3.21. Afterward, Williams and Williamson [142] attempted to compute the shape factors of columns and to assess the influences of these factors on the impulse responses. Also, by evaluating the effectiveness of the scaled distance, it was obtained that the column failure changed from the global minor cracks under large-scaled distance explosion to brittle shear failures at the column base as shown in Figure 3.22.

From the design viewpoint of RC columns under blast loads with the intent to prevent shear failures as concerned in Refs. [143, 144], the use of continuous spiral reinforcements instead of discrete hoops was recommended due to their better

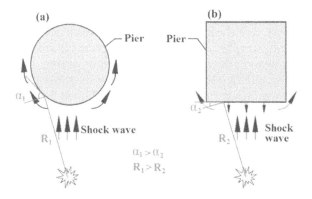

Figure 3.19 Spreading of blast waves around the columns with (a) circular and (b) square cross-sectional shapes [72].

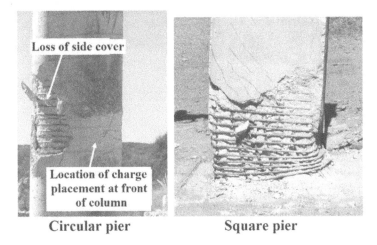

Figure 3.20 Spalling damage modes of RC columns with different cross-sectional shapes under close-in blast loads [22].

performance against shear failures. Moreover, the negative influences of the longitudinal reinforcement splicing were concluded on the blast resistance of columns. Winget et al. [143] concluded that the geometry and clearance parameters can significantly affect the blast loads and the responses of the bridge. It was also revealed that the pier failure mode would be changed from localized spallation to global shear failures at the supports with increasing the size of piers (i.e., height and width dimensions). In addition, more severe damages in the bridge deck were observed by Winget et al. [143] when it was subjected to a below-deck detonation in which the blast pressure confined between the girders, did not have enough space to escape in the air.

As the FE numerical study of bridge columns under blast loads, Yi et al. [68] proposed a new hybrid blast load (HBL) from the combination of two common numerical methods in modeling blast loads in LS-DYNA including: (i) the LBE in which

Figure 3.21 Shear failures of RC columns with different cross-sectional shapes under close-in blast loads [71].

(a) Flexural damages   (b) Shear failure

Figure 3.22 Failure modes of circular RC columns under different blast loads with (a) large standoff and (b) small standoff [71].

blast load is applied directly to the surface of the target structure and (ii) arbitrary Lagrangian-Eulerian (ALE) method in which blast pressure is applied through air or medium. It was claimed that the proposed method solved the drawbacks of the two previous methods. According to these drawbacks, (i) the blast load applied to the shock-front surface elements of the structure using the LBE method is stopped after the erosion of these elements, however, (ii) the FE medium to carry the blast pressure on the surface of structure using ALE method would be very time consuming for large-distance explosions. Afterward, the damage and failure behaviors of RC bridge columns under blast loads were categorized in six levels by Yi et al. [145] as given in Table 3.12.

The influences of different parameters including the cross-sectional shape of the pier, reinforcement arrangement, and the location of explosive charge were evaluated by Hu et al. [72] on the failure behaviors of RC bridge piers under close-in explosions. Similar to the results concluded by Williams and Williamson [65], the blast resistances provided by the square piers were about three times larger than those of circular piers under the same detonations. Besides, it was found that the increase of the detonation height from the pier base led to expansion of the spall damage area due to the increase of effective incidence area of the blast wave on the pier as shown in Figure 3.23. Compared to the negligible effects of transverse rebars, the longitudinal rebars had substantial effects on the damage mitigation of piers and especially in preventing shear failures.

To mitigate the risk of breaches and concrete spalling, Fujikura et al. [144] experimentally investigated the blast resistance of CFST bridge piers under blast loads as an effective strengthening approach. The capability of steel tubes in the protection of core concrete against brittle damages and direct shear failures was observed by increasing the ductility of columns. Thereafter, the blast resistance of CFST columns analytically investigated by Fujikura and Bruneau [147] using SDOF and FRP-based dynamic analyses considering viscose. The importance of considering a shape factor

Table 3.12 Damage levels of RC bridge piers under blast loads classified by Yi et al. [145] compared to those from the literature

Damage level	Damages from FE simulations done by Yi et al. [145]	Corresponding damages from the literature
Eroding of pier bottom concrete		[71]
Shearing of a pier from the footing		[56]
Rebar severance		[146]

(Continued)

Table 3.12 Damage levels of RC bridge piers under blast loads classified by Yi et al. [145] compared to those from the literature (Continued)

Damage level	Damages from FE simulations done by Yi et al. [145]	Corresponding damages from the literature
Breakage of pier		[22]
Spalling of the concrete surface		[23]
Plastic hinge formation		[144]

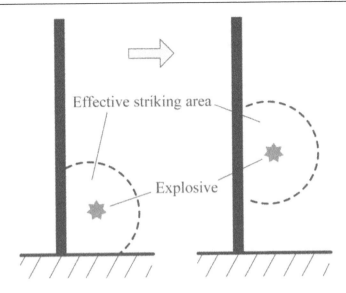

Figure 3.23 Influence of the position of explosion center on the effective striking area [72].

in the simplified analyses was obtained, which reduced the blast impulses, especially on circular columns. Furthermore, a comparative study experimentally was carried out to evaluate the blast resistance of non-ductile RC columns and ductile steel jacketing columns by Fujikura and Bruneau [147]. Unlike capacities behaviors of CFST columns under blast loads observed by Fujikura et al. [144], utilizing steel jackets did not lead to flexural failure modes. Although the columns tended to fail in direct shear at the base, steel jackets protected the columns against spalling damages.

Simplified models adopted by analytical techniques for studying the effects of blast loads on structures may involve different modeling aspects including (i) blast pressure modeling by equivalent static or dynamic analyses, (ii) modeling the interaction between structure and blast loading by using coupled or uncoupled analyses, (iii) discretization of the structure by SDOF or MDOF models, (iv) modeling the material nonlinearity by using elastic or inelastic models, and (v) modeling the geometric nonlinearity by linear or nonlinear models. The advantages of utilizing these analytical modeling aspects their disadvantages have been reviewed by Winget et al. [143]. Among them, most of the blast-resistance studies and design codes of structures utilize uncoupled analyses to model the interaction between the structures and dynamic blast loads to capture more conservative and reasonable responses [143]. Table 3.13 presents a summary of analytical studies on blast responses of RC columns and piers considering different aforementioned modeling aspects associated with analytical options.

Despite the aforementioned researches focused on the blast responses of isolated structural components of girder bridges under explosion, some research works studied the behaviors of entire cable-stayed bridge structures when their different components are subjected to blast loads. Tang and Hao [66, 67] numerically studied blast responses, damage states, and progressive collapse of different components of a cable-stayed concrete bridge under close-in explosions. It was concluded that although the bridge deck components suffered severe localized spalling damages relative to the

Table 3.13 A summary of modeling aspects considered by previous analytical studies

Study	Modeling aspect	Analytical option	Remarks and descriptions
Astarlioglu and Krauthammer [103]	• Blast pressures	• Dynamic analysis	• Consideration of time-dependent idealized blast load
	• Discretization of structure	• SDOF model	• Equivalent SDOF system under idealized dynamic load function
	• Material nonlinearity	• Inelastic model	• Consideration of dynamic strain rate factors of materials
Astarlioglu et al. [53]	• Blast pressures	• Dynamic analysis	• Idealized triangular pulses
	• Discretization of structure	• SDOF model	• Considering the effects of flexural, diagonal shear, and tension membrane behaviors
El-Dakhakhni et al. [25]	• Blast pressures	• Dynamic analysis	• Idealized triangular pulses
	• Discretization of structure	• MDOF model	• Considering the additional moments due to the presence of axial load
	• Material nonlinearity	• Inelastic model	• Considering the high strain rate effects of materials
	• Geometry nonlinearity	• Elastic model	• Linear elastic beams by modeling moment-curvature relationship
Burrell et al. [26]	• Blast pressures	• Dynamic analysis	• Time-varying pressure
	• Discretization of structure	• SDOF model	• Using a lumped inelasticity approach considering confinement and tension behaviors of SFRC
	• Material nonlinearity	• Inelastic model	• Considering strain rate effects of SFRC and steel materials
	• Geometry nonlinearity	• Elastic model	• Linear elastic beams with nonlinear rotational springs at supports and mid-span
Jacques et al. [27]	• Blast pressures	• Dynamic analysis	• Uniformly distributed pressure
	• Discretization of structure	• SDOF model	• Using a lumped inelasticity approach
	• Material nonlinearity	• Inelastic model	• Considering strain rate effects of GFRP and steel materials
	• Geometry nonlinearity	• Elastic model	• Linear elastic beam with nonlinear rotational springs at mid-span
Fujikura and Bruneau [147]	• Blast pressures	• Dynamic analysis	• Equivalent uniform impulse
	• Discretization of structure	• SDOF model	• Using a lumped inelasticity approach considering damping effects and load-mass factors
	• Material nonlinearity	• Inelastic model	• Considering strain rate effects of materials
	• Geometry nonlinearity	• Elastic model	• Equivalent elastic stiffness for cross-section

bridge dimensions, they did not cause the collapse of the whole bridge. However, by observing complete failure in the cross-section of the main pier and back span of the bridge, these components were identified as the most critical components that could force the bridge to collapse. Pan et al. [73] found that girder bridges experienced more severe damages when they were subjected to the below-deck explosion which resulted in significant damages due to large upward forces and deformation. Also, it was observed the bridge tended to whole collapse due to the failure of the main pier under detonation close to the bridge tower. Moreover, the locations around the main piers were identified by Son and Lee [148] as the most critical locations for blast loading scenarios. In addition, the importance of studying the responses of bridge piers during the phase of after blast loading phase was recommended, which can include the large deflections.

Table 3.14 lists a summary of the influences of various parameters on the blast resistance of RC bridge piers. It is found that the majority of the studies focused on the effectiveness of the pier geometry and cross-sectional shape parameter. However, very limited numbers of research works in the literature have investigated the influences of loading-related parameters, and especially no work can be found studying the effects of the axial load levels on blast responses of piers. Therefore, it is extremely suggested for future works to evaluate the sensitivity levels of the axial load parameter to the stability and failure modes of various types of bridge piers under explosions.

Table 3.14 Influence of various parameters on the blast response of bridge piers

Study	Analysis	Parameter	Effectiveness
Williams and Williamson [65, 142]	Experimental and Numerical	• Pier geometry	• More severe damages generated in square-shaped piers than circular-shaped piers
Hao and Tang [67]	Numerical	• Scaled distance	• Minimum scaled distances of 1.20 m/$k^{1/3}$ and 1.33 m/$k^{1/3}$ for the tower and pier to prevent the bridge collapse
Liu et al. [69]	Numerical	• Transverse reinforcement	• Significant positive
Williamson et al. [70, 71]	Experimental	• Pier geometry	• More resistance provided by circular columns than square columns by reducing blast impulses
		• Type of transverse reinforcement	• Better performance of continuous spiral than discrete hoops
		• Reinforcement ratio	• Positive by increasing the column ductility and shear capacity
		• Longitudinal splices	• Negative leading to the breach
Hu et al. [72]	Numerical	• Pier geometry	• Circular piers provide more blast resistances than square piers
		• Longitudinal reinforcement	• Significant positive on the pier shear capacity
		• Transverse reinforcement	• Marginal effects on blast resistances of piers when longitudinal rebars are effectively arranged at the bottom ends
		• Location of detonation	• Piers suffer more severe spallation in the concrete cover with increasing the height of the explosive location
Winget et al. [143]	Numerical	• Below-deck geometry of the bridge	• Significant effect on the blast loads due to confinement effects leading to severe damages
		• Pier clearance	
		• Pier size	• Changing the pier failure modes from localized spallation to global shear failure with increasing pier size
Fujikura et al. [144]	Experimental	• Steel jackets	• Prevent brittle damages and direct shear failures by increasing the ductility of the columns
Fujikura and Bruneau [147]	Experimental	• Steel jackets	• Ineffective to increase the ductility of columns under blast loads

3.5 Chapter review

This chapter presented a comprehensive state-of-the-art review on the responses and failure behaviors of various RC structures subjected to blast loads. The loading and response mechanisms and the damage states of concrete structures such as columns used in bridge piers and framed buildings, and isolated structural members such as beams and slabs under explosions were comprehensively reviewed based on the findings of previous works. In addition, different analysis techniques adopted by the previous studies and the current design codes including simplified analytical, FE simulation, and experimental methods were introduced. It was found the current design codes mainly present simplified approaches in the prediction of blast loads and the design of structures against explosions. Although these analytical methods could predict the overall flexural and ductile responses of structures at some acceptable levels, they are not able to capture the brittle damage modes, localized spalling, and shear failure behaviors of concrete structures. Therefore, it can be concluded that the vast majority of the current design codes consider only far-field detonation effects leading to flexural failure modes in target structures.

Moreover, by reviewing the effectiveness of various parameters on the blast responses of RC columns, no unanimous results were concluded from the evaluation of the axial load level and the column height parameters. However, considering the axial load in the service levels mainly resulted in positive influences. Therefore, more investigations are needed to explore the influences of these parameters. Besides, the majority of previous research works studying blast responses of bridge piers focused on the influences of pier geometry and the cross-sectional shape parameters. It was mainly concluded that circular piers demonstrated more effective performance and resistances against blast waves than square and rectangular piers. In addition, in the protective design of RC beams using high-strength composite materials, it was revealed that utilizing high-ratio longitudinal reinforcements may lead to diagonal shear failures originating from the supports in RC beams.

References

1. Smith PJ. Terrorism in Asia: a persistent challenge sustained by ideological, physical, and criminal enablers. In *Handbook of Asian Criminology*. Springer, New York, NY, 2013, pp. 147–164.
2. Williamson EB, Winget DG. Risk management and design of critical bridges for terrorist attacks. *J Bridge Eng* 2005;10(1):96–106.
3. FHWA. Recommendations for bridge and tunnel security. Blue Ribbon Panel Report, Report No. FHWA-IF-03-036, AASHTO, Federal Highway Administration, Washington, DC, 2003.
4. Ngo TD. Behaviour of high strength concrete subject to impulsive loading. Doctoral Dissertation, Department of Civil and Environment Engineering, The University of Melbourne, Melbourne, Australia, 2005.
5. TM-5-855-1. Fundamentals of protective design for conventional weapons. In *Technical Manual*. US Department of the Army, Washington, DC, 1986.
6. Kennedy WD. *Explosions and explosives in air*. NDRC, Washington, DC, 1946.
7. ASCE. Blast protection of buildings. In *ASCE SEI 59-11*. American Society of Civil Engineers, Reston, VA, 2011.
8. Gel'fandetal B, Voskoboinikov I, Khomik S. Recording the position of a blast-wave front in air. *Combust Explosion Shock Waves* 2004;40(6):734–736.

9. US Department of Defense. Structures to resist the effects of accidental explosions. Report No. UFC 3-340-02, United Facilities Criteria, Washington, DC, 2008.

10. CSA. Design and assessment of buildings subjected to blast loads. CSA 850-12, Canadian Standards Association, Mississauga, Ontario, Canada, 2012.

11. Krauthammer T, Astarlioglu S, Blasko J, Soh TB, Ng PH. Pressure–impulse diagrams for the behavior assessment of structural components. *Int J Impact Eng* 2008;35:771–783.

12. FEMA 427. *Primer for design of commercial buildings to mitigate terrorist attacks.* Federal Emergency Management Agency, Washington, DC, 2003.

13. FEMA 428. Explosive blast. In *Risk Management Series.* US department of Homeland Security, Washington, DC, 2004.

14. Zhang F, Wu C, Zhao X-L, Heidarpour A, Li Z. Experimental and numerical study of blast resistance of square CFDST columns with steel-fibre reinforced concrete. *Eng Struct* 2016;149:50–63.

15. Krauthammer T. *Modern protective structures.* CRC Press, Boca Raton, FL, 2008.

16. TM-5-1300. *Design of structures to resist the effects of accidental explosions. Technical manual.* US Department of the Army, Washington, DC, 1990.

17. US Department of Defense. Design and analysis of hardened structures to conventional weapons effects. UFC 3-340-01, Unified Facilities Criteria, Washington, DC, 2002.

18. US Department of Defense. DoD minimum antiterrorism standards for buildings. Report No. UFC 4-010-01, Unified Facilities Criteria, Washington, DC, 2012.

19. US General Services Administration (GSA). *ISC security design criteria for new federal office buildings and major modernization projects.* Washington, DC, 2003.

20. ASCE. *Design of blast resistant buildings in petrochemical facilities.* American Society of Civil Engineers, Reston, VA, 1997.

21. ASCE. *Minimum design loads for buildings and other structures (7-10).* American Society of Civil Engineers, Reston, VA, 2010.

22. Williamson EB, Bayrak O, Williams GD, Davis CE, Marchand KA, McKay AE, Kulicki J, Wassef W. Blast-resistant highway bridges: design and detailing guidelines. NCHRP Rep. No. 645, National Cooperative Highway Research Program, Washington, DC, 2010.

23. NCHRP 12-72. *Blast-resistant highway bridges: design and detailing guidelines.* National Cooperative Highway Research Program, Washington, DC, 2005.

24. Davis C, Sammarco E, Williamson E. Bridge security design manual. Report No. FHWA-HIF-17-032, Federal Highway Administration, Washington, DC, 2017.

25. El-Dakhakhni WW, Mekky WF, Changiz-Rezaei SH. Vulnerability screening and capacity assessment of reinforced concrete columns subjected to blast. *J Perform Constr Facil* 2009;23(5):353–365.

26. Burrell RP, Aoude H, Saatcioglu M. Response of SFRC columns under blast loads. *J Struct Eng* 2014;141(9):04014209.

27. Jacques E, Lloyd A, Imbeau P, Palermo D, Quek J. GFRP-retrofitted reinforced concrete columns subjected to simulated blast loading. *J Struct Eng* 2015;141(11):04015028.

28. ABAQUS. Analysis user's manual. Version 6.10. Volume I to VI. ABAQUS Inc., an Dassault Systémes, Providence, RI, 2010.

29. LS-DYNA 971. Livermore Software Technology Corporation. Livermore, CA, 2015.

30. AUTODYN. Interactive non-linear dynamic analysis software, version 12, user's manual. SAS IP, Inc., Canonsburg, 2009.

31. LSTC. LS–DYNA keyword user's manual, Ver. 971. Livermore Software Technology Corporation, Livermore, CA, 2016.

32. Tabatabaei ZS, Volz JS. A comparison between three different blast methods in LS-DYNA: LBE, MM-ALE, coupling of LBE and MM-ALE. In *12th International LS-DYNA User Conference*, Detroit, MI, 2012.

33. Abedini M, Mutalib AA, Raman SN, Akhlaghi E. Modeling the effects of high strain rate loading on RC columns using arbitrary Lagrangian Eulerian (ALE) technique. *Rev Int Metodos Numer Calc Diseno Ing* 2018;34(1):1–10.

34. Zhang D, Yao S, Lu F, Chen X, Lin G, Wang W, Lin Y. Experimental study on scaling of RC beams under close-in blast loading. *Eng Fail Aanl* 2013;33:497–504.
35. Aoude H, Dagenais FP, Burrell RP, Saatcioglu M. Behavior of ultra-high performance fiber reinforced concrete columns under blast loading. *Int J Impact Eng* 2015;80:185–202.
36. Hegemier GA, Seible F, Rodriguez-Nikl T, Arnett K. Blast mitigation of critical infrastructure components and systems. In *FIB Proceedings of the 2nd International Congress*, Laussane, Switzerland, 2006.
37. Zhang XH, Zhang CW, Duan ZD. Numerical simulation on shock waves generated by explosive mixture gas from large nuclear blast load generator based on equivalent-energy principles. *J Explosion Shock Waves* 2014;34(1):80–86.
38. Zhang XH, Zhang ZD, Duan ZD, Zhang CW. Analysis for dynamic response and failure process of reinforced concrete beam under blast load. *J Northeast Forestry Univ* 2009;37(4):50–53.
39. Zhang XH, Zhang CW, Duan ZD. Numerical simulation on impact responses and failure modes of steel frame structural columns subject to blast loads. *J Shenyang Jianzhu Univ* 2009;25(4):656–662.
40. Zhang XH, Duan ZD, Zhang CW. Shock responses of steel frame structure near the ground explosion. *Earthq Eng Eng Vibr* 2009;29(4):70–76.
41. Zhang C, Wang W, Zhang X, Lu S. Large scale blast emulator based explosive gas loading methods for structures and recent advances in experimental studies. In *Proceedings of the 8th International Conference on Shock and Impact Loads on Structures*, 2009:769–780.
42. Gebbeken N, Ruppert M. A new material model for concrete in high-dynamic hydrocode simulations. *Arch Appl Mech* 2000;70(7):463–478.
43. McVay MK. Spall damage of concrete structures. No. WES/TR/SL-88-22, Army Engineer Waterways Experiment Station Vicksburg MS Structures LAB, Vicksburg, MS, 1988.
44. Yao SJ, Zhang D, Lu FY, Wang W, Chen XG. Damage features and dynamic response of RC beams under blast. *Eng Fail Aanl* 2016;62:103–111.
45. Magnusson J, Hallgren M, Ansell A. Air-blast-loaded, high-strength concrete beams. Part I: Experimental investigation. *Mag Concr Res* 2010;62(2):127–136.
46. Yan B, Liu F, Song D, Jiang Z. Numerical study on damage mechanism of RC beams under close-in blast loading. *Eng Fail Aanl* 2015;51:9–19.
47. Li J, Hao H. A two-step numerical method for efficient analysis of structural response to blast load. *Int J Prot Struct* 2011;2(1):103–126.
48. Li J, Hao H. Influence of brittle shear damage on accuracy of the two-step method in prediction of structural response to blast loads. *Int J Impact Eng* 2013;54:217–231.
49. Liu Y, Yan JB, Huang FL. Behavior of reinforced concrete beams and columns subjected to blast loading. *Defense Tech* 2018;14(5):550–559.
50. Nagata M, Beppu M, Ichino H, Matsuzawa R. A fundamental investigation of reinforced concrete beams subjected to close-in explosion. *Int J Protect Struct* 2018;9(2):174–198.
51. Li Y, Algassem O, Aoude H. Response of high-strength reinforced concrete beams under shock-tube induced blast loading. *Constr Build Mater* 2018;189:420–437.
52. Bao X, Li B. Residual strength of blast damaged reinforced concrete columns. *Int J Impact Eng* 2010;37(3):295–308.
53. Astarlioglu S, Krauthammer T, Morency D, Tran TP. Behavior of reinforced concrete columns under combined effects of axial and blast-induced transverse loads. *Eng Struct* 2013;55:26–34.
54. Li J, Hao H. Numerical study of concrete spall damage to blast loads. *Int J Impact Eng* 2014;68:41–55.
55. Shi Y, Hao H, Li ZX. Numerical derivation of pressure–impulse diagrams for prediction of RC column damage to blast loads. *Int J Impact Eng* 2008;35(11):1213–1227.
56. Oswald CJ. Prediction of injuries to building occupants from column failure and progressive collapse with the BICADS computer program. In *Proceedings of ASCE Structures Congress and Exposition*, Reston, VA, 2005, pp. 1–22.

57. Yuan C, Chen W, Pham TM, Hao H, Jian C, Shi Y. Strain rate effect on interfacial bond behaviour between BFRP sheets and steel fibre reinforced concrete. *Compos Part B Eng* 2019;174:107032.

58. Thiagarajan G, Kadambi AV, Robert S, Johnson CF. Experimental and finite element analysis of doubly reinforced concrete slabs subjected to blast loads. *Int J Impact Eng* 2015;75:162–173.

59. Foglar M, Kovar M. Conclusions from experimental testing of blast resistance of FRC and RC bridge decks. *Int J Impact Eng* 2013;59:18–28.

60. Wu C, Nurwidayati R, Oehlers DJ. Fragmentation from spallation of RC slabs due to airblast loads. *Int J Impact Eng* 2009;36(12):1371–1376.

61. Wu C, Oehlers DJ, Rebentrost M, Leach J, Whittaker AS. Blast testing of ultra-high performance fibre and FRP-retrofitted concrete slabs. *Eng Struct* 2009;31(9):2060–2069.

62. Jayasooriya R, Thambiratnam DP, Perera NJ, Kosse V. Blast and residual capacity analysis of reinforced concrete framed buildings. *Eng Struct* 2011;33(12):3483–3495.

63. Nguyen K, Navaratnam S, Mendis P, Zhang K, Barnett J, Wang H. Fire safety of composites in prefabricated buildings: from fibre reinforced polymer to textile reinforced concrete. *Compos Part B Eng* 2020;187:107815.

64. Yuan S, Hao H, Zong Z, Li J. A study of RC bridge columns under contact explosion. *Int J Impact Eng* 2017;109:378–390.

65. Williams GD, Williamson EB. Response of reinforced concrete bridge columns subjected to blast loads. *J Struct Eng* 2011;137(9):903–913.

66. Tang EK, Hao H. Numerical simulation of a cable-stayed bridge response to blast loads, Part I: Model development and response calculations. *Eng Struct* 2010;32(10):3180–3192.

67. Hao H, Tang EK. Numerical simulation of a cable-stayed bridge response to blast loads, Part II: Damage prediction and FRP strengthening. *Eng Struct* 2010;32(10):3193–3205.

68. Yi Z, Agrawal AK, Ettouney M, Alampalli S. Blast load effects on highway bridges. I: Modeling and blast load effects. *J Bridge Eng* 2014;19(4):04013023.

69. Liu H, Torres DM, Agrawal AK, Yi Z, Liu G. Simplified blast-load effects on the column and bent beam of highway bridges. *J Bridge Eng* 2015;20(10):06015001.

70. Williamson EB, Bayrak O, Davis C, Williams GD. Performance of bridge columns subjected to blast loads. I: Experimental program. *J Bridge Eng* 2011;16(6):693–702.

71. Williamson EB, Bayrak O, Davis C, Daniel Williams G. Performance of bridge columns subjected to blast loads. II: Results and recommendations. *J Bridge Eng* 2011;16(6):703–710.

72. Hu ZJ, Wu L, Zhang YF, Sun LZ. Dynamic responses of concrete piers under close-in blast loading. *Int J Damage Mech* 2016;25(8):1235–1254.

73. Pan Y, Ventura CE, Cheung MM. Performance of highway bridges subjected to blast loads. *Eng Struct* 2017;151:788–801.

74. Yu R, Zhang D, Chen L, Yan H. Non-dimensional pressure–impulse diagrams for blast-loaded reinforced concrete beam columns referred to different failure modes. *Adv Struct Eng* 2018;21(14):2114–2129.

75. Fallah AS, Louca LA. Pressure–impulse diagrams for elastic–plastic-hardening and softening single-degree-of-freedom models subjected to blast loading. *Int J Impact Eng* 2007;34:823–842.

76. Li QM, Meng H. Pressure-impulse diagram for blast loads based on dimensional analysis and single-degree-of-freedom model. *J Eng Mech* 2002;128(1):87–92.

77. Mutalib AA, Hao H. Development of P-I diagrams for FRP strengthened RC columns. *Int J Impact Eng* 2011;38(5):290–304.

78. Jayasooriya R, Thambiratnam DP, Perera NJ. Blast response and safety evaluation of a composite column for use as key element in structural systems. *Eng Struct* 2014;61:31–43.

79. Kyei C, Braimah A. Effects of transverse reinforcement spacing on the response of reinforced concrete columns subjected to blast loading. *Eng Struct* 2017;142:148–164.

80. Cui J, Shi Y, Li ZX, Chen L. Failure analysis and damage assessment of RC columns under close-in explosions. *J Perform Construct Facil* 2015;29(5):B4015003.

81. Thiagarajan G, Rahimzadeh R, Kundu A. Study of pressure–impulse diagrams for reinforced concrete columns using finite element analysis. *Int J Protect Struct* 2013;4:485–504.
82. Abedini M, Mutalib AA, Raman SN, Alipour R, Akhlaghi E. Pressure–impulse (P–I) diagrams for reinforced concrete (RC) structures: a review. *Arch Comput Method E* 2019;26(3):733–767.
83. Abedini M, Mutalib AA. Investigation into damage criterion and failure modes of RC structures when subjected to extreme dynamic loads. *Arch Comput Method E* 2019;27:1–15.
84. Huang Z, Zhang X, Yang C. Experimental and numerical studies on the bending collapse of multi-cell aluminum/CFRP hybrid tubes. *Compos Part B Eng* 2020;181:107527.
85. Wu KC, Li B, Tsai KC. Residual axial compression capacity of localized blast-damaged RC columns. *Int J Impact Eng* 2011;38(1):29–40.
86. Siba F. Near-field explosion effects on reinforced concrete columns: an experimental investigation. Doctoral Dissertation, Carleton University, Ottawa, Canada, 2014.
87. Wesevich J, Oswald C. *Empirical based concrete masonry pressure–impulse diagrams for varying degrees of damage.* American Society of Civil Engineers, Reston, VA; New York, NY, 2005, pp. 2083–2094.
88. Liu L, Zong ZH, Li MH. Numerical study of damage modes and assessment of circular RC pier under noncontact explosions. *J Bridge Eng* 2018;23(9):04018061.
89. Yan QS. Damage assessment of subway station columns subjected to blast loadings. *Int J Struct Stab Dyn* 2018;18(03):1850034.
90. Zhang F, Wu C, Wang H, Zhou Y. Numerical simulation of concrete filled steel tube columns against blast loads. *Thin Wall Struct* 2015;92:82–92.
91. Wang H, Wu C, Zhang F, Fang Q, Xiang H, Li P, Li Z, Zhou Y, Zhang Y, Li J. Experimental study of large-sized concrete filled steel tube columns under blast load. *Constr Build Mater* 2017;134:131–141.
92. Zhang J, Jiang S, Chen B, Li C, Qin H. Numerical study of damage modes and damage assessment of CFST columns under blast loading. *Shock Vibr* 2016;2016:1–12.
93. Zhang F, Wu C, Zhao XL, Xiang H, Li ZX, Fang Q, Liu Z, Zhang Y, Heidarpour A, Packer JA. Experimental study of CFDST columns infilled with UHPC under close-range blast loading. *Int J Impact Eng* 2016;93:184–195.
94. Li M, Zong Z, Liu L, Lou F. Experimental and numerical study on damage mechanism of CFDST bridge columns subjected to contact explosion. *Eng Struct* 2018;159:265–276.
95. Li M, Zong Z, Hao H, Zhang X, Lin J, Xie G. Experimental and numerical study on the behaviour of CFDST columns subjected to close-in blast loading. *Eng Struct* 2019;185:203–220.
96. Wu KC, Li B, Tsai KC. The effects of explosive mass ratio on residual compressive capacity of contact blast damaged composite columns. *J Constr Steel Res* 2011;67(4):602–612.
97. Codina R, Ambrosini D, de Borbón F. Alternatives to prevent the failure of RC members under close-in blast loadings. *Eng Fail Anal* 2016;60:96–106.
98. Abada M, Ibrahim A. Hybrid multi-cell thin-walled tubes for energy absorption applications: blast shielding and crashworthiness. *Compos Part B Eng* 2020;183:107720.
99. Liu L, Zong Z, Gao C, Yuan S, Lou F. Experimental and numerical study of CFRP protective RC piers under contact explosion. *Compos Struct* 2020;234:111658.
100. Li J, Wu C, Hao H, Liu Z. Post-blast capacity of ultra-high performance concrete columns. *Eng Struct* 2017;134:289–302.
101. Li J, Wu C. Damage evaluation of ultra-high performance concrete columns after blast loads. *Int J Protect Struct* 2018;9(1):44–64.
102. Beppu M, Kataoka S, Ichino H, Musha H. Failure characteristics of UHPFRC panels subjected to projectile impact. *Compos Part B Eng* 2020;182:107505.
103. Astarlioglu S, Krauthammer T. Response of normal-strength and ultra-high-performance fiber-reinforced concrete columns to idealized blast loads. *Eng Struct* 2014;61:1–12.
104. Wang W, Wu C, Li J. Numerical simulation of hybrid FRP-concrete-steel double-skin tubular columns under close-range blast loading. *J Compos Construct* 2018;22(5):04018036.

105. Qasrawi Y, Heffernan PJ, Fam A. Performance of concrete-filled FRP tubes under field close-in blast loading. *J Compos Construct* 2015;19(4):04014067.
106. Rodriguez-Nikl T, Lee CS, Hegemier GA, Seible F. Experimental performance of concrete columns with composite jackets under blast loading. *J Struct Eng* 2012;138(1):81–89.
107. Qasrawi Y, Heffernan PJ, Fam A. Numerical modeling of concrete-filled FRP tubes' dynamic behavior under blast and impact loading. *J Struct Eng* 2016;142(2):04015106.
108. Crawford JE. Chapter 17. Retrofit of structural components and systems. In *Handbook for Blast-Resistant Design of Buildings*. Edited by Dusenberry DO. John Wiley & Sons, Inc., Hoboken, NJ, 2010.
109. Dragos J, Wu C. Interaction between direct shear and flexural responses for blast loaded one-way reinforced concrete slabs using a finite element model. *Eng Struct* 2014;72:193–202.
110. Low HY, Hao H. Reliability analysis of reinforced concrete slabs under explosive loading. *Struct Saf* 2001;23(2):157–178.
111. Low HY, Hao H. Reliability analysis of direct shear and flexural failure modes of RC slabs under explosive loading. *Eng Struct* 2002;24(2):189–198.
112. Silva PF, Lu B. Blast resistance capacity of reinforced concrete slabs. *J Struct Eng* 2009;135(6):708–716.
113. Zhou XQ, Kuznetsov VA, Hao H, Waschl J. Numerical prediction of concrete slab response to blast loading. *Int J Impact Eng* 2008;35(10):1186–1200.
114. Xu K, Lu Y. Numerical simulation study of spallation in reinforced concrete plates subjected to blast loading. *Comput Struct* 2006;84(5–6):431–438.
115. Jia H, Yu L, Wu G. Damage assessment of two-way bending RC slabs subjected to blast loadings. *Sci World J* 2014;2014:1–12.
116. Tai YS, Chu TL, Hu HT, Wu JY. Dynamic response of a reinforced concrete slab subjected to air blast load. *Thero Appl Fract Mech* 2011;56(3):140–147.
117. Gargano A, Das R, Mouritz AP. Finite element modelling of the explosive blast response of carbon fibre-polymer laminates. *Compos Part B Eng* 2019;177:107412.
118. Wang W, Zhang D, Lu F, Wang SC, Tang F. Experimental study on scaling the explosion resistance of a one-way square reinforced concrete slab under a close-in blast loading. *Int J Impact Eng* 2012;49:158–164.
119. Wang W, Zhang D, Lu F, Wang S-C, Tang F. Experimental study and numerical simulation of the damage mode of a square reinforced concrete slab under close-in explosion. *Eng Fail Anal* 2013;27:41–51
120. Li J, Wu C, Hao H. An experimental and numerical study of reinforced ultra-high performance concrete slabs under blast loads. *Mater Des* 2015;82:64–76.
121. Li J, Wu C, Hao H. Investigation of ultra-high performance concrete slab and normal strength concrete slab under contact explosion. *Eng Struct* 2015;102:395–408.
122. Ohtsu M, Uddin FA, Tong W, Murakami K. Dynamics of spall failure in fiber reinforced concrete due to blasting. *Constr Build Mater* 2007;21(3):511–518.
123. Zhao Z, Zhang B, Jin F, Zhou J, Chen H, Fan H. BFRP reinforcing hierarchical stiffened SMC protective structure. *Compos Part B Eng* 2019;168:195–203.
124. Meng Q, Wu C, Su Y, Li J, Liu J, Pang J. Experimental and numerical investigation of blast resistant capacity of high performance geopolymer concrete panels. *Compos Part B Eng* 2019;171:9–19.
125. Li N, Shi C, Zhang Z, Wang H, Liu Y. A review on mixture design methods for geopolymer concrete. *Compos Part B Eng* 2019;178:107490.
126. Mao L, Barnett S, Begg D, Schleyer G, Wight G. Numerical simulation of ultra high performance fibre reinforced concrete panel subjected to blast loading. *Int J Impact Eng* 2014;64:91–100.
127. Foglar M, Hajek R, Fladr J, Pachman J, Stoller J. Full-scale experimental testing of the blast resistance of HPFRC and UHPFRC bridge decks. *Constr Build Mater* 2017;145:588–601.

128. Chernin L, Vilnay M, Shufrin I. Blast dynamics of beam-columns via analytical approach. *Int J Mech Sci* 2016;106:331–345.
129. Chen W, Hao H, Chen S. Numerical analysis of prestressed reinforced concrete beam subjected to blast loading. *Mater Des* (1980–2015) 2015;65:662–674.
130. Zhang C, Gholipour G, Mousavi AA. Nonlinear dynamic behavior of simply-supported RC beams subjected to combined impact-blast loading. *Eng Struct* 2019;181:124–142.
131. Gholipour G, Zhang C, Mousavi AA. Loading rate effects on the responses of simply supported RC beams subjected to the combination of impact and blast loads. *Eng Struct* 2019;201:109837.
132. Rao B, Chen L, Fang Q, Hong J, Liu ZX, Xiang HB. Dynamic responses of reinforced concrete beams under double-end-initiated close-in explosion. *Def Technol* 2018;14(5):527–539.
133. Yoo DY, Banthia N. Mechanical and structural behaviors of ultra-high-performance fiber-reinforced concrete subjected to impact and blast. *Constr Build Mater* 2017;149:416–431.
134. Chen W, Pham TM, Sichembe H, Chen L, Hao H. Experimental study of flexural behaviour of RC beams strengthened by longitudinal and U-shaped basalt FRP sheet. *Compos Part B Eng* 2018;134:114–126.
135. Zhang X, Shi Y, Li ZX. Experimental study on the tensile behavior of unidirectional and plain weave CFRP laminates under different strain rates. *Compos Part B Eng* 2019;164:524–536.
136. Gholipour G, Zhang C, Mousavi AA. Effects of axial load on nonlinear response of RC columns subjected to lateral impact load: ship-pier collision. *Eng Fail Anal* 2018;91:397–418.
137. Gholipour G, Zhang C, Li M. Effects of soil–pile interaction on the response of bridge pier to barge collision using energy distribution method. *Struct Infrastruct E* 2018;14(11):1520–1534.
138. Gholipour G, Zhang C, Mousavi AA. Analysis of girder bridge pier subjected to barge collision considering the superstructure interactions: the case study of a multiple-pier bridge system. *Struct Infrastruct E* 2018;15(3):392–412.
139. Gholipour G, Zhang C, Mousavi AA. Reliability analysis of girder bridge piers subjected to barge collisions. *Struct Infrastruct E* 2018;15(9):1200–1220.
140. Gholipour G, Zhang C, Mousavi AA. Nonlinear numerical analysis and progressive damage assessment of a cable-stayed bridge pier subjected to ship collision. *Mar Struct* 2020;69:102662.
141. Cao R, El-Tawil S, Agrawal AK, Xu X, Wong W. Behavior and design of bridge piers subjected to heavy truck collision. *J Bridge Eng* 2019;24(7):04019057.
142. Williams GD, Williamson EB. Procedure for predicting blast loads acting on bridge columns. *J Bridge Eng* 2011;17(3):490–499.
143. Winget DG, Marchand KA, Williamson EB. Analysis and design of critical bridges subjected to blast loads. *J Struct Eng* 2005;131(8):1243–1255.
144. Fujikura S, Bruneau M, Lopez-Garcia D. Experimental investigation of multihazard resistant bridge piers having concrete-filled steel tube under blast loading. *J Bridge Eng* 2008;13(6):586–594.
145. Yi Z, Agrawal AK, Ettouney M, Alampalli S. Blast load effects on highway bridges. II: Failure modes and multihazard correlations. *J Bridge Eng* 2013;19(4):04013024.
146. Matthews T, Elwood KJ, Hwang SJ. Explosive testing to evaluate dynamic amplification during gravity load redistribution for reinforced concrete frames. In *Proceeding of 2007 ASCE Structures Congress, ASCE*, Reston, VA, 2007.
147. Fujikura S, Bruneau M. Dynamic analysis of multihazard-resistant bridge piers having concrete-filled steel tube under blast loading. *J Bridge Eng* 2011;17(2):249–258.
148. Son J, Lee HJ. Performance of cable-stayed bridge pylons subjected to blast loading. *Eng Struct* 2011;33:1133–1148.
149. Shiravand MR, Parvanehro P. Numerical study on damage mechanism of post-tensioned concrete box bridges under close-in deck explosion. *Eng Fail Anal* 2017;81:103–116.

Chapter 4

Nonlinear dynamic analysis of RC columns subjected to lateral impact loads

4.1 Introduction

Columns as the axial load carrying structural members in bridge piers, buildings, and offshore structures are inherently at the risk of lateral impact loads during the vessel or vehicle collision events. Lateral impact loads applying to the columns may lead to severe structural damages and even collapse of the whole structure. Hence, the effects of various loading-related and structural-related parameters on the failure behavior of the columns against the impact loads should be carefully considered in their structural design. There are many experimental and numerical studies in the literature on the investigation of structural members including beams and columns subjected to lateral impact loads during the vessel, vehicle, or object collisions. As the experimental studies, some of the researchers have conducted the dropping mass impact tests on reinforced concrete (RC) beams [1–5] and RC columns [6–8]. Also, some of the previous works have investigated the impact responses of steel tubes [9–13] and the effects of different strengthening materials of beams [5, 14] and columns [15, 16] against the impact loads. As the numerical studies, many efforts have been done to explore the effects of different parameters such as material type, cross-section type, impact velocity, impact weight, impact location, the shape of impactor, axial load, boundary conditions, impacted member's inertia, and impact loading rate on the dynamic responses and failure modes of structural members including beams [2, 15–20], and columns [11–13, 21–24].

Since the columns are axially restrained members in civil infrastructures, investigating axial load effects on the response and failure modes of the columns and determining their relationships with structural response parameters during the impact event, is very important. Most of the aforementioned studies focused on the behavior of beams in which the axial load is not considered. However, the limited numbers of previous studies considered the effects of axial load on the response of steel tubes [11–13, 21, 25, 26], and RC columns [15, 22] to impact loads. Alam et al. [21] conducted a parametric study on concrete-filled steel tubular (CFST) columns strengthened with the carbon fiber reinforced polymer (CFRP) subjected to simplified mass-spring vehicle model by varying impact velocity, impacting object mass, axial loading range, vehicle stiffness, and CFRP bond length. In their work, the axial load was applied statically at the end of the column in a separate analysis step. Besides, it was declared in [12] that the axial load has a key role in the identification of the failure mechanism of tubular steel columns. It was observed that the residual displacements of CFRP strengthened tubular columns increase with the increase of axial load. As the numerical finite

DOI: 10.1201/9781003262343-4

element (FE) studies, there are some FE simulations of the vehicle collisions [27–34] and vessel collisions [35–40] with bridge piers considering the bridge superstructure.

Some of the previous studies on the vessel-bridge pier collisions found that the bridge superstructure has a significant effect on the distribution of impact load throughout the bridge pier [35], the value of internal forces, and the structural failure modes [36, 41] by utilizing an efficient dynamic analysis technique named coupled vessel to impact analysis (CVIA) on a wide range of bridge configurations. Based on the experimental and numerical parametric studies carried out by Sha and Hao [37] in which the pier superstructure modeled by a concrete block is placed on the top of the pier, it was claimed that although the bridge superstructure affects the pier displacements, it has no significant effect on the impact force.

Despite some limited previous studies on the response of steel columns to impact load with considering axial load effects, the investigation of axial load effects on the dynamic distribution of internal forces, the location of plastic hinges, and failure modes of concrete columns subjected to different impact loading rates and impact locations, are the research gaps of previous works. As an extension to investigate the plastic hinge location formed in laterally impacted beams without considering axial load effects in the works done by [18, 19], this chapter proposes an improved method to estimate the location of plastic hinges in an axially loaded concrete column subjected to lateral impact load. While several researchers have considered the presence of bridge superstructure during the vehicle and vessel collisions with concrete bridge piers, the effects of bridge superstructure axial load on the nonlinear dynamic responses and failure modes of bridge pier have not been investigated in previous works. In this chapter, first, a numerical parametric study is carried out on an axially loaded column subjected to lateral impact loading scenarios by varying axial load ratio, impact velocity (V), and the impact location. The sensitivity of the column impact responses, including peak impact force, displacement, distribution of internal forces, the location of the plastic hinge, and failure modes of the column, is assessed to the different ranges of axial load, impact velocity, and impact location. As a case study in the realistic applications, the FE numerical simulations of ship collision with a concrete cable-stayed bridge pier in the cases with and without superstructure are developed in LS-DYNA [42] to prove the influences of axial load related to the bridge superstructure inertia on the impact responses and failure modes of the impacted pier.

4.2 FE simulation of impact tests on RC columns

In this section, a series of drop-weight impact tests are numerically carried out on an axially loaded RC column. To validate the FE model of the RC column developed in LS-DYNA, the results from FE simulation are compared with those from experimental tests previously conducted by Feyerabend [43]. Afterward, the effects of column axial load on the impact force, column mid-span displacement, the distribution of internal forces, and the location of the plastic hinge are assessed for the mid-span impact scenarios. Then, a comparative study of axial load effects on the impact responses and failure modes of RC columns subjected to lateral impact loading in different locations is accomplished. Subsequently, a parametric study is carried out to obtain the relationships between the dynamic peak impact force, peak flexural moment, plastic hinge location, and axial load ratio with the variability of impact velocity (V) for the mid-span impact scenarios.

Feyerabend [43] carried out many impact tests on the axially loaded RC columns in a horizontal position subjected to transverse dropping mass impact loads, as shown in Figure 4.1. In these tests, a dropping mass of 1,140 kg with a velocity of 4.5 m/s impacted at the mid-span of an RC column with a span length of 4.0 m and the square cross-section with dimensions of 0.3 m × 0.3 m. The diameters of longitudinal and transverse reinforcements are 28 and 12 mm, respectively. One end of the column was restrained against the axial movement and another end was free to move and a mass of 20,000 kg was attached to this end. The initial axial load was applied to the column using external prestressing bars. The dropping mass has a cylindrical shape with a diameter of 100 mm.

In LS-DYNA, the column concrete and the dropping mass are modeled using eight-node constant stress solid elements, and both longitudinal and transverse reinforcements are modeled using Hughes-Liu beam elements. From a study on the mesh size of the FEs used in the model, it was found that results are converged when the mesh size is 20 mm for the column concrete and reinforcements. Selecting a mesh size lesser than this range has a negligible influence on the computational accuracy of the results, while it can be caused to significantly increase the computational time. A nonlinear material model named MAT_CSCM_CONCRETE (MAT_159) [44], which is a continuous surface cap model, is selected to represent the nonlinear damage behavior of the column concrete with considering strain rate effects. Also, an elastic-plastic material model named MAT_PIECEWISE_LINEAR_PLASTICITY

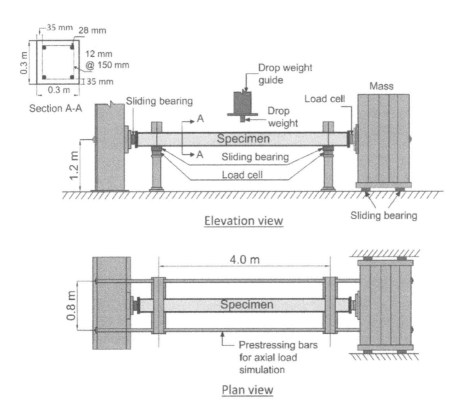

Figure 4.1 Drop-weight impact test setup conducted by Feyerabend [43].

Table 4.1 Material properties used for the column and dropping mass in experiment impact test [43]

Model	Parameter	Value
Column concrete	Mass density (kg/m³)	2,400
	Unconfined compression strength (MPa)	50
	Elements erosion (ERODE)	1.08
	Maximum aggregate size (mm)	19
	Rate effects	Turn on
Column longitudinal reinforcement	Mass density (kg/m³)	7,850
	Young's modulus (GPa)	210
	Poisson's ratio	0.3
	Yield stress (MPa)	523
	Failure strain	0.15
	C_{strain} parameter	40.4
	P_{strain} parameter	5.0
Column transverse reinforcement	Yield stress (MPa)	507
Drop-weight	Mass density (kg/m³)	9,020
	Young's modulus (GPa)	200
	Poisson's ratio	0.3

(MAT_024) considering strain rate effects for the column reinforcement and a rigid material model named MAT_RIGID (MAT_020) for drop-weight are selected in LS-DYNA [44]. The material properties of the column and the dropping mass are tabulated in Table 4.1. To perform the impact scenario in LS-DYNA, a contact algorithm named AUTOMATIC_SURFACE_TO_SURFACE is defined between the column (as the master part) and the dropping mass (as the slave part) with a Coulomb friction coefficient of 0.3 to take into account the sliding forces during the contact. According to LS-DYNA user manual [44], it is advised that the mesh size of the slave part should be smaller than that of the master part. Thus, a mesh size of 10 mm is selected for the dropping mass which is lesser than that of column concrete. In addition, the coupling between the beam concrete and reinforcements is modeled using a keyword named CONSTRAINED_LAGRANGE_IN_SOLID in LS-DYNA. The axial load is applied to the FE model of the column as a uniform surface pressure on the column cross-sectional area same as that procedure used in [22]. This procedure conducts a separate dynamic relaxation analysis using CONTROL_DYNAMIC_RELAXATION keyword in LS-DYNA on the column before the transient impact analysis stage. Figure 4.2 shows the FE model of drop-weight test on the column with considering axial load.

4.3 Drop weight impact of axially loaded RC columns

To validate the FE models of drop-weight impact test on the axially loaded column, the FE simulation results including impact force and column displacement at the column mid-span are compared with those from the experimental tests conducted by Feyerabend [43] for an impact scenario with an impact velocity of 4.5 m/s as shown in Figure 4.3a and b. It is observed that FE results including the peak impact force, the impact duration time, the time of occurrence of the peaks, the peak displacement, and the residual displacements are in good agreement with those of the experimental test.

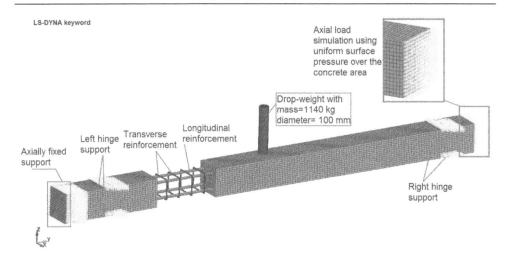

Figure 4.2 FE modeling of drop-weight impact test in LS-DYNA.

4.4 Nonlinear dynamic analysis of axially loaded columns subjected to lateral impact

4.4.1 Mid-span impact

In this section, the effects of axial load on the impact responses of the column including impact force, displacement response, and the distribution of internal forces throughout the column are evaluated. To this end, a series of numerical drop-weight impact tests at the column mid-span with the variability of axial load ranges from 0.1 to 0.8 of axial load carrying capacity of the column are carried out in LS-DYNA. The axial load carrying capacity of the column (P_N) is calculated by Equation (4.1) according to ACI–314 [45].

$$P_N = \beta_1 f_c'(A_c - A_s) + f_y A_s \tag{4.1}$$

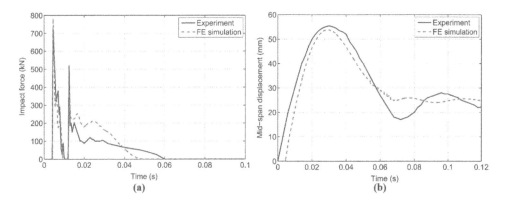

Figure 4.3 Comparison of FE simulation results with experimental test results: (a) impact force results and (b) displacement at the mid-span of the column.

where f_c' and f_y are the concrete compressive strength and rebar yield stress, respectively. A_c and A_s are column cross-sectional area and the reinforcement rebar area, respectively. β_1 is 0.69 for concrete with a grade of $C50$ according to ACI–318 [45].

Figure 4.4a–f shows the comparison of impact responses with the variability of axial load ratio for the scenarios with low-, middle-, and high-impact loading rates with

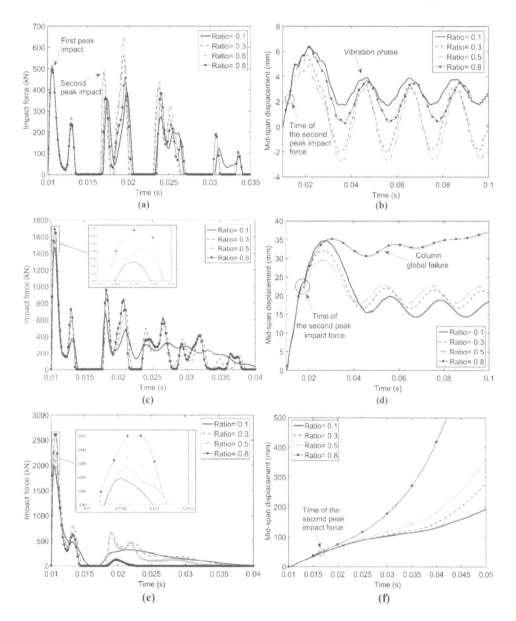

Figure 4.4 Comparison of the column impact responses with the variability of column axial load for different impact loading rates: (a) impact force for $V = 1$ m/s, (b) displacement at the column mid-span $V = 1$ m/s, (c) impact force for $V = 5$ m/s, (d) displacement at the column mid-span $V = 5$ m/s, (e) impact force for $V = 10$ m/s, and (f) displacement at the column mid-span $V = 10$ m/s.

impact velocities of 1 m/s, 5 m/s, and 10 m/s, respectively. Figure 4.4a and b presents the column impact responses when the impact rate is low with an impact velocity of 1 m/s. It is seen that although the first impact impulse (i.e., the first peak impact) is not significantly sensitive for different axial load ratios, the following peak impact forces significantly increase with the increase of axial load ratio except that scenario with an axial ratio of 0.8. According to the previous studies [46], the first impact impulse is not affected by the global stiffness of the impacted structure. Besides, since in low impact rates the column tends to a global response, the following impulses are more affected by additional global-flexure stiffness provided by the axial load on the column than the first impulse. For a high axial ratio of 0.8, it is seen that the value of impact force decreases, and the column displacement increases after the time of second peak impact because of losing the column flexural strength and the formation of the plastic hinge at the column mid-span. However, for the cases with axial load ratios from 0.1 to 0.5, it is seen that the peak of the first mid-span displacement (i.e., the first global response) of the column decreases with the increase of axial load ratio because of the positive influence of axial load on the flexural stiffness of the column. In addition, for these ranges of axial load, it is observed that the increase of ratio causes the increase of the amplitude of cyclic vibrations at the column mid-span during the vibration phase. In other words, although the increase of axial load ratio has a positive influence on the global stiffness of the column at the first cycle of the global response, it plays a negative role during the vibration phase of column response that can cause the formation of possible plastic hinges during this stage.

With the increase of impact velocity to the middle and high ranges of 5 and 10 m/s in Figure 4.4c and e, respectively, it is observed that the value of the first peak impact force is more sensitive than the following peaks because the column tends to a local response more than global response. For these cases, the local response of the structure depends on the first impact impulse, structure inertia, and the stress wave propagation in the structure. For following impulses, because of formation of plastic hinge immediately after the first impulse for those cases with an axial ratio more than 0.3, the axial load has a negative influence.

For a middle impact velocity of 5 m/s, the axial load has no significant effect on the residual displacements of the column between the ratios 0.1 and 0.5 because of increasing the lateral resistance of the column and the degradation of the column vibrations in the vibration phase of the response. Besides, it is observed in Figure 4.4d that although the peak displacement at the mid-span decreases with the increase of axial load ratios from 0.1 to 0.5, that case with a ratio of 0.8 gives the greatest value of displacement because of column failure and large displacement after the time of the second peak impact. In addition, it is observed that residual displacement of the column increases in proportion to the increase of axial load ratios between 0.1 and 0.5. This is because of the permanent damages and plastic hinges in the column with the increase of axial load ratio. However, these vibration responses could not be seen for the high rate impact scenario in Figure 4.4f because of forming the plastic hinge at the column impact zone at the early stage of the column response immediately after the first impulse.

The dynamic distribution of internal forces throughout the structural members is significantly different from those of static analysis [2, 18]. This is because of considering the effects of structure's inertia during dynamic analysis which is not considered in the static analysis. Figure 4.5 shows the theoretical distributions of the internal forces from static and dynamic analyses of a beam subjected to a concentrate impact load (F)

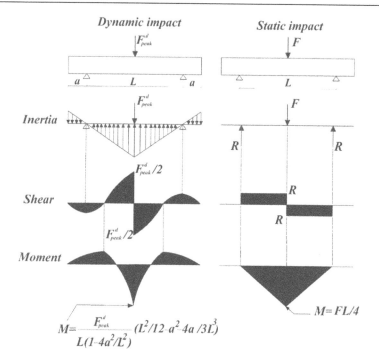

Figure 4.5 Theoretical diagrams of distribution of internal forces in static and dynamic analyses.

at the mid-span. Many studies in the literature [1, 2, 14, 19, 47, 48] have considered that the distribution of inertia is linear throughout the impacted structure.

Figure 4.6a–c shows the distribution of internal forces including inertial force, shear force, and flexural moment throughout the column for different axial load ratios at the initial peak time of impact force about 0.005 s. Also, Figure 4.6d shows the distribution of flexural moment after the formation of plastic hinges. It is observed that the peak of all internal forces increase with increasing axial load ratio. In Figure 4.6a, it is seen that the distance of the stationary point (where the inertial force of the column is zero [18]) from the column mid-span increases with increasing axial load ratio. In addition, the peaks of sagging and hogging moments increase with increasing axial load ratio as shown in Figure 4.6c. Figure 4.6d shows the distribution of flexural moments through-out the column after the time of the second peak impact during the residual displace-ments when the permanent damages are formed in the column. It should be noted that the plastic hinges are formed when the flexural moments reach the dynamic moment capacity of the column cross-section presented by the black line in Figure 4.6c and d. From Figure 4.6d, it can be found that the distance between the locations of plastic hinges formed at the two sides of the column mid-span called effective length accord-ing to [19], decreases with the increase of axial load ratio.

4.4.2 Plastic hinge locations

Some studies exist in the literature which investigated the plastic hinge location in RC beams subjected to lateral impact loads. Cotsovos et al. [19] proposed an effective length on RC beams which represents the distance between the locations of plastic

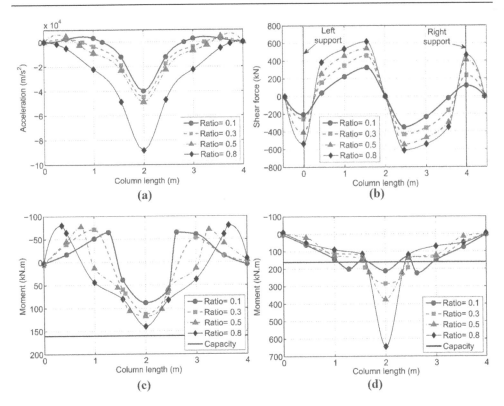

Figure 4.6 Comparing the distribution of internal forces throughout the column for different axial load ratios: (a) acceleration (inertial per unit mass of the column length), (b) shear force, (c) flexural moment at the peak time of impact force (t = 0.005 s), and (d) flexural moment after the formation of plastic hinges.

hinges formed at the two sides of the beam mid-span. This length was determined based on the stress wave propagation concepts in concrete structures subjected to impact loads. According to this approach, the stress waves produced by impact loads in the beams do not necessarily reach the supports during a very short impact duration time and the plastic hinges are formed when the cracks appear on the upper surface of the beam. In this method, the beam's inertia was not taken into account to estimate the effective length and the distribution of flexural moment throughout the beam. Figure 4.7a shows the diagram for moment distribution assumed by Cotsovos et al. [19] without considering the beam's inertia during high rates of impact loading. Therefore, in this chapter, the effective length estimated by Cotsovos et al [19]. which is called the static effective length (L_{eff}^{static}) which is calculated by Equation (4.2) based on the stress wave propagation concepts:

$$L_{eff}^{static} = 2v_w \Delta t_c \tag{4.2}$$

where v_w and Δt_c are the wave velocity and required time to reach the cracks at the upper face of the beam in static analysis, respectively, which are obtained as follows:

$$v_w = \sqrt{\frac{G}{\rho}} \tag{4.3}$$

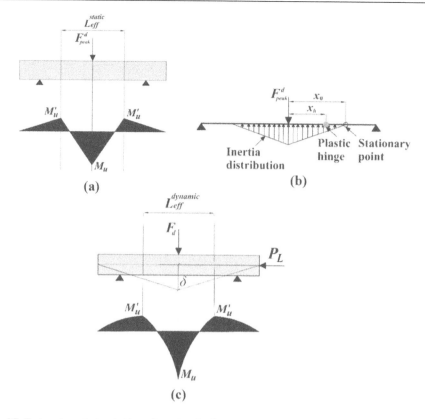

(a)

(b)

(c)

Figure 4.7 Estimation of plastic hinge location for beams subjected to impact load: (a) static analysis proposed by [19], (b) dynamic analysis proposed by [18], and (c) proposed in the present study with taking into account of axial load and the structure's inertia.

where G and ρ are the shear modulus and mass density of concrete, respectively.

$$\Delta t_c = \sqrt{\frac{4 M_{cr}}{\dot{F} . v_w}} \qquad (4.4)$$

where M_{cr} is the cracking moment which is obtained by Equation (4.5). $\dot{F}$ is the loading rate equal to $F_d / \Delta t_c$ in which F_d is the dynamic impact force at the time of reaching cracks to the upper face of the beam.

$$M_{cr} = \frac{f_t \, bh^2}{6} \qquad (4.5)$$

where f_t is the tensile strength of concrete, b and h are the width and height of the beam cross-section, respectively.

In the proposed method by Cotsovos et al. [19], by assuming the fixed ends for the beam in L_{eff}^{static}, the flexural moment equilibrium equation can be written as follows:

$$M_u = \frac{F_{peak}^d \cdot L_{eff}^{static}}{4} - M_u' \qquad (4.6)$$

where M_u is the positive plastic moment at the mid-span which was assumed equal to the flexural capacity of the beam cross section. M_u' is the negative plastic moment which was assumed equal to the cracking moment M_{cr} or $0.3M_u$. F^d_{peak} is the peak of dynamic impact force.

According to another approach proposed by Pham and Hao [18], it is possible that the plastic hinges are formed in a limited length of the beam between the stationary points where the beam's accelerations and inertia are zero as shown in Figure 4.7b. The distance between this stationary point and the beam mid-span is calculated by Equation (4.7):

$$x_0 = \frac{G_m}{m}\left(\frac{1}{2\alpha} - 2\right) \tag{4.7}$$

where x_0 is the distance between the stationary point and the beam mid-span, G_m is the impactor weight, m is the beam's mass per unit length, and α is the ratio of impactor kinetic energy E_k to the absorbed energy ΔE by the structure ($\alpha = E_k/\Delta E$).

Consequently, the location of plastic hinge formation (x_h) at the peak time of impact force is calculated according to Equation (4.8) proposed by Pham and Hao [18]:

$$x_h = x_0 - \sqrt[3]{\frac{6M_0\ x_0^2}{F^d_{peak}}} \tag{4.8}$$

where the M_0 is the plastic moment of the column which remains constant after the formation of plastic hinge during the residual response of the structure, and F^d_{peak} is the dynamic peak impact force.

Unlike the two aforementioned approaches in which the location of plastic hinge was estimated at peak time of impact force at the early stage of impact loading, this chapter proposes an improved method to estimate the location of final plastic hinges formed in the axially loaded RC columns subjected to lateral impact loads at the mid-span by considering the column's inertia as shown in Figure 4.7c. In other words, the proposed method in this study improves the method proposed by Cotsovos et al. [19] into the dynamic analysis by considering the column's inertia and axial load. In this method, the axial load (P_L) applied to the column end affects the flexural moment equilibrium when the column has a plastic displacement (δ) at the mid-span at the formation time of plastic hinge. Therefore, with neglecting the overhang length (a) of the column as given in Figure 4.5, the flexural moment equilibrium equation for a dynamic approach can be re-written as follows:

$$M_u = \frac{F_d \cdot L^{dynamic}_{eff}}{12} - M_u' + P_L \cdot \delta \tag{4.9}$$

where M_u is the positive plastic moment which is assumed equal to the flexural moment capacity of the column, M_u' is the negative plastic moment which is assumed equal to $0.3M_u$, F_d is the dynamic impact force at the formation time of plastic hinge, and $L^{dynamic}_{eff}$ is the distance between the final plastic hinges at the two sides of the column mid-span.

Figure 4.8 shows the failure modes and the location of the plastic hinge in the column subjected to a middle rate impact loading with an impact velocity of 5 m/s at the mid-span with the variability of axial load ratio (i.e., P_L/P_N which is the ratio of the axial load applied, P_L, to the axial load capacity of column, P_N). It is observed that

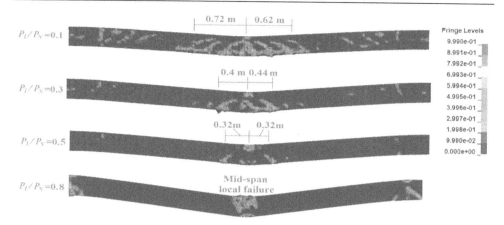

Figure 4.8 Comparison of the column failure modes and the location of the plastic hinges with the variability of axial load ratios.

the failure mode of the column changes from a global failure mode for a low axial load ratio 0.1 to a local failure mode in the high ratio of 0.8. In other words, the plastic hinges are prone to form near the column mid-span because of stress concentration at the column mid-span in proportion to the increase of axial load.

To determine the plastic hinge locations from the observations in FE simulations, the dynamic interactions of the column axial load versus the flexural moment (P–M) are compared with the P–M interaction diagram for the column capacity plotted according to ACI-318 [45] at critical zones throughout the column as shown in Figure 4.9a–d. It is seen that the severity of the damages increases with the increasing axial load ratio. Figure 4.10a shows the sensitivity of plastic hinge location to the axial load ratio obtained from the calculation method proposed in the present study by Equation (4.9) (*Cal*), from the observations in FE simulations by checking the P–M interaction diagrams (*Obs*), and from the calculation methods proposed by the previous works [18, 19]. It is revealed that the distance of plastic hinge formation from the column mid-span [i.e., the half of effective length, $L_{eff}^{dynamic}$ in Equation (4.9)] decreases with the increase of axial load ratio for all methods. Also, it is seen that the calculation and the observation results from the FE simulations are in good agreement. Besides, the results estimated by Pham and Hao [18] are more consistent with the results from the proposed method in the present study rather than those estimations proposed by Cotsovos et al. [19]. This is because of taking into account the column's inertia effects in both proposed methods in the present study and the work done by Pham and Hao [18]. Figure 4.10b shows the comparison of plastic hinge locations in the column subjected to different impact velocities and axial load ratios between the observations on FE simulations (*Obs*) and from the calculation method proposed in Equation (4.9) (*Cal*). It is found that the plastic hinge location is not only affected by the axial load ratio but also is sensitive to the impact velocity. It is observed that the distance of the plastic hinge location from the column mid-span decreases with increasing impact velocity. Also, Figure 4.10b demonstrates that the estimated results from the calculation and observation methods are in good agreement.

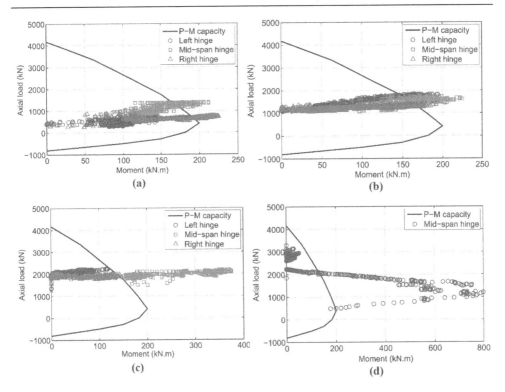

Figure 4.9 Determination of plastic hinges locations using the *P–M* interaction diagrams of the column and from the observations in FE simulations for different axial load ratios: (a) $P_L/P_N = 0.1$, (b) $P_L/P_N = 0.3$, (c) $P_L/P_N = 0.5$, (d) $P_L/P_N = 0.8$.

4.4.3 Different impact locations

In this section, the effects of axial load on the failure modes and the impact responses of the RC column subjected to lateral impact loads at low elevations are evaluated. The failure modes of the column with an impact elevation of 0.5 m above the column base for the middle and high impact velocities of 5 and 10 m/s

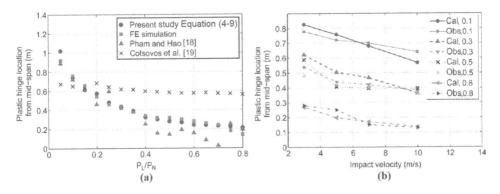

Figure 4.10 Comparing the sensitivity of plastic hinge location between the different methods for (a) axial load variation and (b) impact velocity variation.

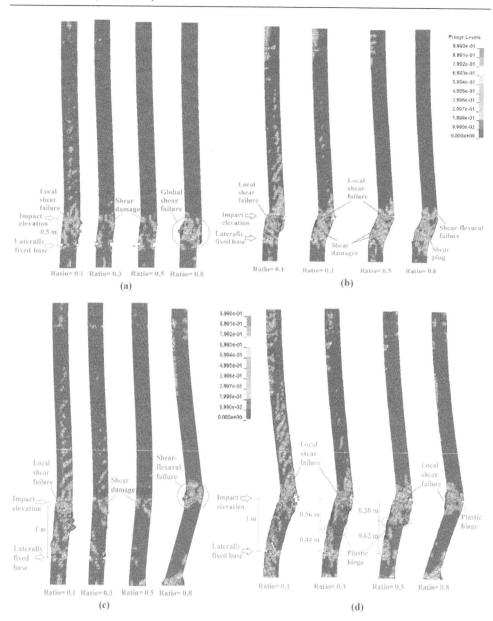

Figure 4.11 Comparing the column failure modes with the variability of axial load ratio subjected to different impact elevations (H_i) and different impact velocities (V). (a) H_i=0.5 m, V=5 m/s; (b) H_i= 0.5 m, V = 10 m/s; (c) H_i = 1 m, V = 5 m/s; (d) H_i = 1 m, V = 10 m/s.

are shown in Figure 4.11a and b, respectively. It is observed that the increase of axial load ratio causes the change of local shear failure mode of the column at the impact elevation to a global shear failure associated with the shear plug damage originated from the column base. In other words, the severity of local shear failure decreases at the impact elevation and it increases at the column base with the increase of axial load ratio. This is because of increasing the trapped axial stresses at the base of the column in proportion to the increase of axial load ratio. As shown in Figure 4.11b, for the impact scenario with a high impact velocity of 10 m/s, the column suffers a shear-flexural failure mode because of the formation of a plastic hinge at the impact elevation for high ratios of axial load. However, the formation of the plastic hinge at this elevation is not seen for the impact scenarios with a lesser impact velocity (i.e., 5 m/s) as shown in Figure 4.11a. For this impact rate, the column is prone to a global shear failure mode. Also, the increase of the column lateral shear resistance for the middle ratios of axial load (i.e., 0.3 and 0.5) is observed for both high and middle impact rates in Figure 4.11a and b.

Figure 4.11c and d shows the column failure modes and the location of plastic hinges for middle and high impact rates with impact velocities of 5 and 10 m/s, respectively, when the impact elevation is 1 m above the column base. It can be seen that the distance of plastic hinges from the impact point decreases with the increase of axial load ratio for a high impact rate in Figure 4.11d. Accordingly, for a high axial load ratio of 0.8, the column suffers a local shear-flexural failure at the impact elevation through the formation of a plastic hinge at this elevation. However, for the cases with the lesser ratios (i.e., the ratios from 0.1 to 0.5), the column suffers a global shear-flexural failure mode. This combined failure mode includes the occurrence of a local shear failure at the early stage of the column response and the formation of plastic hinges around the impact point at the following phase of the column response. For the middle impact rate scenarios, the axial load plays a positive role in the column shear resistance as shown in Figure 4.11c.

Figure 4.12a–h shows the comparison of the impact force and the displacement results of the column subjected to the middle and high rate impacts with the impact velocities of 5 and 10 m/s, respectively, at low impact elevations of 1 and 0.5 m from the column base with the variability of axial load ratio. By comparing the impact force results as given in Figure 4.12a, c, e, and g for the impact scenarios at low elevations with those from the mid-span impact scenario in Figure 4.4c and e, it can be found that the value of the first peak impact force increases with the decrease of impact elevation. Also, it is observed that the increase of the first peak impact force is more significant when both impact rate and axial load ratio are in high ranges. Besides, with decreasing impact elevation, the value of the following impulses observed for the column mid-span impact scenario decreases because of the degradation of the column vibrations during its vibration phase.

In Figure 4.12b and f, it is seen that the first peak of displacement decreases with the increase of axial load ratio which is like that behavior observed for the mid-span impact scenario for a middle impact rate (also see Figure 4.4d). However, the residual displacements present different behaviors. Unlike the mid-span impact, the column demonstrates an extra resistance against the lateral impact force for the axial load ratios of 0.3 and 0.5. Therefore, the column has a lesser residual displacement in these ranges of the axial load for the impact scenarios at the low elevations. However, these behaviors are not observed for those impact scenarios with high impact rates in Figure 4.12d and h because of losing the column shear and flexural strength at the early stage of its response immediately after the first peak impact force.

Recently, a comprehensive study was done by Aghdamy et al. [13] on the influence of different loading-related and structural-related parameters including impactor's mass, initial impact velocity, impact location, initial contact area, and compressive axial load on the response of concrete-filled double-skin steel tube columns subjected

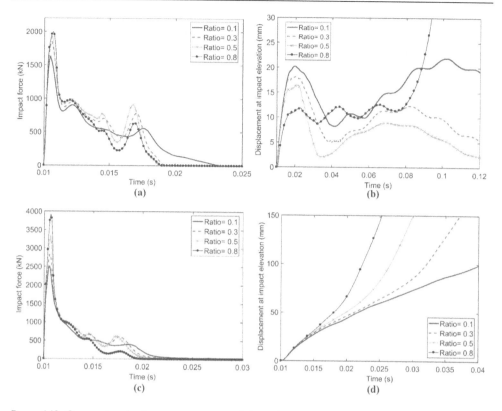

Figure 4.12 Comparing the impact responses of the column with the variability of axial load ratio subjected to different impact elevations (H_i) and different impact velocities (V). (a) Impact force for $H_i = 0.5$ m and $V = 5$ m/s; (b) Displacement for $H_i = 0.5$ m and $V = 5$ m/s; (c) Impact force for $H_i = 0.5$ m and $V = 10$ m/s; (d) Displacement for $H_i = 0.5$ m and $V = 10$ m/s.

to lateral impact load t. However, in the present study, it is found that the increase of axial load significantly increases the first peak impact force, particularly when the impact loading rate is relatively high. Also, in the work done by Aghdamy et al. [13], it was claimed that the influence of the impact location is negligible on the initial peak impact force. While from the results of this study, it is found that the impact location has a significant influence on the initial peak impact force and the failure modes of the column.

4.4.4 Parametric analysis

The sensitivity of the different column response parameters to the mid-span impact loading scenarios with the variability of axial load is assessed. Afterward, the empirical relationships between the column impact responses (including the peak impact force, peak flexural moment at the column mid-span, and the location of the plastic hinge from the column mid-span) and the axial load ratio (P_L/P_N) are quantified by varying impact velocities (V). Other parameters related to the column cross-sectional material, dimensions, and the weight of the dropping mass are assumed to be constant. As shown in Figure 4.13a and b, it is revealed that the peak of the impact force

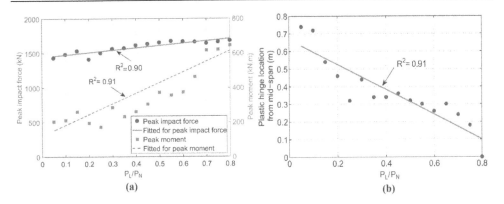

Figure 4.13 Sensitivity of the column impact response parameters to the variability of the axial load ratio: (a) peak of impact force and flexural moment and (b) plastic hinge location from the column mid-span.

and the flexural moment significantly increase at the column mid-span with increasing axial load ratio with the high correlations, while the distance of the plastic hinge from the column mid-span decreases with increasing axial load ratio.

To quantify the relationships between the column impact responses [i.e., peak impact force (F_{max}), peak flexural moment (M_{max}), and the distance of plastic hinge location from the mid-span (x_h)] and the axial load ratio (P_L/P_N) with the variability of impact velocity (V), the relative parameters of the impact responses are defined as follows:

$$\bar{F}_{max} = \frac{F_{max}}{F_{max0}} \tag{4.10a}$$

$$\bar{M}_{max} = \frac{M_{max}}{M_{max0}} \tag{4.10b}$$

$$\bar{x}_h = \frac{x_h}{x_{h0}} \tag{4.10c}$$

where $\bar{F}_{max}$, $\bar{M}_{max}$, and $\bar{x}_h$ are the relative peak impact force, peak flexural moment, and plastic hinge location from the column mid-span, respectively, which vary with the axial load ratio. F_{max0}, M_{max0}, and x_{h0} are the peak impact force, peak flexural moment, and plastic hinge location varying with impact velocity when the axial load ratio is zero.

The relationships between the impact responses (i.e., F_{max0}, M_{max0}, and x_{h0}) and the impact velocities when the axial load ratio is zero, are shown in Figure 4.14a and d. The linear fitting formulas written between these parameters are given in Equations (4.11a)–(4.11c), respectively. Also, the relationships between the relative impact responses ($\bar{F}_{max}$, $\bar{M}_{max}$, and $\bar{x}_h$) and the axial load ratio are shown in Figure 4.14b, c, and e, respectively. Also, the logarithmic and exponential fitting formulas written between these parameters are given in Equations (4.12a)–(4.12c), respectively. From Figure 4.14b, c, and e, it is observed that the value of all relative impact responses increases with a steep slope until an axial load ratio of 0.5.

$$F_{max0} = 129V + 614.2 \tag{4.11a}$$

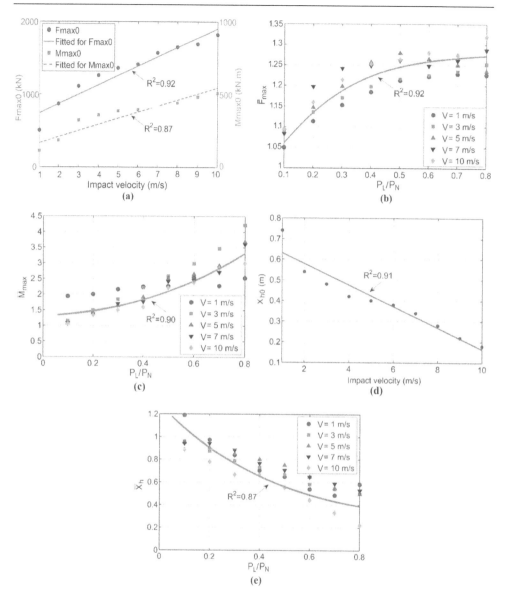

Figure 4.14 Relationships between the column impact responses and the loading-related parameters: (a) F_{max0}, M_{max0} and V, (b) $\bar{F}_{max}$ and P_L/P_N, (c) $\bar{M}_{max}$ and P_L/P_N, (d) x_{h0} and V, and (e) $\bar{x}_h$ and P_L/P_N.

where F_{max0} and V are kN and m/s. respectively.

$$M_{max0} = 39.05V + 145.4 \qquad (4.11b)$$

where M_{max0} and V are kN m and m/s. respectively.

$$x_{h0} = -0.05V + 0.68 \qquad (4.11c)$$

where x_{h0} and V are m and m/s. respectively.

$$\bar{F}_{max} = 0.09\ln(P_L/P_N)+1.29 \tag{4.12a}$$

$$\bar{M}_{max} = 1.84e^{(P_L/P_N)} - 0.75 \tag{4.12b}$$

$$\bar{x}_h = -0.26\ln(P_L/P_N)+0.45 \tag{4.12c}$$

By substituting Equations (4.11a) and (4.12a)into Equation (4.10a), Equations (4.11b) and (4.12b) into Equation (4.10b), and Equations (4.11c) and (4.12c) into Equation (4.10c), F_{max}, M_{max}, and x_h, varying with both impact velocity (V) and axial load ratio (P_L/P_N) parameters can be expressed as follows:

$$F_{max} = (129V+614.2)[0.09\ln(P_L/P_N)+1.29] \tag{4.13a}$$

$$M_{max} = (39.05V+145.4)[1.84e^{(P_L/P_N)} - 0.75] \tag{4.13b}$$

$$x_h = (-0.05V+0.68)[-0.26\ln(P_L/P_N)+0.45] \tag{4.13c}$$

4.5 Case study

In this section, to prove the effects of axial load on the dynamic responses and the failure behaviors of the axially loaded impacted columns in the realistic applications, a numerical study of ship collision with a cable-stayed bridge pier in the cases with and without the superstructure is carried out in LS-DYNA. A high-energy impact scenario of the ship collision on the bridge pier is considered in which the impact characteristics are impact velocity of 5 m/s, ship impact weight of 10,000 tons, and a head-on impact on the piles' cap level.

In this chapter, a cable-stayed Zhanjiang Bay Bridge located in Guangdong province in China is considered as a case study subjected to ship collision. The bridge includes a main cable-stayed bridge with two pylons (i.e., the piers with numbers 3 and 4) and a steel deck with a main span of 480 m as shown in Figure 4.15. The side spans have been built of concrete decks and girders, while the remaining spans adopt the steel box girder. The height of the ship navigation clearance is 48 m, and the least

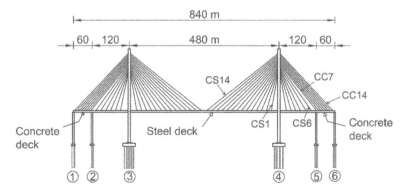

Figure 4.15 Longitudinal view of cable-stayed Zhanjiang Bay Bridge.

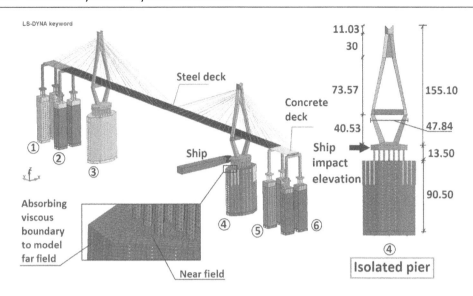

Figure 4.16 FE model of Zhanjiang Bay Bridge in the cases with and without the superstructure (unit in m).

width of clearance is 400 m [49]. Figure 4.16 shows the details of FE modeling of the bridge structure, superstructure, and substructure in LS-DYNA in the cases with and without the superstructure. The height of the main pylons and the piles from the piles' cap surface is 155.1 and 104 m, respectively. The foundation of main piers has 31 friction concrete piles with a diameter of 2.5 m, and the mud-line level is at the depth of 13.5 m lower than piles' cap surface.

In LS-DYNA, the concrete material of all piers and the concrete decks are modeled using 8-node constant stress solid element, reinforcements of the piers and the cables using beam elements, and steel deck plates using four-node shell elements. The soil-structure interaction (SSI) is modeled using continuum model of the soil behavior in the near field using solid elements, while the viscous dashpot elements are used as the absorbing boundary to prevent the reflection of stress waves from the far field. Because of limited information about the soil layer characteristics and the ratio of reinforcements, the reasonable assumptions are considered in this study. The damping coefficient of viscous damper elements (C_s) attached to a certain partial area (A_p) related to the specific soil layer with the mass density (ρ_s) is calculated as follows:

$$C_s = \rho_s.A_p \tag{4.14}$$

The concrete of piers with grade 50 (*C*50) is modeled using a continuous surface cap model (CSCM) named MAT_CSCM_CONCRETE (MAT_159) in LS-DYNA to capture the damage behaviors and important mechanical features of the concrete material such as the pressure hardening, strain hardening, and third stress invariant-dependent deviatoric plane [50]. The steel beam elements used for the reinforcements are modeled by the MAT_PLASTIC_KINEMATIC (MAT_003) similar to that model used for ship bow steel plates, and an elastic-plastic material by considering the strain rate effects named MAT_PIECEWISE_LINEAR _PLASTICITY (MAT_24) is used for the steel deck. In addition, steel cables are modeled by truss beam elements with an elastic material model named MAT_CABLE_DISCRETE_BEAM (MAT_71). Furthermore, the nonlinearity of soil behavior surrounding the piles is modeled using an elastic-plastic constitutive model named MAT_MOHR_COULUMB (MAT_173). Owing to the limited information on soil strength characteristics, those recommended parameters in the literature [9, 51, 52] are utilized in this study. The material properties used for bridge and ship components are tabulated in Table 4.2.

In this study, a container ship with a weight of 10,000 tons is considered as the impacting vessel on the piles' cap level of pier-4 as shown in Figure 4.16. Figure 4.17 shows the details of the FE model of the ship including the internal stiffeners, outer plates in the bow portion which are modeled using four-node shell elements, and the stern portion is modeled using eight-node solid element in LS-DYNA. According to the previous studies in the literature [53, 54], the ratio of length to thickness of the FE elements used for the ship bow is selected in the range of 8–10 to capture the converge results. Totally, 39,292 and 48,816 elements are used for internal stiffener plates and outer plates in ship bow portion, respectively. A series of numerical mesh size convergence tests were carried out for FE model of the ship to reach the reliable ship-pier collision results. In these tests, the ship bow collides with a rigid wall with different mesh sizes as shown in Figure 4.18. It was found that impacted wall with a

Table 4.2 Material properties used for bridge simulations in LS-DYNA

Member	Material model	Parameters	Values
Pier concrete	MAT_CSCM_CONCRETE	Mass density	2,400 kg/m³
		Unconfined compression strength	50.0 MPa
		Elements erosion (ERODE)	1.08
		Maximum aggregate size (Dagg)	19.0 mm
		Rate effects	Turn on
Steel reinforcement bars	MAT_PLASTIC_ KINEMATIC	Mass density	7,865 kg/m³
		Young's modulus	2.1×10^5 MPa
		Poisson's ratio	0.27
		Yield stress	310 MPa
		Failure strain	0.35
		C_{strain} parameter	40.4
		P_{strain} parameter	5.0
Steel deck	MAT_PIECEWISE_ LINEAR _PLASTICITY	Mass density	7,850 kg/m³
		Young's modulus	2.1×10^5 MPa
		Poisson's ratio	0.3
		Yield stress	350 MPa
		Failure strain	0.15
Steel cables	MAT_CABLE_DISCRETE_ BEAM	Mass density	7,850 kg/m³
		Young's modulus	2.1×10^5 MPa
		Initial tensile force	Vary from 2.1 to 4.9 MPa
Soil in near field	MAT_MOHR_COULUMB	Mass density	1,890 kg/m³
		Shear modulus	6.32 MPa
		Poisson's ratio	0.25
		Angle of friction	32°
		Value of cohesion	11.0 kPa

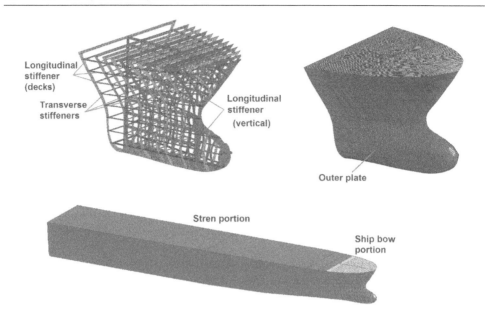

Figure 4.17 FE model of container ship used in this study.

mesh size of 200 mm can capture the converge results. Therefore, the mesh size of 200 mm is selected for FE model of the impacted pier at the impact zone on the piles' cap and for the pier columns. The calibration of contact force versus the crushing depth of the ship bow during the collision with a rigid wall with a mesh size of 200 mm is shown in Figure 4.18a and b. The results from the present study are compared with those from the previous work [55, 56] and the standard codes [57–59]. The ratio of longitudinal reinforcements for all piers is considered around 2–4%. By considering the minimum requirements according to standard codes [45], the space between the

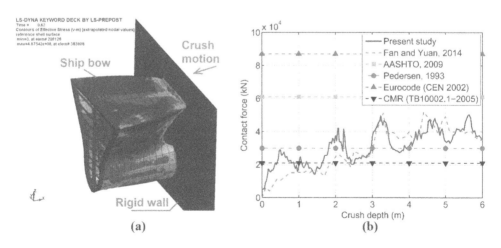

Figure 4.18 Calibration of FE model of ship bow: (a) ship-rigid wall convergence tests and (b) comparing the contact force versus crushing depth between the present study and the previous works.

Table 4.3 Contact algorithms defined in LS-DYNA

Contact algorithm	Between the components of
AUTOMATIC_NODES_TO_SURFACE	Piles and soil
AUTOMATIC_SINGLE_ SURFACE	Ship bow internal plates and outer plates
AUTOMATIC_SURFACE_TO_SURFACE	Ship bow and pier
	Deck and piers
TIDE_NODES_TO_SURFACE	Steel deck and concrete decks
TIDE_ SURFACE _TO_SURFACE	Soil layers

transverse reinforcements is 300 and 150 mm in the columns and the beam member, respectively. In order to simulate the interaction between the concrete and reinforcements, a coupling algorithm called CONSTRAINED_LAGRANGE_IN_SOLID is used in LS-DYNA. To simulate the contact and interactions between the separate components and to prevent from any penetrations of the elements, several contact algorithms are used in LS-DYNA, which are tabulated in Table 4.3.

To simulate the ship-bridge pier collision with considering the superstructure axial load, two loading stages, (i) stress initialization before ship-bridge collision and (ii) ship collision stage as the transient loading stage, are assigned in LS-DYNA. In the first stage, the whole bridge is subjected to the gravity load under self-weight and the prestress tensile forces are applied to the cables using a dynamic relaxation method with a control keyword named CONTROL_ DYNAMIC_ RELAXATION in LS-DYNA. Also, a global damping 5% was adopted to reach an equilibrium state during the dynamic relaxation stage. The gravity load is applied to bridge superstructure during both relaxation and transient states. In Figure 4.19a and b, the results from the dynamic relaxation analysis stage are shown. It is obtained that bridge deck reaches an equilibrium state after 0.1 s in dynamic relaxation with the prestress tensile forces of cables from 2.1 to 4.9 MN. The cable identifications (ID) are presented in Figure 4.15 in which cables attached to the steel deck have an ID of CS and those attached to the concrete decks have an ID of CC. Since the arrangement of the cables is symmetric on the bridge, the cables forces are shown in Figure 4.19a only for the half-length of the bridge.

Figure 4.20a and b shows the comparison of contact force and lateral displacement of the impacted pier (pier-4), respectively, between the cases with and without the

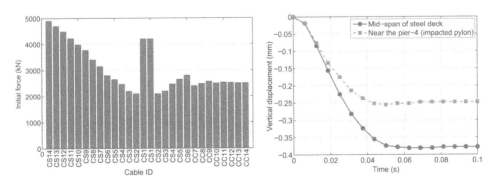

Figure 4.19 Results from the stress initialization stage: (a) initial forces of the cables and (b) vertical displacements of the bridge deck during dynamic relaxation stage.

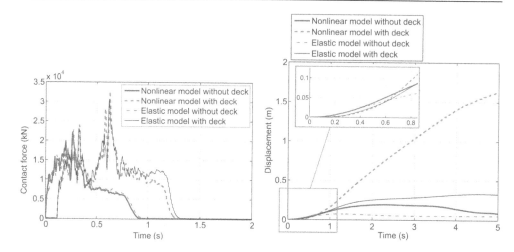

Figure 4.20 Comparing the pier responses between the cases with and without superstructure: (a) contact force and (b) displacement at the impact level (piles' cap level).

superstructure for both nonlinear and linear elastic material models for the pier. In Figure 4.20a, it is observed that the values of peak impact force and the impact duration time for the case with superstructure are greater than those of isolated model of the pier without superstructure. Also, it is obtained that the severity of ship impact on the pier superstructure significantly increases at the second peak time of impact force about 0.7 s after the collision. Because the superstructure's inertia causes the increase of lateral resistance of the pier. In Figure 4.20b, although the pier demonstrates a lesser displacement for that case with superstructure than that of the isolated pier before the second peak time at 0.7 s, the axial load of the superstructure causes the increase of the residual displacements after this time because of losing the pier strength and the progressive damages.

Figure 4.21a and b shows the comparison of the pier inertial force and the substructure forces (including the shear force at the piles' head level and the total soil lateral

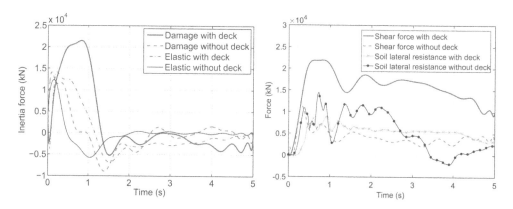

Figure 4.21 Comparison of inertial force and substructure forces of the pier between the models with and without the superstructure: (a) inertial force and (b) shear force at the piles' head, and the total soil lateral resistance.

resistance surrounding the piles), respectively, between the cases with and without the superstructure. The greater peak inertial force against the ship impact force is seen for the pier model in the presence of the superstructure rather than to that of the isolated model. Also, some phase discrepancies in the pier inertia responses are observed between the models with and without superstructure after rebounding the ship from the pier after the time 1.0 s. From Figure 4.21b, it is observed that the superstructure axial load causes not only the increase of inertial force of the pier but also the shear force at the piles' head. In contrast, the peak soil lateral force surrounding the piles decreases in the presence of the superstructure because of decreasing the pier lateral deflection. In addition, the comparison of the distribution of the internal forces including the inertial force (acceleration-related) and the flexural moments throughout the height of impacted pier from the piles' tip level at the peak time of impact force are shown in Figure 4.22a and b, respectively, between the cases with and without the superstructure. It is observed that both peak inertial force and peak flexural moment at the impact elevation on the piles' cap level and the piles' head (specified level by the black line) have greater values in the presence of superstructure than that of the isolated model.

In order to evaluate the effects of axial load on the failure mode and damage state of the pier, the situation and curvature of the plastic hinges formed in the pier columns are assessed for both cases with and without superstructure (deck) as shown in Figure 4.23a and b, respectively. The plastic hinges formed at the base of top columns in the front side of the impacting ship and in the back side named PH_1 and PH_2, respectively, are considered to study their curvature using their down and upper angles named a_1 and a_2, respectively. The formation of extra plastic hinges is seen in the slender top columns of the pier in the isolated model comparing with the pier model in the presence of the superstructure. Since the isolated is not restrained at the top boundaries, the impact load can be easily transferred upward the pier. Therefore, the pier presents a global response and the location of plastic hinges transfers of upward the pier.

In other words, the pier in the presence of the superstructure is able to redistribute the impact load downward, which leads to the increase in the severity of the local shear failures throughout the pier height. Figure 4.24a and b shows the intensity of the

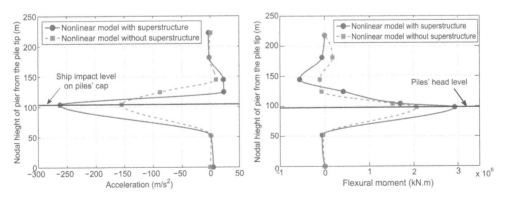

Figure 4.22 Comparing the distribution of internal forces throughout the pier height from the piles' tip level at the peak time of impact force between the cases with and without superstructure: (a) acceleration (inertia related) distribution and (b) flexural moment distribution.

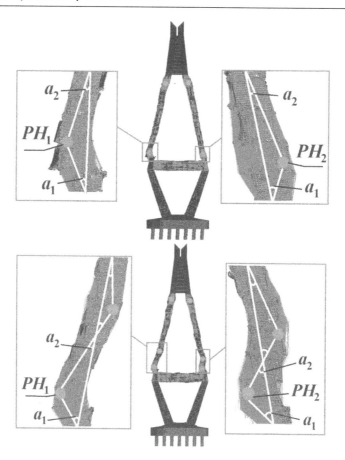

Figure 4.23 Damage states and plastic hinges situations in the column of the impacted pier in cases of (a) with superstructure and (b) without superstructure.

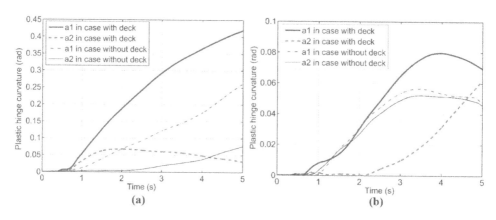

Figure 4.24 Conclusions comparing the curvature of plastic hinges at the base of top column: (a) plastic hinge PH_1 and (b) plastic hinge PH_2.

plastic hinges formed in the pier top columns by assessing the curvature angles. It is observed that curvature angles of the plastic hinges formed in both top columns of the pier in the case with superstructure are greater than those of isolated model.

4.6 Conclusions

In spite of some limited studies on the effects of axial load on the behavior of steel columns, this chapter evaluated the influences of the axial load on the distribution internal forces, the location of plastic hinges, and the failure modes of RC columns subjected to lateral impact load using FE simulations in LS-DYNA. First, the sensitivity of the impact responses and the failure behaviors of an axially loaded column subjected to dropping mass at the column mid-span varying with different axial load ratios were investigated. It was obtained that the increase of axial load ratio causes the increase of peak impact force and the internal forces of the column including the inertial force, the flexural moment, and the shear force. For the scenarios with a middle impact velocity, it was found that the axial load has no significant effect on the residual displacements of the column between the ratios 0.1 and 0.5 because of increasing the lateral resistance of the column and the degradation of the column vibrations in the vibration phase of response. Also, by observing the decrease of the distance of the plastic hinges from the column mid-span, it was found that the column failure mode changes from a global failure mode to local failure in proportion to increase of axial load ratio. From a comparative study on the effects of the axial load on the column impact responses and failure modes with different impact locations at low elevations, it was found that the initial peak impact force increases with the decrease of the impact elevation particularly for high ranges of the impact velocity and axial load ratio. Also, the failure mode of the column changed from a global-flexure mode to the shear-flexural or global shear failure modes with the decrease of the impact elevation.

In addition, a parametric study was carried out to quantify the relationships between the column impact responses (including the peak impact force, the peak flexural moment, and the location of plastic hinges) and the axial load ratio by varying the impact velocity for the column mid-span impact scenarios. It was found that the estimation of plastic hinge location by considering the column's inertia gives more consistent results with those from the observations on FE simulations rather than that estimation method without considering the inertia.

To prove the axial load effects on the response of the axial load carrying structures subjected to lateral impact loads in the realistic applications, a ship-bridge pier collision case study is simulated by FE numerical software LS-DYNA. The ship impact load is applied on the cable-stayed bridge pier in different cases with and without the superstructure. From the results, it was obtained that superstructure axial load increases the impact force and the peak internal forces including the inertial force, flexural moment, and shear force. Also, it was revealed that the pier with superstructure tends to transfer the impact load downward the pier because of redistribution of impact load which leads to the increase in the intensity of the local shear failures.

References

1. Banthia N, Mindess S, Bentur A, Pigeon M. Impact testing of concrete using a drop–weight impact machine. *Exp Mech* 1989;29(1):63–69.

2. Cotsovos DM. A simplified approach for assessing the load-carrying capacity of reinforced concrete beams under concentrated load applied at high rates. *Int J Impact Eng* 2010;37(8):907–917.

3. Saatci S, Vecchio FJ. Nonlinear finite element modeling of reinforced concrete structures under impact loads. *ACI Struct J* 2009;106(5):717–725.

4. Fujikake K, Li B, Soeun S. Impact response of reinforced concrete beam and its analytical evaluation. *J Struct Eng ASCE* 2009;135(8):938–950.

5. Pham TM, Hao H. Behavior of fiber-reinforced polymer-strengthened reinforced concrete beams under static and impact loads. *Int J Protect Struct* 2017;8(1):3–24.

6. Louw JM, Maritz G, Loedolff MJ. RC cantilever columns under lateral impact load: An experimental investigation, Structures under Shock and Impact II, In Proceedings of the 2nd International Conference, Portsmouth, UK, 1992:309–319.

7. Remennikov AM, Kaewuruen S. Impact resistance of reinforced concrete columns: experimental studies and design considerations. In 19th Australasian Conference on the Mechanics of Structures and Materials, Christchurch, New Zealand, 2006:817–824.

8. Bisby L, Ranger M. Axial–flexural interaction in circular FRP-confined reinforced concrete columns. *Constr Build Mater* 2010;24(9):1672–1681.

9. Zeinoddini M, Parke GAR, Harding JE. Axially pre-loaded steel tubes subjected to lateral impacts: an experimental study. *Int J Impact Eng* 2002;27:669–690.

10. Zeinoddini M, Harding JE, Parke GAR. Axially pre-loaded steel tubes subjected to lateral impacts: a numerical simulation. *Int J Impact Eng* 2008;35(11):1267–1279.

11. Al-Thairy H, Wang YC. A numerical study of the behaviour and failure modes of axially compressed steel columns subjected to transverse impact. *Int J Impact Eng* 2011;38(8):732–744.

12. Alam MI, Fawzia S Numerical studies on CFRP strengthened steel columns under transverse impact. *Compos Struct* 2015;120:428–41.

13. Aghdamy S, Thambiratnam DP, Dhanasekar M, Saiedi S. Effects of load-related parameters on the response of concrete-filled double-skin steel tube columns subjected to lateral impact. *J Constr Steel Res* 2017;138:642–662.

14. Pham TM, Hao H. Impact behavior of FRP–strengthened RC beams without stirrups. *J Compos Constr ASCE* 2016;20(4):04016011.

15. Thilakarathna I, Thambiratnam D, Dhanasekar M, Perera N. Shear-critical impact response of biaxially loaded reinforced concrete circular columns. *ACI Struct J* 2013;110(4):565.

16. Huynh L, Foster S, Valipour H, Randall R. High strength and reactive powder concrete columns subjected to impact: experimental investigation. *Constr Build Mater* 2015;78:153–171.

17. Pham TM, Hao H. Effect of the plastic hinge and boundary conditions on the impact behavior of reinforced concrete beams. *Int J Impact Eng* 2017;102:74–85.

18. Pham TM, Hao H. Plastic hinges and inertia forces in RC beams under impact loads. *Int J Impact Eng* 2017;103:1–11.

19. Cotsovos DM, Stathopoulos ND, Zeris CA. Behavior of RC beams subjected to high rates of concentrated loading. *J Struct Eng ASCE* 2008;134(12):18or39–18or51.

20. Guo J, Cai J, Chen W. Inertial effect on RC beam subjected to impact loads. *Int J Struct Stab Dy* 2017;17(04):1750053.

21. Alam MI, Fawzia S, Zhao XL. Numerical investigation of CFRP strengthened full scale CFST columns subjected to vehicular impact. *Eng Struct* 2016;126:292–310.

22. Thilakarathna HM, Thambiratnam DP, Dhanasekar M, Perera N. Numerical simulation of axially loaded concrete columns under transverse impact and vulnerability assessment. *Int J Impact Eng* 2010;37(11):1100–1112.

23. Jiang H, Chorzepa MG. An effective numerical simulation methodology to predict the impact response of pre-stressed concrete members. *Eng Fail Anal* 2015;55:63–78.

24. Madurapperuma MA, Wijeyewickrema AC. Performance of reinforced concrete columns impacted by water-borne shipping containers. *Adv Struct Eng* 2012;15(8):1307–1327.
25. Al-Thairy H, Wang YC. A simplified analytical method for predicting the critical velocity of transverse rigid body impact on steel columns. *Int J Impact Eng* 2013;58:39–54.
26. Makarem FS, Abed F. Nonlinear finite element modeling of dynamic localizations in high strength steel columns under impact. *Int J Impact Eng* 2013;52:47–61.
27. AuYeung S, Alipour A. Evaluation of AASHTO suggested design values for reinforced concrete bridge piers under vehicle collisions. *Transp Res Rec* 2016; 2592:1–8.
28. Abdelkarim OI, ElGawady MA. Design of short reinforced concrete bridge columns under vehicle collision. *Transp Res Rec* 2016;2592:27–37.
29. Agrawal AK, Liu GY, Alampalli S. Effects of truck impacts on bridge piers. *Adv Mat Res* 2013;639:13–25.
30. El-Tawil S, Severino E, Fonseca P. Vehicle collision with bridge piers. *J Bridge Eng ASCE* 2005;10(3):345–353.
31. Abdelkarim OI, ElGawady MA. Performance of bridge piers under vehicle collision. *Eng Struct* 2017; 140:337–352.
32. Abdelkarim OI, ElGawady MA. Performance of hollow-core FRP–concrete–steel bridge columns subjected to vehicle collision. *Eng Struct* 2016;123:517–531.
33. Sharma H, Gardoni P, Hurlebaus S. Performance-based probabilistic capacity models and fragility estimates for RC columns subject to vehicle collision. *Comput Aided Civ Inf* 2015;30(7):555–569.
34. Sharma H, Hurlebaus S, Gardoni P. Performance-based response evaluation of reinforced concrete columns subject to vehicle impact. *Int J Impact Eng* 2012;43:52–62.
35. Davidson MT, Consolazio GR, Getter DJ, Shah FD. Probability of collapse expression for bridges subject to barge collision. *J Bridge Eng ASCE* 2012;18(4):287–296.
36. Getter DJ, Consolazio GR, Davidson MT. Equivalent static analysis method for barge impact–resistant bridge design. *J Bridge Eng ASCE* 2011;16(6):718–727.
37. Sha Y, Hao H. Laboratory tests and numerical simulations of barge impact on circular reinforced concrete piers. *Eng Struct* 2013;46:593–605.
38. Sha Y, Hao H. Nonlinear finite element analysis of barge collision with a single bridge pier. *Eng Struct* 2012;41:63–76.
39. Sha Y, Hao H. Numerical simulation of barge impact on a continuous girder bridge and bridge damage detection. *Int J Protect Struct* 2013;4(1):79–96.
40. Jiang H, Wang J, Chorzepa MG, Zhao J. Numerical investigation of progressive collapse of a multispan continuous bridge subjected to vessel collision. *J Bridge Eng ASCE* 2017;22(5):04017008.
41. Davidson M, Consolazio G, Getter D. Dynamic amplification of pier column internal forces due to barge–bridge collision. *Transp Res Rec* 2010;2172:11–22.
42. LS–DYNA 971, Livermore Software Technology Corporation, Livermore, CA, 2015.
43. Feyerabend M. Hard transverse impacts on steel beams and reinforced concrete beams. PhD Thesis, University of Karlsruhe (TH), Karlsruhe, Germany, 1988 (in German).
44. LS–DYNA keyword user's manual ver. 971: Livermore Software Technology Corporation (LSTC), Livermore, CA, 2016.
45. ACI 318-14, Building code requirements for structural concrete and commentary, American Concrete Institute, Farmington Hills, MI, 2014.
46. Pham TM. Influence of global stiffness and equivalent model on prediction of impact response of RC beams. *Int J Impact Eng* 2018;113:88–97.
47. Saatci S, Vecchio FJ. Effects of shear mechanisms on impact behavior of reinforced concrete beams. *ACI Struct J* 2009;106(1):78–86.
48. Pham TM, Hao H. Prediction of the impact force on reinforced concrete beams from a drop weight. *Adv Struc Eng* 2016;19(11):1710–1722.

49. Cao Y, Luo L, Zhou Y. Ship Collision Protection Device for Zhanjiang Bay Bridge. In *17th Congress of IABSE*, Chicago, FL, 2008.

50. Murray YD. User manual for LS–DYNA concrete material model 159, Report No. FHWA-HRT-05-062. APTEK, Inc., Colorado Springs, CO; 2007.

51. Fan W, Yuan WC. Numerical simulation and analytical modeling of pile-supported structures subjected to ship collisions including soil-structure interaction. *Ocean Eng* 2014;91:11–27.

52. Reese LC, Van Impe WF. *Single Piles and Pile Groups under Lateral Loading*. Taylor & Francis Group PLC, London, UK, 2001.

53. Fan W, Yuan W. Ship bow force–deformation curves for ship–impact demand of bridges considering effect of pile–cap depth. *Shock Vib* 2014;201425:1–19.

54. Alsos HS, Amdahl J. On the resistance of tanker bottom structures during stranding. *Mar Struct* 2007;20(4):218–237.

55. Pedersen PT, Valsgard S, Olsen D, Spangenberg S. Ship impacts: bow collisions. *Int J Impact Eng* 1993;13(2):163–187.

56. Fan W, Yuan W Ship bow force–deformation curves for ship–impact demand of bridges considering effect of pile–cap depth. *Shock Vib* 2014;201425:1–19.

57. AASHTO. *Guide Specifications and Commentary for Vessel Collision Design of Highway Bridges*, 2nd edn. American Association of State Highway and Transportation Officials, Washington, DC, 2009.

58. European Committee for Standardization (CEN), EUROCODE 1-Actions on structures. Parts 1–7: General Actions-Accidental actions due to impact and explosions, 3rd draft, CEN, Brussels, Belgium, 2002.

59. China Ministry of Railways (CMR), General code for design of railway bridges and culverts (TB10002.1–2005), China Railway Press, Beijing, China, 2005 (in Chinese).

Chapter 5

Progressive damage assessment of bridge pier subjected to ship collision

5.1 Introduction

Bridge piers spanning across the navigable coastal channels are potentially at the risk of vessel collisions that can lead to severe damages or even collapse of such structures. Therefore, the structural responses of bridge piers should be carefully designed against the extreme impact loads. An occurrence of catastrophic barge accident on Sunshine Skyway Bridge in the United States in 1980 led to the collapse of one pier and three spans of the bridge with a length of 396 m. Thereafter, many researchers were attracted to study vessel-bridge collision events and to determine the extreme impact loads in design codes. According to AASHTO [1] specifications, vessel collisions caused 31 major bridge collapses and 342 fatalities worldwide from 1960 to 2002. In 2007, the impact of a fully loaded cargo ship to Jiujiang Bridge in China led to the death of nine humans and the collapse of the 200-m bridge superstructure [2].

Although the current bridge design specifications [1] present a static analysis based on the experimental ship-ship impact tests, vessel collisions with bridge structures are dynamic. Hence, many studies were done in the literature to recognize the nonlinear dynamic behaviors of bridges subjected to vessel collisions analytically, numerically, and experimentally. Since the static analysis procedures neglect dynamic amplification effects (e.g., superstructure inertial effects), the calculated collision load and bridge structure demands can be underestimated [3, 4]. Since 2002, a series of full-scale experimental barge collision tests have been carried out on the old St. George Island Bridge. The force-deformation relationships predicted from the numerical simulations a jumbo hopper barge collision with pier columns varying in the shapes and sizes [5] to capture the insights into the barge impact loads captured from the aforementioned full-scale barge-pier collision tests [6]. Afterward, various approaches include analytical [3, 7–10] and finite element (FE) high-resolution techniques [11–17]. In addition, the shock spectrum-based analyses were utilized to compute the impact loads arising from the collision of barge [11, 18] and ship [19–21] vessels with bridge pier structures.

In-line with the study done by Consolazio and Cowan [3] who proposed an analytical method named the coupled vessel to impact analysis (CVIA), Fan et al. [8] improved this method by taking into account the material strain rate effects of the striking vessel to prevent from underestimated responses of bridge structures. Despite many previous works that assume elastic or rigid material models for bridge piers, Sha and Hao [12] utilized a nonlinear damage material model for the FE model of a single reinforced concrete (RC) column subjected to barge collisions in LS-DYNA [22] without considering the strain rate effects.

DOI: 10.1201/9781003262343-5

To assess the vulnerability of an impacted pier realistically, the structural failure modes and the progressive damage behaviors should be carefully studied during the ship collision and free vibration phases of the pier response. Generally, the damage mechanism of concrete structures can be categorized into two different modes: (i) the ductile damages due to global flexural deformations in the tensile zone of structure and (ii) the brittle damages resulting in shear failures that commonly appear as the concrete spalling under impulsive and extreme loads with very short durations. When an RC structure is subjected to a lateral extreme load, the structure likely tends to fail in shear modes shortly after the onset of impact loading. Progressing of such shear damages in axially loaded RC structures such as bridge piers would lead to the structural responses in the plastic ranges and losing of concrete and reinforcement strengths (i.e., the formation of plastic hinges) during the following phases. Hence, considering the nonlinearity of materials such as strain rate effects is very important to capture the realistic behaviors of RC structures under extreme and impulsive loads. Recently, some research works were carried out to develop the nonlinear damage models for the concrete materials based on the experimental impact tests of RC beams [23–26]. Afterward, these models were utilized through high-resolution FE modeling software codes such as LS-DYNA. Although the nonlinearity of concrete materials for girder bridge piers was considered in some previous studies [13, 27], the damage behaviors and the strain rate effects of these materials were not considered. Fan and Yuan [28] numerically evaluated the failure modes of a pile-supported structure adopting different material models under ship collisions considering soil-structure interactions. There exist several works in the literature that study the nonlinear responses of bridge piers under the collision of vehicles [29–32], vessels [14, 33], and shipping objects [34].

Most of the aforementioned works studied the behaviors of girder bridges under barge collisions. Besides, an investigation is made to detect damage behaviors and failure modes of concrete cable-stayed bridge piers subjected to ship collision to rectify the shortcomings of previous studies. Cable-stayed bridges built across the wide rivers and navigable inland waterways are one of the most common types of offshore bridges in China. The main piers (pylons) of such bridges located at the main channel are inherently at the risk of vessel collisions. The pylons are totally different from girder bridge piers geometrically (which are mostly in diamond shapes), and also in terms of the axial load-carrying mechanism through the prestressed cables attached between the pylons and the superstructure. Therefore, the occurrence of severe damages in pylons may lead to the progressive collapse of the whole bridge. Owing to the mentioned discrepancies between the girder and cable-stayed bridge piers, it is expected that such structures would indicate different failure behaviors and damage patterns to vessel collision loads compared to those of girder bridge piers. Hence, it is the motivation of this study to numerically evaluate the process of progressive damages and failure behaviors of the main pier (pylon) of a cable-stayed bridge subjected to a 10,000-deadweight-tonnage (DWT) ship collision in LS-DYNA.

Despite previous research works that studied the influences of the steel material strain rate factor of striking vessels on the impact responses of bridge piers [8, 28], the concrete material strain rate effects on the global responses of the impacted pier are investigated in this chapter using an analytical two-degree-of-freedom (2-DOF) system. To this end, first, the importance of the concrete strain rate factor during the ship-pier collision is assessed by comparing the impact responses between those from the pier FE models with and without considering the material strain rate effects. Then, the concrete material strain rate effects are formulated by developing an analytical model with the 2-DOF system and assuming the pier as an idealized vertical beam with roller-fixed supports. Besides, the damage levels that progress in the impacted pier are categorized using three proposed different damage indices during the simulation time. Consequently, the damage index results are compared with the progressive damage behaviors observed during the ship-pier collision using FE simulation to determine the most efficient approach that represents the damage levels.

5.2 FE modeling and validations

5.2.1 FE model of ship

Figure 5.1 shows the details of the FE model of a typical 10,000-DWT container that ship navigates in inland waterways of China. This ship has 11.94-kg impact mass considering the mass of ship bow and 10% added mass due to the hydrodynamic effects, a total length of 121 m (including 100 m for the rear portion and 21 m for bow portion), a width of 20 m, a molded depth of 12 m, and a design draft of 6 m. More details about the characteristics of this ship can be found in the literature [35]. Since the prediction

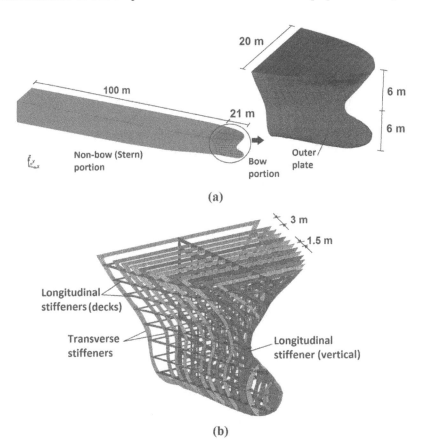

Figure 5.1 FE model of the ship: (a) ship portions, (b) internal stiffeners of bow portion without the outer plates.

of ship impact load profoundly depends on the accuracy of ship bow FE modeling, the configuration of internal stiffeners and the ship bow are accurately modeled to capture the reasonable and realistic results and crush deformations as shown in Figure 5.1b [1, 35, 36].

All the plates of the ship bow portion are modeled by Hughes-Liu shell elements, and the non-bow portion (i.e., stern) is modeled by solid elements. To simulate the local deformations and folding behaviors in the plates of the bow portion, the mesh size of shell elements should be selected sufficiently small. On the other hand, the selection of a very small mesh size can be excessively led to an increase in computational time. According to the previous studies [35, 37], to achieve the converged results, the ratio of the shell element length to the thickness used in the ship bow should be between the ranges of 8 and 10. To meet this criterion, the mesh sizes between 50 and 150 mm are selected for the plates of the ship bow portion. Accordingly, the FE model of the ship bow portion contains 39,292 elements for the internal stiffeners and 48,816 elements for the outer plates. In LS-DYNA, an elastoplastic material model in LS-DYNA named MAT_PLASTIC_KINEMATIC (MAT_003) [38] is utilized for plates of the ship bow that represents the isotropic and kinematic hardening behaviors considering the effects of steel material strain rate. The values of strain rate parameters for steel materials are given in Table 5.1 according to Cowper and Symonds's formulas presented as follows [39]:

$$\frac{\sigma'_d}{\sigma_s} = 1 + \left(\frac{\dot{\varepsilon}}{C_{strain}}\right)^{1/P_{strain}} \tag{5.1}$$

where σ'_d is the dynamic flow stress at uniaxial plastic strain rate $\dot{\varepsilon}$, and σ_s is the static flow stress. C_{strain} and P_{strain} are the constants of steel material.

The non-bow portion of the ship with rather coarse meshes is modeled using a rigid material model MAT_RIGID (MAT_020) to reduce the computation time of the simulations in LS-DYNA. The material properties used for the ship components are depicted in Table 5.1.

To exam the reliability of the ship bow FE model, a rigid wall with 200-mm mesh and moving with a constant velocity of 5 m/s collides with the fixed ship bow portion during a quasi-static analysis in LS-DYNA as discussed in the previous work [14]. The bow impact forces versus the crushing deformations are compared with those calculated by the empirical code formulas [1, 40, 41] and the previous work [35] as illustrated in Figure 5.2. The comparison between the results from the present study and those from the previous work done by Fan and Yuan [35] revealed a good agreement between them. However, the collision forces predicted by AASHTO [1] and Eurocode

Table 5.1 Material parameters used for ship bow

Parameter	Magnitude
Mass density (kg/m³)	7,850
Young's modulus (GPa)	206
Poisson's ratio	0.3
Yield stress (MPa)	235
Tangent modulus (MPa)	885
Failure strain	0.3
Strain rate parameter C_{strain}	40.4
Strain rate parameter P_{strain}	5.0

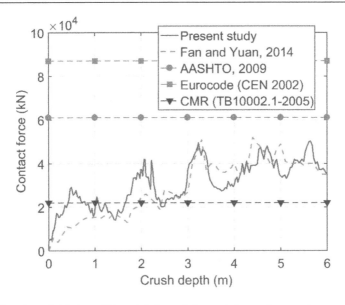

Figure 5.2 Calibration of ship bow FE model through force-deformation results.

[40] are more conservative for design considerations than those calculated by China's bridge design code [41].

5.2.2 FE model of bridge pier

As a case study, the Zhanjiang Bay Bridge located in the Guangdong province of China is selected subjected to ship collision. Figure 5.3 illustrates the detailed FE modeling of the bridge components and the elevations of the impacted RC pylon (pier-4) in which the

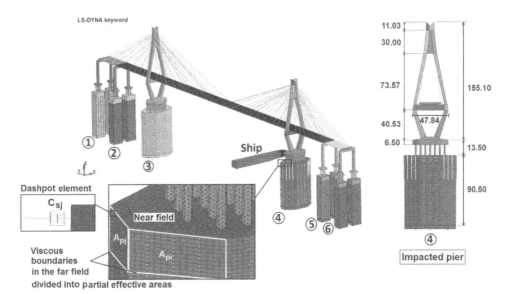

Figure 5.3 Detailed FE model of Zhanjiang Bay Bridge (unit: m).

ship collides at the piles' cap level. Detailed information about this bridge can be found in the referenced works [14, 42]. The FE modeling characteristics of the bridge pier, the superstructure, and the soil surrounding the piles have been described in the previous work [14]. As shown in Figure 5.3, the soil-pile interaction is modeled using a continuum model by solid elements in the near-field, and by discrete elements in the far-field. In the near-field, a contact algorithm named AUTOMATIC_NODES_TO_SURFACE is defined between the pile nodes as the slave part (with finer mesh), and the surrounding soil as the master part with the friction coefficients of 0.26 in LS-DYNA to take into account the frictional forces in the simulation. In the far-field, to prevent the reflection of stress waves from the boundaries of soil continuum model that can cause additional dynamic impulse, viscous dashpot elements with a damping coefficient of, C_s, are adopted attached to the effective partial areas, A_p, of the desired soil layer with a mass density of ρ_s as illustrated in Figure 5.3. Therefore, the damping coefficient of dashpot elements relative to the ith partial area can be calculated as follows:

$$C_{si} = \rho_{si} \cdot A_{pi} \qquad (5.2)$$

In addition, the soil behaviors in the near-field are modeled using an elastic-plastic constitutive material model named MAT_MOHR_COULOMB (MAT_173) [38] that is used by adopting several characteristics of soil such as mass density, shear modulus, Poisson's ratio, angle of friction, and cohesiveness value as given in the previous study [14].

Figure 5.4 shows the details of the cross-sectional dimensions of the pier members, the arrangement of longitudinal and transverse reinforcements used in the pylons. The longitudinal reinforcement ratio for all piers is considered between 2% and 4% and the transverse rebars were spaced between 300 and 150 mm in the columns and the beam, respectively, based on the recommendations by ACI-318 [43]. It should be noted that the properties of the material models and the contact algorithms between the various components of the ship-pier collision system have been utilized in this study similar to those given in the previous study [14]. Also, the nonlinear behaviors of the pier concrete were simulated using a material model named MAT_CSCM_CONCRETE (MAT_159) [38] considering the strain rate effects and representing the isotropic and kinematic hardening behaviors of the concrete. The damages and cracks in this concrete material model are simulated by an element erosion parameter named *ERODE* based on the maximum principal strain of the concrete. The reliability of this model has been proved using a series of impact tests on RC members in the previous works [14, 44, 45] for *ERODE* = 1.08, which means the elimination of the concrete elements from the simulation when their strain reaches 8% of the maximum principal strain.

As the FE validation study of the pier, a series of mesh convergence tests were carried out in LS-DYNA through the collision of a 10,000-DWT ship with a velocity of 5 m/s with different FE models of piles' cap of the main pylon with mesh sizes of 50, 100, and 200 mm. Since the results were very close to each other, it was found that a 200-mm mesh for the pier can give the converged and accurate results with efficient computational time. Also, the energy balance between the ship and the pier components with the hourglass energy less than 10% of the pier absorbed energy as given in the following section proved the convergence of the results. Hence, the FE model of the impacted pier with 200-mm mesh is considered in the following sections of this chapter.

Owing to the substantial influences of bridge superstructures on the dynamic behaviors of impacted columns and piers proven by Gholipour et al. [14, 16], the

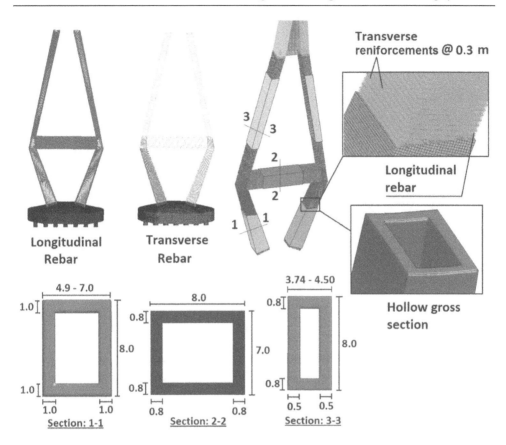

Figure 5.4 Cross-sectional details of the impacted pylon (pier-4) with longitudinal and transverse rebars (unit: m).

vulnerability of the impacted pier is assessed in the presence of the superstructure in this study. To this end, the whole bridge is subjected to self-weight to reach an equilibrium state in 0.1 s using a control keyword in LS-DYNA named CONTROL_ DYNAMIC_ RELAXATION before the onset of the transient analysis of ship impact loading considering 5% of global damping as recommended in the bridge design codes [46]. It is worth noting that the bridge undergoes the gravity load during both stress initialization (i.e., dynamic relaxation analysis) and transient loading phases. The tensile forces generated in the cables of the superstructure after reaching the stable state of the bridge are captured in the rages between 2.1 and 4.9 MN as discussed in the previous study done by Gholipour et al. [14].

5.3 Strain rate effects of materials

The importance of the concrete material strain rate effects is studied in this section. The dynamic strength enhancements due to the material strain rate effects discussed in the previous studies [14, 44] are similarly adopted in this study for the materials of the pier and ship components.

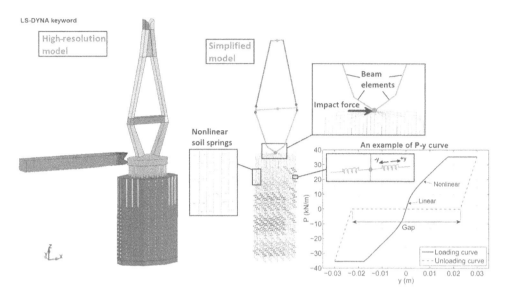

Figure 5.5 High-resolution and simplified models of the impacted pier developed in LS-DYNA.

To recognize the influence of the material strain rate on the global responses of the pier, different FE models are developed in LS-DYNA as shown in Figure 5.5 varying in the modeling approach and the material model. These pier models are: (i) linear elastic high-resolution model (without considering the strain rate effects of both steel and concrete materials), (ii) simplified linear model developed by elastic beam elements attached to mass elements without considering the material strain rate effects, (iii) non-linear high-resolution model with a suffix of "with rate" considering the strain rate effects of both steel and concrete materials, and (iv) nonlinear high-resolution model with a suffix of "without rate" without considering the concrete material strain rate effects. Therefore, the strain rate effects of both concrete and steel reinforcement materials can be assessed through the comparison of the pier impact responses from the linear elastic high-resolution and linear simplified models with those from the nonlinear high-resolution model "with rate". Besides, the concrete material strain rate effects can be explored through the comparison of the pier impact responses from the nonlinear high-resolution model "without rate" with those from the nonlinear high-resolution model "with rate." In the simplified linear model as shown in Figure 5.5, the soil-pile interaction is modeled using nonlinear elastic-plastic spring elements attached to the piles' nodes representing the force-displacement (*P-y*) curves generated according to the methods proposed by Reese et al. [47] for lateral resistance of cohesionless soil using strength characteristics of the desired soil layer such as unit weight, internal friction angle, and subgrade modulus. To this end, the generated loading and unloading curves are applied to the nonlinear material named MAT_SPRING_GENERAL_ NONLINEAR (MAT_S06) used for the soil springs. An example of *P-y* curves composed of loading and unloading (with a slope the same as initial loading slope) curves considering the soli gap model [15] is illustrated in Figure 5.5.

Since the element erosion (i.e., the elimination of elements) can lead to excessive large deflections and overestimate responses of the pier modeled with nonlinear materials, the *ERODE* parameter is set to less than 1.0 [45] to be omitted from the simulation. Besides, since the presence of the superstructure may increase the resistance of the pier against the collision loads [14, 16], the isolated models of the pier are studied in this section to prevent the interference effects of the materials and the superstructure.

The impact force acquired from the high-resolution head-on collision simulation of a 10,000 DWT with 5.0-m/s velocity is applied to the equivalent mass element of the piles' cap in the linear simplified model as shown in Figure 5.5. The displacements of different pier models under similar impact forces are compared in Figure 5.6a–c. It is observed that the displacements of the pier models at the piles' cap and the deck levels follow each other until they reach 1.5 and 0.9 m representing the elastic range of the responses, respectively. Afterward, the pier models with nonlinear materials give greater peak, residual displacements, and larger residual deflections (i.e., the difference of the pier displacements between the piles' cap and the deck level) after reaching the plastic range (also see Figure 5.6c). Also, it is found that the pier nonlinear model "with rate" presents smaller deflections in the plastic ranges during the global response phases rather than the nonlinear model "without rate" due to the resistance enhancements arising from the concrete strain rate effects. The summary of the peak displacement results at various elevations of different pier models is listed in Table 5.2.

Similar to the displacement results, the internal forces and the absorbed energy of the pier structure capture almost the same values during the elastic phase of the pier response until reaching the plastic ranges as shown in Figure 5.7a–c. The greater internal forces generated in the pier nonlinear model "with rate" rather than linear and nonlinear models "without rate" are observed during the following phases of the pier response in Figure 5.7a and b. In Figure 5.7c, although the absorbed energy curve of the pier model with linear-elastic materials demonstrates a turning behavior after following those of nonlinear models until 1.0 s, significant residual energy is observed in the linear model due to global displacement and energy dissipated by the soil. This is because of the large deflection of the pier linear model (more than 1.0 m) as shown in Figure 5.6a, which results in nonlinear plastic behaviors and residual deformations in the surrounding soil as illustrated in Figure 5.5 by *P-y* curves plotted for the soil lateral resistance.

Besides, the impact force-time histories from the ship with the different pier models with and without rate are almost identical in Figure 5.7d. This is because the ship bow deformations govern the impact forces rather than those of the pier. Also, the pier model "with rate" results in less energy due to fewer deformations of the pier rather than those of the model "without rate" as shown in Figure 5.7d. Figure 5.8 demonstrates the energy contributions of the various components during the collision of a 10,000-DWT ship with an impact mass of 11.94 Mkg and 5.0-m/s velocity with the pier nonlinear model "with rate." It is found that the ship bow absorbs about 58% of the system internal energy (equal to 87.0 MJ), which is the largest energy contribution. As such, the internal energy contribution of the pier structure, the soil layers surrounding the piles, and the hourglass energy dissipated during the simulation are 25%, 14.5%, and 2.5%, respectively.

As discussed earlier, it was found that the strain rate parameter can significantly affect the global responses and the internal forces of the impacted pier. Therefore, it is worth formulating and quantifying the effects of this parameter using an analytical 2-DOF system as shown in Figure 5.9.

Based on the study by Fujikake et al. [48], a concrete structure under impact loads may demonstrate local and overall responses in which the overall response mainly

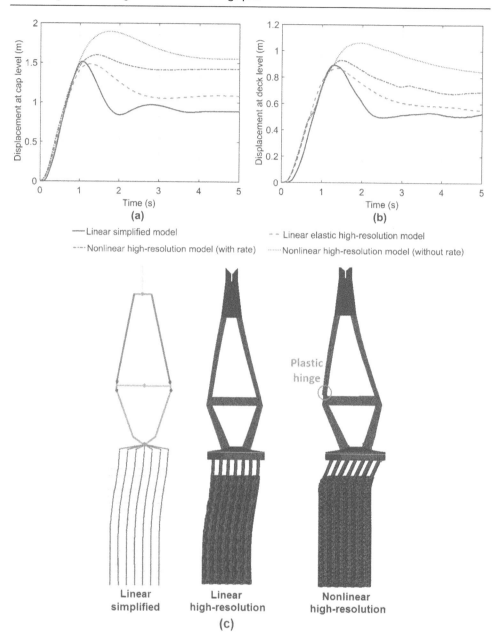

Figure 5.6 Comparing the pier displacements between those from the high-resolution and simpli-fiedmodels:(a)attheimpactlevel(piles'cap),(b)atthedecklevel,and(c)pierresidualdeflections magnified with a scale of 5.0.

depends on the dynamic loading rate effects during the following impact phases rather than the initial impact phase. Figure 5.10a illustrates the proposed 2-DOF system developed by mass elements coupled with discrete elements including the springs and dashpots connected to each other in parallel. The strain rate effects of both ship and pier components are formulated using the dynamic damping coefficients (i.e., C_v

Table 5.2 Pier displacement results from different models of the pier

Pier model	Peak displacement at the impact level (m)	Peak displacement at the deck level (m)	Residual deflection (m)
Linear simplified model	1.51	0.88	0.31
Linear high resolution	1.49	0.85	0.62
Nonlinear high resolution (with rate)	1.60	0.92	0.68
Nonlinear high resolution (without rate)	1.83	1.07	0.80

for the ship and C_{SG} for the pier as shown in Figure 5.9a) that represent the dynamic increase factor of the demand parameter, D $(\dot{\delta})$, generated due to the time derivation of the pier deflections and the ship crush depths $(\dot{\delta})$, as given in the following:

$$D_{(with\ rate)} = D_{(without\ rate)} + D(\dot{\delta}) \tag{5.3}$$

$$D(\dot{\delta}) = (\beta - 1) \cdot D_{(without\ rate)} \tag{5.4}$$

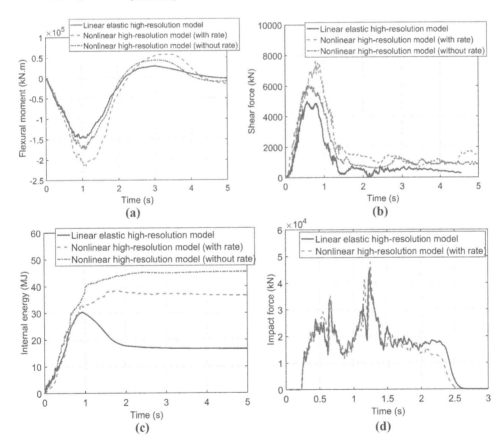

Figure 5.7 Comparison of the pier impact results between those from the linear and nonlinear models: (a) flexural moments at the impact level, (b) shear forces at the impact level, (c) impact forces, and (d) internal energy absorbed by the pier structure.

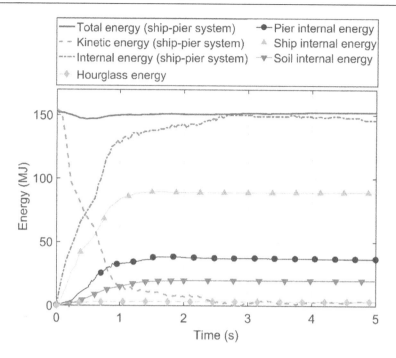

Figure 5.8 Energy contributions of the various components in the ship-pier collision system.

where $D_{(with\,rate)}$ and $D_{(without\,rate)}$ indicate the demands of the ship and pier components with and without considering strain rate effects. β is the dynamic increase factor equal to $D_{(with\,rate)}/D_{(without\,rate)}$.

First, the contributions and influences of local response parameters (k_{SL} and C_{SL}) of the impacted structure in the 2-DOF system are examined by developing several quasi-static impact simulations between the ship and the isolated piles' cap with rigid and nonlinear material models for the concrete as shown in Figure 5.10a. By comparing the contact force results in Figure 5.10b, it is found that local response parameters can

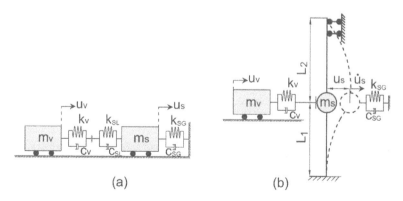

Figure 5.9 2-DOF model of ship-pier impact collision system: (a) a simplified model with mass elements connected to discrete elements and (b) a simplified model with an idealized cantilever-simply supported vertical beam.

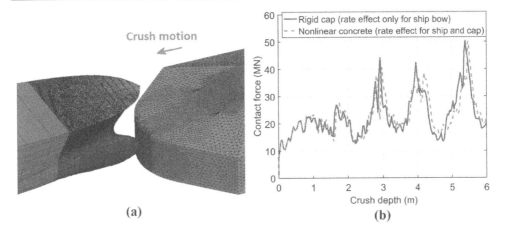

Figure 5.10 (a) Quasi-static simulations of ship-piles' cap collision and (b) comparison between the contact forces and the quasi-static simulations.

be neglected from the simplified model (i.e., k_{SL}, $C_{SL} \rightarrow 0$). Therefore, the equation of motion for the simplified 2-DOF system can be presented as follows:

$$\begin{bmatrix} m_V & 0 \\ 0 & m_S \end{bmatrix} \begin{Bmatrix} \ddot{u}_V \\ \ddot{u}_S \end{Bmatrix} + \begin{bmatrix} C_V & -C_V \\ -C_V & C_V + C_{SG} \end{bmatrix} \begin{Bmatrix} \dot{u}_V \\ \dot{u}_S \end{Bmatrix} + \begin{bmatrix} k_V & -k_V \\ -k_V & k_V + k_{SG} \end{bmatrix} \begin{Bmatrix} u_V \\ u_S \end{Bmatrix} = \begin{Bmatrix} 0 \\ 0 \end{Bmatrix}$$

$$(5.5)$$

where m_V, C_V, and k_V are the vessel mass, damping coefficient relevant to the time derivative of the bow crush depth, and the stiffness of the vessel bow, respectively. m_S, C_{SG}, and k_{SG} are the pier mass, damping relevant to the time derivative of the pier deflection, and the pier stiffness, respectively. u_V, $\dot{u}_V$, and $\ddot{u}_V$ are the vessel, and u_{SG}, $\dot{u}_{SG}$, and $\ddot{u}_{SG}$ are the pier displacement, velocity, and acceleration parameters, respectively.

To solve the equation of motion given in Equation (5.5), the components of the mass, stiffness, and damping matrixes should be quantified. In this study, the equivalent masses of the ship and pier components are assumed as deterministic parameters. Besides, the static stiffness of the ship bow (k_V) can be identified using the force-deformation relationship captured from a quasi-static simulation of ship bow collision with the piles' cap similar to the procedure shown in Figure 5.10b in which the influences of the steel material strain rate used for the plates of the ship bow are not taken into account. The static stiffness of the pier structure (k_{SG}) can be obtained using Equation (5.6) by applying a unit load to the mid-height of the idealized model of the pier laterally as shown in Figure 5.10b as a vertical cantilever-simply supported beam. In this simplified model, since the top of the main pylons is restrained by the cables, the pin support is assumed as the boundary condition at this level. However, it is assumed that the pier is translational and rotational fixed at the base

$$k_{SG} = \frac{1}{96EI} \left(\frac{8L_1^2 L_2^3 (2L + L_2) - 3L_2 L^5}{L^3} \right)$$

$$(5.6)$$

where E is Young's modulus of the pier concrete, and I is the moment of inertia of the pier equivalent cross-section. L and L_2 are the pier total height and the height of the ship impact elevation from the pier top, respectively. According to the design codes [49, 50], a minimum required length equal to the sum of the pile free length (i.e., the pile length above the mudline) and three times the pile diameter is considered the equivalent height of impact elevation (i.e., L_1) from the fixed base of the pier.

According to the study of RC beams under dropping mass impact loads by Fujikake et al. [48], the damping coefficient of the striking object can be written for the ship-pier system as follows:

$$C_V = \xi C_{cr} = 2\xi \sqrt{\frac{m_V m_S}{m_V + m_S}} \cdot k_{dV}$$

(5.7)

where C_{cr} is the critical damping coefficient, and ξ is the damping ratio that is considered equal to 0.2 according to [28] for a 10,000-DWT ship to capture desired results. k_{dV} is the dynamic stiffness of ship bow considering strain rate effects equal to $k_{dV} = \beta_V k_V$ in which β_V is the dynamic increase factor due to the influences of steel material strain rate in the ship bow that can be calculated as follows [28]:

$$\beta_V = 1 + \left(\frac{\dot{u}_{V0}}{2l_{eq}C}\right)^{1/P}$$

(5.8)

where $\dot{u}_{V0}$ is the ship initial velocity, l_{eq} is the equivalent length of the ship that is equal to 3.4 m for 10,000-DWT ships, C and P are the strain rate constants of mild steel materials equal to 40.4 and 5, respectively, as given in Ref. [39].

Besides, the damping coefficient of the pier structure (C_{SG}) can be calculated as follows:

$$C_{SG} = 2\xi \sqrt{m_S \cdot k_{dSG}}$$

(5.9)

where ξ is the damping ratio equal to 0.05 as recommended by Aviram et al. [46] for RC bridges. m_S is the equivalent mass of the pier that can be calculated according to the proposed methods given by Biggs [51] and Krauthammer and Shahriar [52] using the distributed mass along with the pier height, and the shape functions under different phases of the pier deformations from elastic to purely plastic ranges. k_{dSG} is the dynamic stiffness of the pier equal to $k_{dSG} = \beta_S \cdot k_{SG}$, in which β_S is the dynamic increase factor of the pier strength due to the influences of the concrete material strain rate that can be calculated using Equation (5.10) based on the experimental tests done by Guo et al. [26]:

$$\beta_S = \frac{f'_{cd}}{f'_{cs}} = 1 + \frac{9.309 \times 10^{-3} E}{f'_{cs}} \cdot \dot{\varepsilon}^{[1-(-0.817)]}$$

(5.10)

where f'_{cd} and f'_{cs} are the dynamic and static unconfined compressive strength of concert, respectively, E is the static elastic modulus of concrete, and $\dot{\varepsilon}$ is the strain rate of concrete.

From the structural analysis of the idealized vertical beam (Figure 5.9b), the strain rate of the desired cross-section with a curvature of κ and the natural surface depth of z can be calculated as follows:

$$\dot{\varepsilon} = \frac{\partial \varepsilon}{\partial t} = z \frac{\partial \kappa}{\partial t} = z \cdot \dot{\kappa}$$

(5.11)

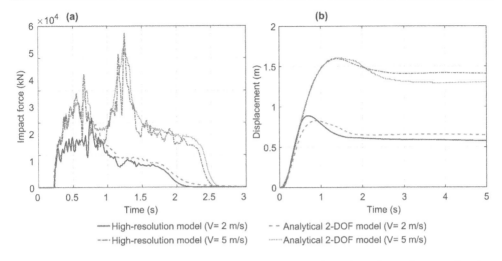

Figure 5.11 Comparison between the impact results of the pier from the high-resolution and analytical analyses of ship-pier collisions: (a) impact forces, (b) displacements at the impact level.

where $\dot{\kappa}$ is the curvature rate at the critical cross-section.

By solving the structural problem of the 2-DOF model given in Figure 5.9b, the relationship between the curvature of the beam cross-section and the deflection at the impact level (i.e., the height of L_1 from the fixed base of the pier) can be written as follows:

$$\kappa = u_{SG} \underbrace{\left(\frac{12L_1^2 L_2 (2L + L_2)}{8L_1^2 L_2^3 (2L + L_2) - 3L_2 L^5} \right)}_{=\lambda_L} \to \dot{\kappa} = \dot{u}_{SG} \cdot \lambda_L \qquad (5.12)$$

where λ_L is a constant parameter with a certain value in terms of deterministic height parameters of the pier.

Consequently, by substituting Equation (5.12) into Equation (5.11), the relationship between the strain rate and the time derivation of the pier deflection can be written as follows:

$$\dot{\varepsilon} = z \cdot \lambda_L \cdot \dot{u}_{SG} \qquad (5.13)$$

The reliability of the proposed 2-DOF system in the prediction of the pier impact responses is assessed by comparing the pier impact responses from the high-resolution and analytical approaches under ship impact scenarios with the impact velocities of 2.0 and 5.0 m/s as illustrated in Figure 5.11a and b. It is revealed that the results from the analytical method, including the peak values of the impact forces and displacements, loading durations, and the residual deformations, are agreed well with those from the high-resolution method.

5.4 Results and discussion

5.4.1 Damage progression characteristics

In this section, first, the damage progression process of the pier under the ship collision is evaluated. Then, the failure modes of the pier with nonlinearity are investigated compared to those from the linear high-resolution model with an elastic material model for the concrete. Finally, three damage indices based on the different parameters are proposed to determine an efficient approach describing the damage progression process of the impacted pier.

Figure 5.12a–g shows the damage progression process of the pier components from the initiation of tensile cracks to the severe damages and cross-sectional failures. In Figure 5.12a, the unconfinement damages appear in the concrete cover of upper columns at 0.2 s after the onset of the ship collision load due to the bonding strength loss of the concrete. Then, the shear cracks are generated at the base and the top zones of the upper columns (slender columns), simultaneous with the initial peak impact force at 0.3 s, as shown in Figure 5.12b. Hereafter, the progression of shear cracks in the upper columns leads to the element erosions and the shear damages in the cross-section of the columns after 0.5 s as shown in Figure 5.12c. In Figure 5.12d, the initiation of shear damages is observed in the lower columns and the piles' head after 0.7 s. Afterward, the progressive shear damages associated with spall damages in the outer surface of the top and bottom zones of the upper columns are observed after 1.0 s as shown in Figure 5.12e, simultaneous with the rebounding of the ship and the initiation of the free vibration phase of the pier response. During the free vibration phase, the plastic hinges are formed at the base and top zones of the upper columns due to the accumulation of the cross-sectional fractures and the strength loss of the reinforcements, which represents the combined shear and flexural failure modes of the pier column as shown in Figure 5.12f and g.

Figure 5.13a and b shows the maximum shear stress and the plastic strain results at checkpoints A and B as identified in Figure 5.12b and d, respectively, during the damage progression process. It is found that when the stresses suddenly decrease, the plastic strain reaches its peak value (i.e., 1.0) named softening behavior. In contrast, the hardening behavior is captured with the occurrence of a sudden increase in the concrete strength along with a sudden decrease in the plastic strain behaviors. The FEs are eroded and eliminated from the simulation when the values of both stress and strain parameters reach zero.

5.4.2 Failure modes of impacted pier

To identify the predominant failure modes on the responses of the impacted pier under ship impact, the shear forces and flexural moments generated at the various cross-section zones of the pier columns are evaluated. Figure 5.14a–d shows that the shear forces reach their peak values at 0.5 s. Besides, this time stage (i.e., 0.5 s) is recognized as the turning point of the flexural moments when the direction of the flexural moments is reversed at various cross-sections of the pier columns as shown in Figure 5.15. Figure 5.16 demonstrates the direction of the flexural moments and the corresponding tensile and compression surfaces of the pier columns before and after the time of 0.5 s.

In addition, the direction of flexural moments illustrated in Figure 5.16 can be proven by investigating the axial forces of the longitudinal reinforcements located on the inner and outer faces of pier columns as shown in Figure 5.17a–d. The positive values imply that the tensile face and the negative values indicate the compression face of the pier columns. As shown in Figure 5.17a and b, the reinforcements lose their strengths through the reach of zero value for the axial forces of the reinforcements around the

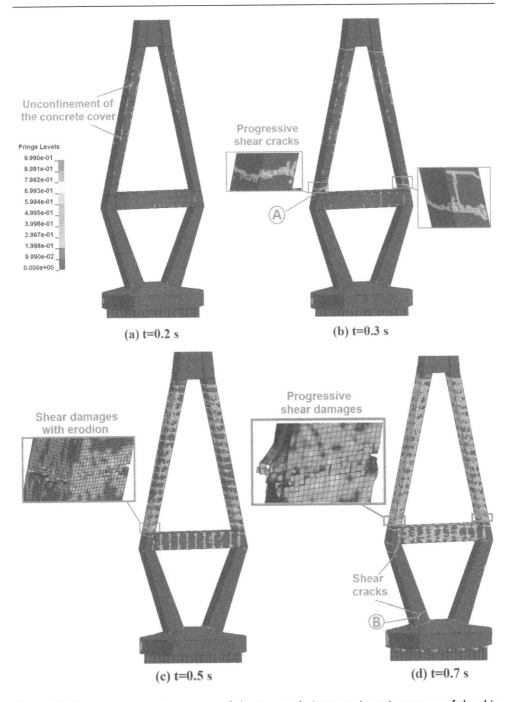

(a) t=0.2 s

(b) t=0.3 s

(c) t=0.5 s

(d) t=0.7 s

Figure 5.12 Damage progressing process of the impacted pier at various time stages of the ship collision in which concrete damages are shown using the contours of effective plastic strain. *(Continued)*

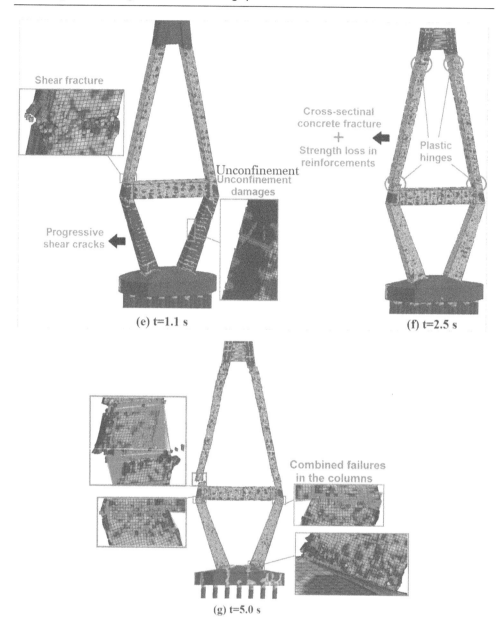

Figure 5.12 (Continued)

time of 1.5 s, which indicates the formation of plastic hinges in the cross-sections of *E-E* and *F-F* at the base of the upper columns.

The predominant failure modes on the responses of the impacted pier are shown in Figure 5.18a–d. It can be seen that after 0.5 s, the shear failure initiates in the upper columns of the pier (Figure 5.18a). Then, the flexural failure mode is associated with the shear failure mode in the bottom columns with the formation of plastic hinges at the base and top zones of the upper columns after 1.0 s as shown in Figure 5.18b (also

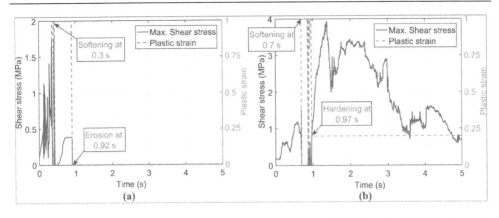

Figure 5.13 Maximum shear stress and the plastic strain results at: (a) checkpoint A, (b) checkpoint B.

see Figure 5.17c and d). With the combination of shear and flexural failure modes after 1.0 s, the progressive damages lead to the accumulation of cross-sectional fractures and the formation of plastic hinges in the column as shown in Figure 5.18c. Hereafter, the pier experiences large curvy deformations and deflections after 2.5 s due to the combined shear-flexural failure modes as shown in Figure 5.18d.

To study the nonlinear effects of concrete material on the global failure modes of the impacted pier, the failure modes of the pier with nonlinearity in the presence of the

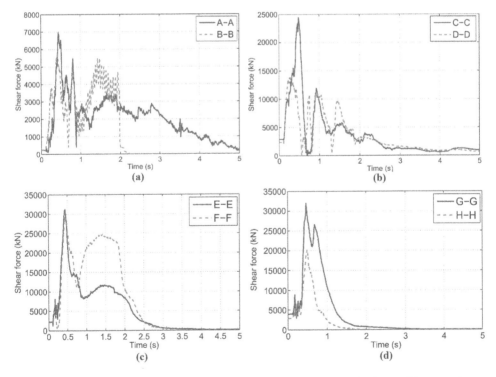

Figure 5.14 Shear forces at various cross-sections of the pier: (a) at the base of lower columns, (b) at the top of lower columns, (c) at the base of upper columns, and (d) at the top of upper columns.

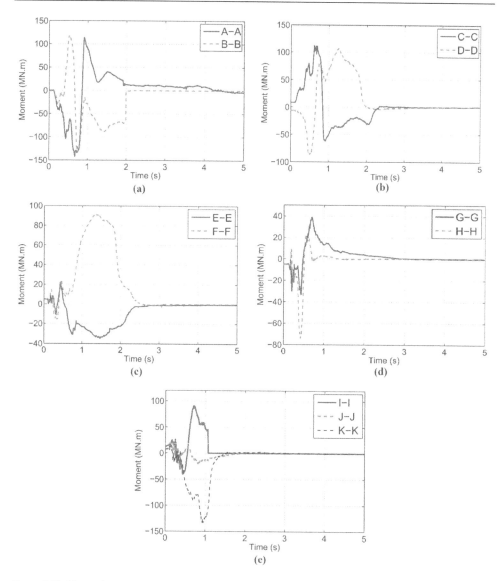

Figure 5.15 Flexural moments at various cross-sections of the pier: (a) at the base of lower columns, (b) at the top of lower columns, (c) at the base of upper columns, (d) at the top of upper columns, and (e) beam sections.

bridge superstructure are compared with those from a linear high-resolution model of the pier adopting an elastic material. In this way, the occurrence times of the peak shears and flexural moments in the pier columns are considered a criterion to determine the predominant failure mode on the responses of the pier.

Figure 5.19a shows the peak shear forces and flexural moments of the linear model of the pier, which occur shortly after rebounding the ship around 1.0 s before the occurrence of the peak displacements at various elevations. In contrast, the maximum internal demands occur earlier than those of the linear model because of the

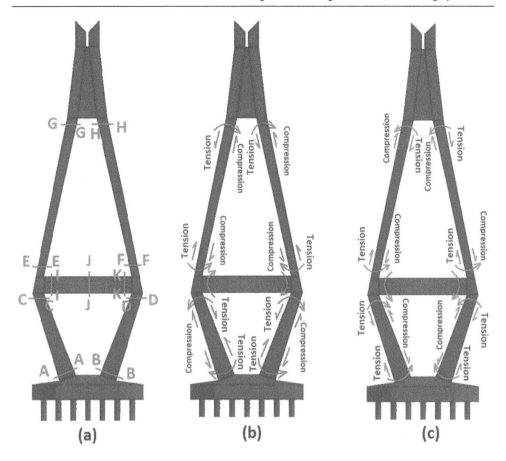

Figure 5.16 Failure modes of the impacted pier at various cross-sections: (a) showing the location of various cross-sections, (b) flexural directions before 0.5 s, and (c) flexural directions after 0.5 s.

overall strength loss of the pier during the damage progressing process as shown in Figure 5.19b. In addition, Figures 5.20 and 5.21 illustrate the mode shapes and the deflections along with the pier height from the mudline level to the level of the cable at the occurrence time of the maximum demands, and at the various time stages of the simulation, respectively. In Figures 5.20 and 5.21a, the peak values of the shear force and the flexural moment at the base of the pier occur around 1.0 s when the deflection of pier base (i.e., at the height of 16.75 m) has value rather than other time stages. In addition, the deflection value at the base elevation of the pier linear model at 1.0 s is larger than that of the nonlinear model. However, the internal forces generated at this elevation of the nonlinear model have larger peak values rather than those of the pier linear model as illustrated in Figure 5.22. Besides, for the nonlinear model, it is found that once the shear failure occurs at the base of upper columns (at the height of 61.75 m) at 1.0 s, the large curvy deformations are generated due to the formation of plastic hinges. Subsequently, during the following time stages between 1.0 and 1.5 s, the value of internal forces generated at these elevations of the nonlinear model are larger than those of the pier linear model due to the large curvy deformations in the upper columns and the piles' head level (see Figures 5.21b and 5.22).

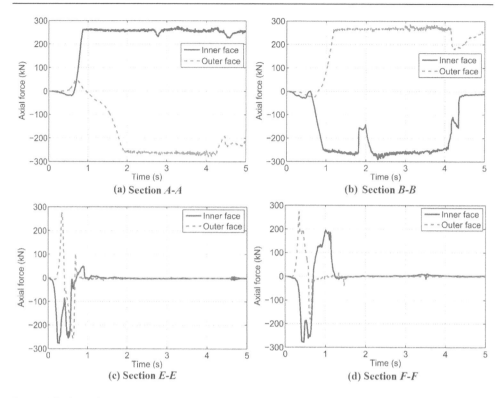

Figure 5.17 Axial forces of the longitudinal reinforcements of the pier at various cross-sections.

5.4.3 Damage indices

Because of limited studies in the literature about the damage index of bridge piers progressing with the time, there is no global and conventional target value in the literature in the definition of damage state levels of such structures under impact load. Therefore, it is attempted in this study to classify the damage levels of the impacted pier based on observations on the FE simulations (see Figure 5.12) and find a proper and reasonable approach evaluating the damage index comparing with those observations. Three damage indices calculating through different approaches are proposed to evaluate the damage levels of impacted pier progressing during a ship collision. These damage indices are based on the pier deflection, the eroded internal energy due to the elimination of the pier concrete FEs, and the axial load-carrying capacity of the columns.

5.4.3.1 Damage index based on the pier deflection

In this method, the ratio of the maximum pier deflection at the critical elevation (i.e., the piles' cap elevation as the ship impact level) of the pier model with nonlinear concrete material (i.e., damaged) to that of pier model with linear elastic (i.e., undamaged) material is considered the damage criterion, and the damage index can be calculated as follows:

$$DI_{(def)} = 1 - \frac{D_{damaged}}{D_{undamaged}} \tag{5.14}$$

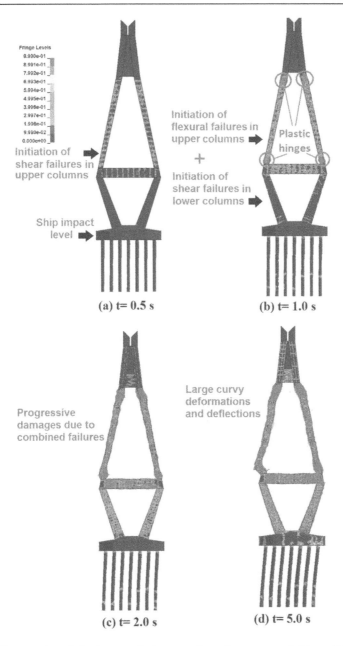

Figure 5.18 Failure modes of the impacted pier at various time stages of ship collision (magnified deflections with a 5-time scale factor).

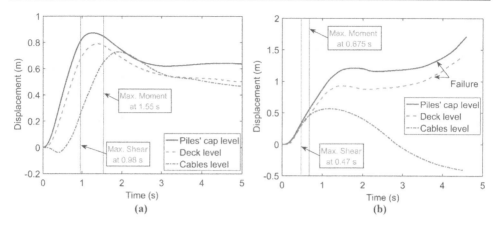

Figure 5.19 Displacement of the pier in the presence of the deck at various elevations subjected to ship collision with an impact velocity of 5.0 m/s for: (a) the linear elastic high-resolution model, and (b) the nonlinear high-resolution model.

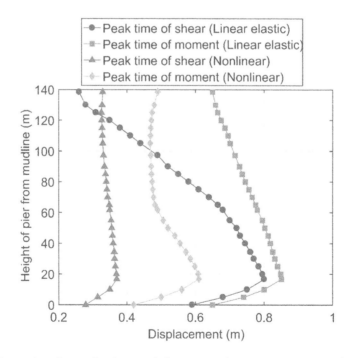

Figure 5.20 Comparing the mode shapes of the pier in the occurrence time of the maximum demands.

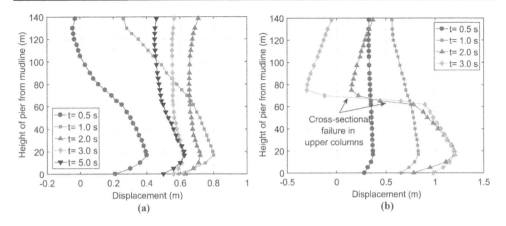

Figure 5.21 Mode shapes of the impacted pier in: (a) linear model, (b) nonlinear model.

where $DI_{(def)}$ is the deflection-based damage index, $D_{damaged}$ is the displacement of the pier with a nonlinear material model, and $D_{undamaged}$ is the displacement of the pier with an elastic material model.

5.4.3.2 Damage index based on the eroded internal energy

This approach evaluates the ratio of eroded internal energy of the pier concrete owing to the elimination of FEs to the total internal energy of the pier structure during the simulation as follows:

$$DI_{(eng)} = \frac{E_{eroded}}{E_{total}} \tag{5.15}$$

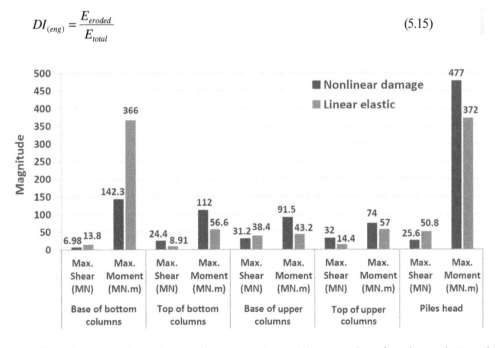

Figure 5.22 Comparing the peak value of maximum demands between those from linear elastic and nonlinear high-resolution models of the impacted pier.

where $DI_{(eng)}$ is the energy-based damage index, E_{eroded} is the eroded internal energy that represents the lost internal energy of the pier structure due to the element erosion during the collision simulation, and E_{total} is the total internal energy of the pier structure with the nonlinear material model.

5.4.3.3 Damage index based on the axial load-carrying capacity

The ratio of the axial load-carrying capacity of the pier columns captured during the collision simulation to their design axial load capacities is considered the base of this method. The total damage index value is calculated by averaging of subset damage indices from the various cross-sections of the pier columns as follows:

$$DI_{(cap)} = \frac{\sum_{i=1}^{m} \sum_{j=1}^{n} DI_{ij}}{m \cdot n} \tag{5.16}$$

where DI_{ij} is the subset damage indices from the various damaged cross-sections of the pier columns that are calculated using Equation (5.17), m and n are the numbers of pier columns and the damaged cross-sections, respectively

$$DI_{ij} = 1 - \frac{(P_L)_{ij}}{(P_N)_{ij}} \tag{5.17}$$

where P_N and P_L are the axial load-carrying capacity of the column and the axial load applied to the pier columns at various cross-sectional zones, respectively. The axial load-carrying capacity of the column is calculated according to ACI-318 [43] as follows:

$$P_N = \beta_1 f_c'(A_c - A_s) + f_y A_s \tag{5.18}$$

where f_c' is the compressive strength of concrete, and f_y is the yield stress of reinforcements, respectively. A_c and A_s are the area of the column cross-section and longitudinal reinforcements, respectively. β_1 is considered equal to 0.69 for C50 concrete ($f_c' = 50$ MPa) based on ACI-318 [43].

In Figure 5.23, damage indices calculated by the three aforementioned methods are compared during the simulation time of 5.0 s. It is obvious that with the progress of the damage level, the value of damage indices is prone to 1.0. In addition, it is observed that the damage indices based on the internal energy [$DI_{(eng)}$] and the axial load-carrying capacity [$DI_{(cap)}$] are activated earlier than the deflection-based method [$DI_{(def)}$] around 0.4 s after the onset of ship collision. Since the upper columns of the pier have a relatively smaller buckling stiffness (slender columns) than base columns, they would suffer more severe shear and flexural (plastic hinges) damages. As shown in Figures 5.12 and 5.18, the formations of plastic hinges and shear fractures are observed in the upper columns earlier than base columns. Hence, after 0.7 s, the damage index based on the axial load capacity of the upper column increases with a steeper slope and higher values than that based on the capacities of the base columns.

Based on the observations in Figure 5.12 on the progressive damage process, between the times of 1.0 and 2.0 s, the top columns lose their shear, flexural, and axial load-carrying capacities through the formation of shear fractures, plastic hinges

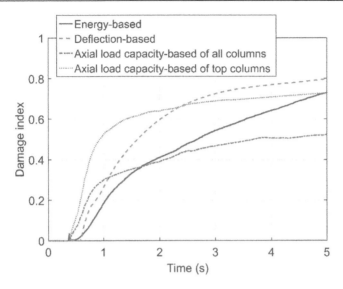

Figure 5.23 Comparison of damage indices from the different methods.

at the bottom and uppers zones that cause the high-level damage state throughout the pier structure. This high-level damage state can be properly represented by the deformation-based damage index and the damage index based on the axial load-carrying capacity of the upper columns as shown in Figure 5.23, which results in damage indices higher than 0.6 after 2.0 s. However, the results from $DI_{(eng)}$ and $DI_{(cap)}$ of the pier total columns capture significantly smaller damage indices that are not in good agreement with those from the observations.

From the observations in Figure 5.12a–g, the damage levels progressing in the pier can be classified as follows:

- Low-level damage ($DI = 0$–0.2): the initiation of shear damages between 0.3 and 0.75 s after a collision in upper columns.
- Medium-level damage ($DI = 0.2$–0.5): the progression of shear damages and the initiation of flexural damages (i.e., the formation of plastic hinges) between 1.0 and 2.0 s, which leads to the cross-sectional fracture along with curvy deformations in the upper columns.
- High-level damage ($DI = 0.5$–0.8): the accumulation of progressive damages arising from the combined shear and flexural failures, which leads to large curvy deformations and deflections between 2.0 and 5.0 s.
- Collapse ($DI = 0.8$–1.0): the pier collapses that may lead to the collapse of the whole bridge.

5.5 Conclusions

Despite many studies on barge collisions with girder bridge piers in the literature, this chapter investigated the damage progressing process and failure modes of a concrete cable-stayed bridge pier subjected to a container 10,000-DWT ship collision by FE simulations in LS-DYNA.

The main findings of this study from the FE simulations can be drawn as follows:

- The importance of the pier concrete strain rate effects on the global responses of the impacted pier was evaluated by comparing of the internal forces and the impact responses generated in the nonlinear and linear models of the pier. The enhancements in the internal forces during the global phase of the pier response were captured by adopting a nonlinear material model considering the concrete strain rate effects rather than those of a linear model.
- The strain rate effects were formulated using an analytical 2-DOF system for the ship-bridge collision system. Despite significant influences of the materials' strain rate on the global responses of the pier, the local responses were slightly affected by this parameter.
- The proposed analytical 2-DOF model captured the efficient results that agreed well with those from the high-resolution simulations. Since an idealized vertical beam with an equivalent cross-section was utilized for the pier in the 2-DOF system, the proposed analytical approach can be generalized for concrete bridge piers varying in the geometry and dimension parameters.
- By evaluating the damage progression process of the impacted pier from the minor cracks to the severe damages and cross-sectional fractures during the ship collision and free vibration stages, the time of 0.5 s was recognized as a key time stage in the determination of the pier predominant failure modes.
- By comparing the damage progression process observed from the FE simulations and those from damage indices, the pier deflection-based method gave a more efficient approach rather than other methods.

References

1. AASHTO. *Guide specifications and commentary for vessel collision design of highway bridges*, 2nd edn. American Association of State Highway and Transportation Officials, Washington, DC, 2009.
2. http://news.sohu.com/20070615/n250593101.shtml, 2007.
3. Consolazio GR, Cowan DR. Numerically efficient dynamic analysis of barge collisions with bridge piers. *J Struct Eng–ASCE* 2005;131(2131):1256–1266.
4. Davidson MT, Consolazio GR, Getter D. Dynamic amplification of pier column internal forces due to barge-bridge collision. *Transp Res Rec* 2010;2172:11–22.
5. Consolazio GR, Cowan DR. Nonlinear analysis of barge crush behavior and its relationship to impact resistant bridge design. *Comput Struct* 2003;81:547–557.
6. Consolazio GR, Cowan DR, Biggs A, Cook RA, Ansley M, Bollmann HT. Part 2: Design of foundations and structures: full-scale experimental measurement of barge impact loads on bridge piers. *Transp Res Rec* 2005;1936:79–93.
7. Yuan P, Harik IE, Davidson MT. Multi-barge flotilla impact forces on bridges. Report No. KTC-08-13/SPR261-03-2F, Kentucky Transportation Center College of Engineering, University of Kentucky, Lexington, KY, 2008.
8. Fan W, Yuan W, Zhi Y, Fan Q. Dynamic demand of bridge structure subjected to vessel impact using simplified interaction model. *J Bridge Eng–ASCE* 2011;16(1):117–126.
9. Sha Y, Hao H. A simplified approach for predicting bridge pier responses subjected to barge impact loading. *Adv Struct Eng* 2014;17(1):11–23.
10. Wang W, Morgenthal G. Dynamic analyses of square RC pier column subjected to barge impact using efficient models. *Eng Struct* 2017;151:20–32.

11. Yuan P, Harik IE. Equivalent barge and flotilla impact forces on bridge piers. *J Bridge Eng* 2009;15:523–532.

12. Sha Y, Hao H. Nonlinear finite element analysis of barge collision with a single bridge pier. *Eng Struct* 2012;41:63–76.

13. Davidson MT, Consolazio GR, Getter DJ, Shah FD. Probability of collapse expression for bridges subject to barge collision. *J Bridge Eng–ASCE* 2012;18(4):287–296.

14. Gholipour G, Zhang C, Mousavi AA. Effects of axial load on nonlinear response of RC columns subjected to lateral impact load: ship–pier collision. *Eng Fail Anal* 2018;91:397–418.

15. Gholipour G, Zhang C, Li M. Effects of soil–pile interaction on the response of bridge pier to barge collision using energy distribution method. *Struct Infrastruct Eng* 2018;14(11):1520–1534.

16. Gholipour G, Zhang C, Mousavi AA. Analysis of girder bridge pier subjected to barge collision considering the superstructure interactions: the case study of a multiple-pier bridge system. *Struct Infrastruct Eng* 2019;15(3):392–412.

17. Gholipour G, Zhang C, Mousavi AA. Reliability analysis of girder bridge piers subjected to barge collisions. *Struct Infrastruct Eng* 2019;15(9):1200–1220.

18. Getter DJ, Consolazio GR, Davidson MT. Equivalent static analysis method for barge impact-resistant bridge design. *J Bridge Eng–ASCE* 2011;16(6):718–727.

19. Fan W, Liu Y, Liu B, Guo W. Dynamic ship-impact load on bridge structures emphasizing shock spectrum approximation. *J Bridge Eng–ASCE* 2016;21(10):04016057.

20. Fan W, Yuan WC. Shock spectrum analysis method for dynamic demand of bridge structures subjected to barge collisions. *Comput Struct* 2012;90:1–12.

21. Fan W, Zhang Y, Liu B. Modal combination rule for shock spectrum analysis of bridge structures subjected to barge collisions. *J Eng Mech –ASCE* 2016;142(2):04015083.

22. LS-DYNA 971. *Livermore Software Technology Corporation*, Livermore, CA, 2015.

23. Murray YD, Abu-Odeh AY, Bligh RP. Evaluation of LS-DYNA concrete material model 159. Report No. FHWA-HRT-05-063, APTEK Inc., Colorado Springs, CO, 2007.

24. Jiang H, Wang X, He S. Numerical simulation of impact tests on reinforced concrete beams. *Mater Des* 2012;39:111–120.

25. Jiang H, Zhao J. Calibration of the continuous surface cap model for concrete. *Finite Elem Anal Des* 2015;97:1–19.

26. Guo J, Cai J, Chen W. Inertial effect on RC beam subjected to impact loads. *Int J Struct Stab Dyn* 2017;17(4):1750053.

27. Lu YE, Zhang LM. Progressive collapse of a drilled-shaft bridge foundation under vessel impact. *Ocean Eng* 2013;66:101–112.

28. Fan W, Yuan WC. Numerical simulation and analytical modeling of pile supported structures subjected to ship collisions including soil–structure interaction. *Ocean Eng* 2014;91:11–27.

29. Abdelkarim OI, ElGawady MA. Performance of hollow-core FRP–concrete–steel bridge columns subjected to vehicle collision. *Eng Struct* 2016;123:517–531.

30. Abdelkarim OI, ElGawady MA. Performance of bridge piers under vehicle collision. *Eng Struct* 2017;140:337–352.

31. Fan W, Xu X, Zhang Z, Shao X. Performance and sensitivity analysis of UHPFRC-strengthened bridge columns subjected to vehicle collisions. *Eng Struct* 2018;173:251–268.

32. Do TV, Pham TM, Hao H. Dynamic responses and failure modes of bridge columns under vehicle collision. *Eng Struct* 2018;156:243–259.

33. Jiang H, Wang J, Chorzepa MG, Zhao J. Numerical investigation of progressive collapse of a multispan continuous bridge subjected to vessel collision. *J Bridge Eng–ASCE* 2017;22(5):04017008.

34. Madurapperuma MA, Wijeyewickrema AC. Response of reinforced concrete columns impacted by tsunami dispersed 20 and 40 shipping containers. *Eng Struct* 2013;56:1631–1644.

35. Fan W, Yuan W. Ship bow force-deformation curves for ship-impact demand of bridges considering effect of pile-cap depth. *Shock Vib* 2014;2014:1–19.

36. Consolazio GR, Davidson MT, Cowan DR. Barge bow force-deformation relationships for barge–bridge collision analysis. *Transp Res Rec* 2009;2131:3–14.

37. Alsos HS, Amdahl J. On the resistance of tanker bottom structures during stranding. *Mar Struct* 2007;20(4):218–237.

38. LSTC. LS-DYNA keyword user's manual, version 971. Livermore Software Technology Corporation, Livermore, CA, 2016.

39. Jones N. *Structural impact*. Cambridge University Press, Cambridge, UK, 2011.

40. European Committee for Standardization (CEN). Eurocode 1 – Actions on structures. Parts 1–7: General actions–accidental actions due to impact and explosions, 3rd draft, CEN, Brussels, Belgium, 2002.

41. China Ministry of Railways (CMR). *General code for design of railway bridges and culverts (TB10002.1-2005)*. China Railway Press, Beijing, China, 2005 (in Chinese).

42. Cao Y, Luo L, Zhou Y. Ship collision protection device for Zhanjiang Bay Bridge. In 17th Congress of IABSE, Creating and Renewing Urban Structures, International Association for Bridge and Structural Engineering Chicago, 2008.

43. ACI 318-14. Building code requirements for structural concrete and commentary. American Concrete Institute, Farmington Hills, MI, 2014.

44. Zhang C, Gholipour G, Mousavi AA. Nonlinear dynamic behavior of simply-supported RC beams subjected to combined impact-blast loading. *Eng Struct* 2018;181:124–142.

45. Murray YD. User manual for LS-DYNA concrete material model 159. Report No. FHWA–HRT–05–062, APTEK Inc., Colorado Springs, CO, 2007.

46. Aviram A, Mackie KR, Stojadinović B. Guidelines for nonlinear analysis of bridge structures in California. Report No. UCB/PEER 2008/03, Pacific Earthquake Engineering Research Center, University of California Berkeley, CA, 2008.

47. Reese LC, Cox WR, Koop FD. Analysis of laterally locked piles in sand. In Fifth Annual Offshore Technology Conference, Houston, Texas, Paper No. OTC 2080 (GESA Report No. D-75-9), 1974.

48. Fujikake K, Li B, Soeun S. Impact response of reinforced concrete beam and its analytical evaluation. *J Struct Eng–ASCE* 2009;135(8):938–950.

49. API. Recommended practice for planning designing and constructing fixed offshore platforms, recommended practice 2A-WSD, 21st edn. American Petroleum Institute, Dallas, TX, 2000.

50. Ghosn M, Fred M, Jian W. Design of highway bridges for extreme events. Report No. 489, Transportation Research Board, Washington, DC, 2003.

51. Biggs JM. *Introduction to structural dynamics*. McGraw-Hill, New York, NY, 1964.

52. Krauthammer T, Shahriar SA. A computational method for evaluating modular prefabricated structural element for rapid construction of facilities, barriers, and revetments to resist modern conventional weapons effects. Report No. ESL-TR-87-60. Engineering and Services Laboratory Air Force Engineering and Services Center, Tyndall Air Force Base, FL, 1988.

Chapter 6

Simply-supported RC beams subjected to combined impact-blast loads

6.1 Introduction

With the increase of terrorist attacks in recent years, many researchers have focused on the study of dynamic responses and failure behaviors of reinforced concrete (RC) structures subjected to extreme impulsive loads (e.g., impact and blast loads) to propose the essential recommendations for the anti-terrorist design of such structures. Understanding the mechanism of applying impulsive load on RC structures plays the most key role in the recognition of the consequent damage patterns and failure modes. Many studies exist in the literature on the investigation of RC structures and components, including beams, columns, slabs, framed buildings, and bridge piers subjected to impact or blast loads analytically [1–5], numerically [6–10], and experimentally [4, 11–13].

While many previous works evaluated the blast responses of RC columns [13–18], slabs [19–22], framed buildings [23, 24], bridge piers [10, 25–33], there is only a limited number of studies on the investigation of RC beams subjected to blast loads experimentally [11, 12, 34] and numerically [6, 35, 36]. RC beams are flexural-critical members that are commonly used to attach slabs in RC structures. Although some applications of isolated RC beams can be found in real life such as parking ramps, high-rise buildings, and girder highway bridges, designing these components against extreme dynamic loadings such as impact and blast loads is very important because their failure may cause the progressive collapse of the whole structure. Zhang et al. [11] conducted an experimental study on the damage patterns of RC beams under close-in blast loading. It was observed that the beam spallation length increases with the decrease of explosive scaled distance and the beam is prone to fail in flexural failure mode. Yao et al. [12] studied experimentally and numerically the damage levels of RC beams with the variability of stirrup ratio subjected to different explosive charges. It was found that the severity of the damage decreases with increasing of stirrup ratio and reducing the space between the stirrups. In addition to the experimental works, Yan et al. [6] numerically investigated the damage mechanism of RC beams subjected to close-in blast loading. The increase of spallation length at the bottom cover and extending of damage in the beam depth was observed by propagating of main flexural cracks. Also, it was found that the spall damages are mainly generated due to the reflected tensile stresses from the beam bottom surface and the main cracks in the depth of beam are produced by meeting up of flexural cracks from the bottom and compressive cracks from the top of the beam. Li and Hao [35] proposed a two-step numerical method to analyze the structural responses of RC beams to blast loads in the loading and free vibration phases.

DOI: 10.1201/9781003262343-6

Also, the accuracy of the proposed method in the prediction of damage states, stresses, displacement responses, and shear failure [36] of RC beams was evaluated in comparison with those from the direct blast simulation.

Despite limited investigations on RC beams subjected to blast loads, many studies exist in the literature assessing the impact responses of RC beams. Because of fundamental differences between the mechanisms of applying impact and blast loads, a structure may present different behaviors and failure modes under each of these loads. When a structure subjects to an impact loading, the impact stresses propagate from the impact point toward the supports and farther zones as the longitudinal and shear waves. While the blast loads apply as the distributed load with a higher rate than that of concentrated impact load on the structure surface, the blast loadings could be led to more impulsive loadings with a shorter duration time than impact loadings.

Fujikake et al. [4] experimentally investigated the responses and damage patterns of RC beams with different ratios of longitudinal reinforcement subjected to dropping mass impact loads with different impact heights. Besides, an analytical deflection-based damage index (DI) was proposed to assess the damage levels of RC beams. Two local and global response phases were declared for RC beams under low-rate impact loads. The local response is dominant in the impact zone in a very short time after the onset of impact load, while the global response phase is dominant at the free vibration stage, which extremely depends on the loading rate. Pham et al. [37] found that the contact stiffness during an actual impact event on RC beams has a significant influence on the peak value and the duration of impact loading. Also, the influence of the global stiffness of RC beams is numerically and analytically investigated by Pham and Hao [38] on the impact behaviors and responses. It was revealed that although the global response mode arising from the global stiffness dominates the beam behavior at the free vibration stage, it is not properly effective at the early impact impulse stage. Therefore, the failure behavior of RC beams extremely depends on the impact duration. Cotsovos et al. [9] numerically studied the dynamic response of RC beams subjected to a high-rate impact loading. The increase of the beam load-carrying capacity was concluded with increasing in the loading rate that results in the reduction of the distance between the plastic hinges formed at both sides of the mid-span named effective length. In line with this study, Pham and Hao [8] proposed a method to estimate the position of plastic hinges formed in RC beams based on the theory of stress wave propagation. Afterwards, the effects of various factors, including plastic hinges, boundary conditions, impact velocity, projectile weight, and concrete strength, on the impact behavior of RC beams are numerically examined in Ref. [7].

Since the beams are flexure-resistant components, reaching their ultimate shear strength earlier than the flexural strength is more likely. Hence, some researches have focused on the evaluation of the shear response mechanism of RC beams cited as follows. Saatci and Vecchio [39] studied the velocity of impact force propagation from the impact point to the farther boundaries of RC beams. It was found that this velocity is slower than the velocities in longitudinal and shear waves. Also, Isaac et al. [40] concluded that the velocity of impact force propagation is significantly different from the shear wave phenomenon in RC beams. Accordingly, it was obtained from Ref. [41] that the probability of the occurrence of shear failure increases with increasing impact loading rates.

Despite the aforementioned previous works, RC structures are likely subjected to the combination of impact and blast loads simultaneously or sequentially during a possible terrorist activity or accidental event. Such combined loading scenarios could occur during the impact of heavy objects with a relatively low velocity such as vehicles [42, 43] and vessels [44–47] or rather lighter objects with a high velocity such as missiles and projectiles, which carry the explosive materials, could be led to the subsequent blast loading after the onset of impact loading on civil infrastructures (e.g., bridges, framed buildings, etc.) and structural RC components (e.g., beams, columns). Besides, the blast load can be applied on the structure before impact loading during other possible combined loading scenarios such as impacting of fragments or species of exploded objects and projectiles with high velocities, or collision of errant vehicles or heavier objects with lower velocities moving due to the explosion waves after reaching of blast loading on the target structure. Therefore, this chapter aims to numerically evaluate the dynamic responses and failure behaviors of the RC beams subjected to combined impact-blast loading that has not been reported in the literature. Investigating the effects of impact loading stresses on the blast responses and damage patterns of simply-supported RC beams is the main contribution of this chapter. To this end, different combined loading scenarios on RC beams are considered in this study by varying the sequence of applying loads and the time lag between the initiations of impact and blast loads. Also, the vulnerability of RC beams under combined loadings to some structural parameters, including beam depth, span length, and configuration of longitudinal and transverse reinforcements, is assessed through a proposed DI based on the residual flexural capacity of the beam cross-section. Since there is no experimental work in the literature investigating RC beams under combined impact-blast loads, the finite element models of RC beams developed by LS-DYNA [48] are validated using the comparison of finite element (FE) simulation results and those from the experimental tests done by Fujikake et al. [4] and Zhang et al. [11] under the impact and blast loads, respectively. In this chapter, an RC beam with a sufficient flexural-shear resistance provided by the reinforcements named $S2222$ considered by Fujikake et al. [4] is selected to study under combined impact-blast loads with a middle-rate impact velocity of 6.86 m/s and middle-range blast energy with a scaled distance of 0.25 m/kg$^{1/3}$ applied on the beam mid-span.

6.2 Theoretical background

In this section, the theory of impact and blast loadings applied to RC beams is presented to realize the similarities and differences between the applying load mechanisms and the structural responses. According to Hao [49], loads applied on structures can be categorized into three types based on their duration in comparison with the natural vibration period of the impacted structure. These types are (i) quasi-static loading in which the structure reaches its maximum response before ending of the impact duration; (ii) dynamic loading in which the structure reaches its maximum response almost at the same time with ending of the impact duration; (iii) impulsive loading in which impact duration ends before reaching the structure its maximum response. These different types of impact loading could result in different failure behaviors and responses in RC beams. Based on previous studies in the literature, different possible failure modes of simply-supported RC beams with middle-range flexural and shear resistances subjected to impact or blast loads are presented as follows.

The response of an impacted RC structure may include local and global phases. Figure 6.1a shows the propagation of compressive stress waves from the location of

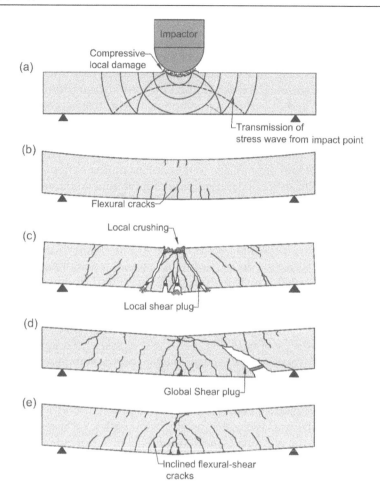

Figure 6.1 Theoretical response of RC beams subjected to impact loading: (a) load applying mechanism and local response, (b) global flexural failure, (c) local shear failure, (d) global shear failure, (e) global shear-flexural failure.

concentrated impact load on the beam to farther boundaries and the reflection of tensile waves from the free boundaries at the bottom (tension zone). Also, the localized compressive damage in the top cover is observed which is extremely dependent on the impact intensity, the confinement, and interlocking (bond) resistance between the concrete and reinforcements in the compression zone. Figure 6.1b–d shows the different failure behaviors and crack patterns observed in previous studies for the impacted simply-supported beams with middle-range flexural and shear resistance provided by the reinforcements. These behaviors profoundly depend on the impact loading rate on the beam. When the beam is subjected to a low-rate impact and relatively long duration at the mid-span, it experiences a relatively ductile response and the flexural failure initiates with the formation of vertical flexural cracks from the tension zone (i.e., bottom surface) as shown in Figure 6.1b [4, 50]. In this case, the failure of the beam in flexure mode through the formation of a plastic hinge at the mid-span is possible if the magnitude and the duration of impact load are rather high. In contrast, when a beam with a middle-range flexural and shear resistances is subjected to a high-rate impact loading with a rather short duration, it suffers a local shear plug initiated at the impact point and propagated downward the beam at an approximate angle of 45° on both sides of the mid-span as shown in Figure 6.1c [9, 51]. Besides, if the beam would have a low-shear strength, the occurrence of a global brittle shear plug originating from the supports is possibly subject to a middle-rate impact loading as shown in Figure 6.1d [50]. Besides, when the beam with a middle-range shear and flexural strength is under a middle-rate impact loading, a global flexural-shear failure mode is observed as shown in Figure 6.1e. In this failure mode, the damages initiate with the appearance of vertical flexural cracks at the tension zone and propagating with the formation of inclined shear cracks in the beam depth. This type of failure behavior commonly occurs when the beam is under dynamic-type impact loading. Besides, the increase of flexural cracks generated at the top surface due to the negative flexural moments (hogging moments) is observed with increasing impact loading rate. Because the stress waves produced in the beams subjected to a rather high-rate impact loading do not necessarily reach the supports during a very short impact duration that causes the formation of flexural cracks on the upper surface of the beam [52].

Unlike concentrated impact loads, which are applied to the limited area, the blast load is applied on the beam surface as a uniformly distributed load and the shock wavefront pressure as shown in Figure 6.2a. Localized damages of RC beams subjected to air blast load, including the spalling at the beam bottom (tensile zone) and the compressive damage at the beam top, are presented in Figure 6.2a. When the compressive shock wave reaches the beam top surface facing the detonation, it can cause local compressive damages if the blast energy is sufficiently high with a rather close standoff distance of the explosive. With propagating of the compressive shock wave within the beam depth and reaching the free bottom surface (i.e., free boundary condition), it reflects as a tensile wave that causes the tensile damages and the concrete spall damage and fragmentation at the beam bottom surface (see Figure 6.2a). Therefore, the level of spall damage generated in the beam subjected to a blast loading extremely depends on the shear and bond resistances and the confinement of the beam concrete provided by the reinforcements in the trapped impulse zone and particularly at the beam bottom cover. Also, the concrete spallation depends on the stress wave shapes and the boundary conditions influencing the characteristics (compression or tension waves) of the reflected stress wave. A classification of the concrete spallation damage into the three categories from low-level spallation with observing of a few cracks to high-level spallation with the generation of the breach and hollow damages in the beam depth was presented in the previous studies [53]. The level of these localized compressive and spall damages is also dependent on the energy of blast loading, which is commonly identified using a scaled distance of explosive charge from the structure as given follows:

$$Z = \frac{R}{W_{TNT}^{1/3}} \tag{6.1}$$

where R is the distance of the explosive from the beam, and W_{TNT} is the weight of the explosive.

Although the local spallation damage may not cause the overall failures, it can be significantly led to losing the load-carrying capacity of the beam at the trapped

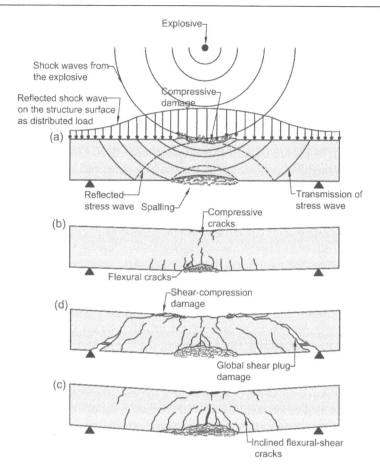

Figure 6.2 Theoretical responses of RC beams subjected to air blast loading: (a) local responses, (b) global flexural failure, (c) global shear failure, (d) global flexural-shear failure.

impulse zone. Besides the localized damages, RC beams relative to their shear and flexural resistances could indicate different global flexural, shear, or flexure-shear failures subjected to different intensities and duration of blast loading. Like the response of simply-supported RC beams under impact loading, RC beams could experience a common global flexural response as shown in Figure 6.2b, when the ratio of blast duration to the natural vibration period of the beam is rather long [35]. In this type of failure mode, the beam may suffer a plastic hinge subjected to high-energy blast loading (i.e., high-weight explosive charge) with losing of concrete strength and yielding of the reinforcements. However, RC beams subjected to blast loading with a very short duration could experience diagonal or direct shear failures as shown in Figure 6.2c depending on the boundary conditions and intensity of blast loading [18, 35, 36]. Also, between the flexural and shear failures, the occurrence of flexural-shear failure is possible as shown in Figure 6.2d, when the beam is subjected to the middle range of blast intensity and duration.

The load applying mechanism is one of the most significant differences between the impact and blast loads which are applied to the structure as the concentrated and uniformly distributed loading, respectively. This discrepancy causes the different shear failures in the beam when it is under the impulsive level of impact and blast loadings with a very short duration. In other words, unlike the global brittle shear failures started directly or diagonally from the beam supports due to the impulsive blast loading, the beam subjected to a high-rate impact loading (i.e., impulsive) suffers a local shear plug around the impact zone [41, 51] (see Figures 6.1c and 6.2c). Since the spalling damage has a brittle nature that results in high-rate impacts and impulsive blast loadings [16], the accumulation of this localized damage with global shear and flexural-shear failures is more possible than ductile flexural response resulted from low-rate loadings.

6.3 FE modeling of RC beam

Since no available experimental work exists in the literature on the investigation of RC beams subjected to the combination of impact and blast loads, the reliability of FE models of RC beams under impact and blast loadings developed in LS-DYNA is separately examined by comparing FE simulation results with those from the experimental tests conducted by Fujikake et al. [4] and Zhang et al. [11], respectively.

As an FE validation study of RC beams subjected to blast loading, a fix-support beam subjected to air blast loading with a standoff distance of 0.4 m is shown in Figure 6.3a, which was experimentally conducted by Zhang et al. [11]. The beam has a square cross-section with dimensions of 0.1 m × 0.1 m, a span length of 1.1 m, a concrete cover with a depth of 20 mm. Besides, the diameter of transverse and longitudinal reinforcements in the compression and tension sides is 6 mm, and the spacing of stirrups is 60 mm. The deflection of the beam clamped at the steel frame was measured by a cluster of needles fixed in a barrel filled with fine sand. Figure 6.3b shows the FE model of the beam developed in LS-DYNA in which the beam concrete is modeled using 8-node constant stress solid element, the end supports using tetrahedral solid elements, and the reinforcements using Hughes-Liu beam elements.

Besides, as the FE validation study of RC beams under impact loading, a simply-supported RC beam with an identification of S2222 subjected to a dropping mass of 400 kg experimentally conducted by Fujikake et al. [4] as shown in Figure 6.4a is considered in this chapter. The FE model of the beam developed in LS-DYNA is illustrated in Figure 6.4b which has the cross-sectional dimensions of 0.25 m in depth, 0.15 m in width, 1.7 m in length, the cover depth of 40 mm, two numbers of longitudinal reinforcements with a diameter of 22 mm in the compression and tension sides, and transverse reinforcements with a diameter of 10 mm spaced 75 mm along the beam. The beam concrete and reinforcements are modeled with a similar property to those of the beam considered under blast loading. According to the literature [54], utilizing a node-sharing mesh method can lead to the creation of unnecessarily small element sizes that may significantly cause the increase of the computational time of FE simulation. To solve this problem, a coupling method named CONSTRAINED_LAGRANGE_IN_SOLID in LS-DYNA is utilized with frictionless sliding between the reinforcement and concrete elements that permits a uniform meshing throughout the beam. In addition, this constraint method provides the independence of the concrete meshing on the location of reinforcements that assists the feasibility of parametric studies. Also, a contact algorithm named AUTOMATIC_SURFACE_TO_SURFACE between the RC beam as the master part and the dropping mass as the

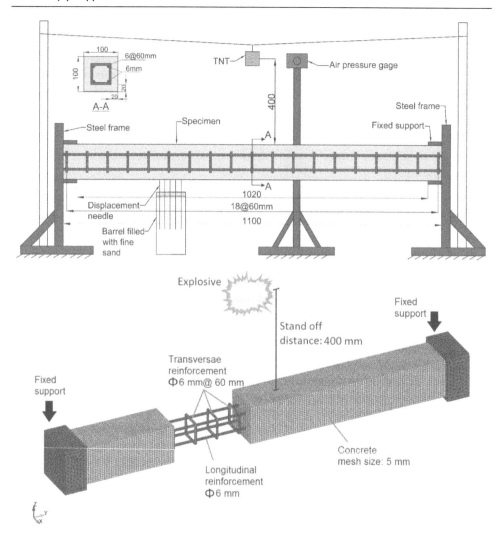

Figure 6.3 Blast testing of RC beam: (a) experimental setup done by Zhang et al. [11] (units in mm) and (b) numerical FE simulation in LS-DYNA by the present study.

slave part is defined in which Coulomb friction coefficients are considered equal to 0.3 to take into account the sliding force and energy during the contact. Moreover, this contact algorithm is used between the FE models of the beam and the supports for both impact and blast loading studies.

6.3.1 Material model

The concrete beams under blast and impact loadings have an unconfined compression strength of 40.45 and 42 MPa, respectively. For both of these studies, the concrete behavior of the beams is modeled using a non-linear material model named MAT_CSCM_CONCRETE (MAT_159) in LS-DYNA [55] which is a continuous surface

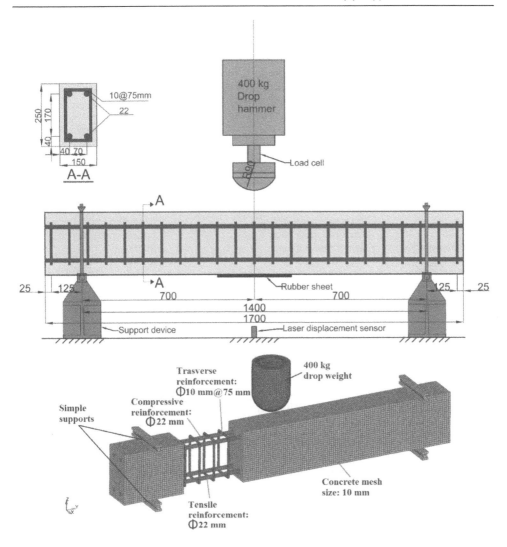

Figure 6.4 Dropping mass impact test on the RC beam: (a) experimental setup done by Fujikake et al. [4] (units in mm) and (b) numerical FE simulation developed in the present study using LS-DYNA.

cap model with considering strain rate effects. The feasibility of utilizing this model through three input specifications of the concrete material including the unconfined compression strength (f_c'), the element erosion parameter (*ERODE*), and the maximum aggregate size (*Dagg*) related to the fracture energy parameter is one of the main advantages of this model. In this model, the element erosion is modeled based on the maximum principal strain (*ERODE*). According to the literature [54, 56], the damage (i.e., the element erosion herein) occurs when the value of ERODE is set more than 1.0 in LS-DYNA. Based on comprehensive research done on this model in the literature [54], the *ERODE* ranges between 1.05 and 1.1 were recommended for RC structures under extreme loads such as impact loading. From a series of FE simulations of

Table 6.1 Material properties used for RC beams under blast and impact loadings

Model	Parameter	Case study	Value
Concrete	Mass density (kg/m³)	Impact and blast	2,440
	Unconfined compression strength (MPa)	Blast	40.45
	Elements erosion (ERODE)	Impact	42
	Maximum aggregate size (mm)	Impact and blast	1.08
	Rate effects	Impact and blast	10
		Impact and blast	Turn on
Rebar (longitudinal and transverse)	Mass density (kg/m³)	Impact and blast	7,800
	Young's modulus (GPa)	Impact and blast	210
	Yield stress (MPa)	Blast	395
	Poisson's ratio	Impact and blast	0.3
	Failure strain	Impact and blast	0.15
	Strain rate parameter C_{strain}	Impact and blast	40.4
	Strain rate parameter P_{strain}	Impact and blast	5.0
Longitudinal rebar	Yield stress (MPa)	Impact	418
Transverse rebar	Yield stress (MPa)	Impact	295

dropping mass and blast loadings on RC beams, a value of 1.08 gave the failure behaviors and responses in good agreement compared with those from the experimental tests that existed in the literature. This means that the element erosion occurs when the value of the maximum principal strain of the concrete exceeds a value of 0.08. Also, the aggregate size parameter affects the softening behavior of the damage formulation and the fracture energy of the concrete material. According to the literature [54, 57], considering the sizes between 8 and 32 mm is recommended as the maximum aggregate size for the concrete with the unconfined compression strengths between 20 and 58 MPa. In this chapter, for the concrete with f_c' around 40 MPa, a size of 10 mm is assigned as the maximum aggregate size to capture a high compressive strength for the concrete.

The yield strength of all the reinforcements in the beam under blast loading is 395 MPa, and for the beam under impact loading, this parameter has the values of 418 and 295 MPa for the longitudinal and transverse reinforcements, respectively. Also, Young's modulus is 210 GPa for all reinforcements. An elastic-plastic material model considering strain rate effects named MAT_PIECEWISE_LINEAR_PLASTICITY (MAT_024) is selected in LS-DYNA to model the behaviors of the reinforcements. This model provides the isotropic and kinematic hardening plasticity behaviors for the reinforcements. Moreover, a rigid material model named MAT_RIGID (MAT_020) is selected for the dropping mass and all supports in both studies. The properties of material models used for RC beams under blast and impact loadings are given in Table 6.1.

6.3.2 Strain rate effect

Both concrete and steel materials in RC structures under high-rate dynamic loads indicate a higher strength due to their high strain rate effects than those under quasi-static conditions. In this chapter, the strength enhancement is defined by the dynamic increase factor (*DIF*) at a given strain rate of concrete and steel materials.

The *DIF* of concrete compressive strength (*CDIF*) has been provided by CEB Code 1990 [57] as follows:

$$CDIF = \frac{f_{cd}}{f_{cs}} = \begin{cases} \left(\dfrac{\dot{\varepsilon}_{cd}}{\dot{\varepsilon}_{cs}}\right)^{1.026\alpha} & \text{for } \dot{\varepsilon}_{cd} \leq 30 \text{ s}^{-1} \\[3mm] \gamma\left(\dfrac{\dot{\varepsilon}_{cd}}{\dot{\varepsilon}_{cs}}\right)^{1/3} & \text{for } \dot{\varepsilon}_{cd} > 30 \text{ s}^{-1} \end{cases} \quad (6.2)$$

where f_{cd} is the dynamic compressive strength at strain rate $\dot{\varepsilon}_{cd}$; f_{cs} is the static compressive strength at $\dot{\varepsilon}_{cs}$; the dynamic strain rate $\dot{\varepsilon}_{cd}$ is in the range of 3×10^{-5} to 3×10^2 s^{-1}, and the static strain rate $\dot{\varepsilon}_{cs}$ is equal to 3×10^{-5} s^{-1}; $\alpha = 1/(5 + 9 f_{cs}/f_{c0})$ in which f_{c0} is 10 MPa; log $\gamma = 6.156\alpha - 2$.

Besides, the DIF of concrete tensile strength (*TDIF*) is defined according to Malvar and Crawford [58], as follows:

$$TDIF = \frac{f_{td}}{f_{ts}} = \begin{cases} \left(\dfrac{\dot{\varepsilon}_{td}}{\dot{\varepsilon}_{ts}}\right)^{\delta} & \text{for } \dot{\varepsilon}_{td} \leq 1 \text{ s}^{-1} \\[3mm] \beta\left(\dfrac{\dot{\varepsilon}_{td}}{\dot{\varepsilon}_{ts}}\right)^{1/3} & \text{for } \dot{\varepsilon}_{td} > 1 \text{ s}^{-1} \end{cases} \quad (6.3)$$

where f_{td} is the dynamic compressive strength at strain rate $\dot{\varepsilon}_{td}$, f_{cs} is the static compressive strength at $\dot{\varepsilon}_{ts}$, the dynamic strain rate $\dot{\varepsilon}_{td}$ is in the range of 10^{-6} to 160 s^{-1}, and the static strain rate $\dot{\varepsilon}_{ts}$ is equal to 10^{-6} s^{-1}, $\delta = 1/(1 + 8 f_{cs}/f_{c0})$ in which f_{c0} is 10 MPa; log $\beta = 6\delta - 2$.

For steel reinforcements, the *DIF* relationship is defined by Malvar [59] as follows:

$$DIF = \left(\frac{\dot{\varepsilon}}{10^{-4}}\right)^{\eta} \quad (6.4)$$

where $\dot{\varepsilon}$ is the strain rate of steel; η for the yield stress f_y in MPa is $\eta_{fy} = 0.074 - 0.04$ $f_y/60$; and for the ultimate stress is $\eta_{fu} = 0.019 - 0.009 f_y/60$.

6.3.3 Validation of FE models under impact and blast loads

From the mesh convergence tests on the RC beam under blast loading, it was obtained that the FE simulation results are in good agreement with experimental results when the mesh size ratio of beam concrete elements to the beam depth is 0.05 (i.e., the mesh size is 5 mm). Besides, selecting a smaller mesh size than 5 mm for the concrete caused an inefficient increase in the simulation computational time. Therefore, considering the same element size for the reinforcements, the total number of elements of the beam is 89,640 (i.e., 88,000 for the concrete and 1,640 for the reinforcements) for the RC beam under blast loading. Two different blast loading scenarios varied for scale distances with values of 0.57 m/kg$^{1/3}$ (in which the explosive charge weight is 0.36 kg for the scenario named B2-1) and 0.4 m/kg$^{1/3}$ (in which explosive charge weight is 1.0 kg for the scenario named B2-4) are considered to validate the FE model of RC beam. The blast load is modeled by a function named LOAD_BLAST_ENHANCED based

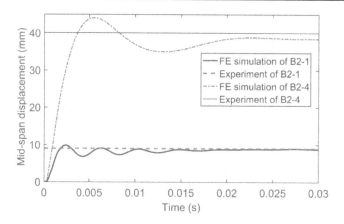

Figure 6.5 Comparing the beam displacements at the mid-span between those from the FE simulations and the experimental tests.

on the empirical relationships obtained from the blasting tests as reported in the literature [60]. This function reduces the computational time by eliminating the shock wave propagation in the air from the simulation. In this method, only the interaction of shock wave with the structure is modeled by assigning a shock-front surface on the structure through the LOAD_BLAST_SEGMENT_SET command in LS-DYNA. Since RC beams are not commonly used in the lower elevations of the structure near the ground, the amplification of the initial shock wave due to the interaction with the ground surface is neglected in this study. The reliability and applicability of this function have been proven in the literature [61]. Figure 6.5 shows the comparison of maximum displacements at the mid-span of the RC beam between those from the FE simulations and experimental tests for the two aforementioned blast loading scenarios (i.e., B2-1 and B2-4). According to the experimental work done by Zhang et al. [11], the maximum deflection value for B2-1 is 9 mm and for B2-4 is 40 mm. Compared to these values, the peak and residual displacements of the beam from the FE simulations of B2-1 are 9.83 and 8.86 mm, and for B2-4 are 44 and 38.4 mm, respectively.

Figure 6.6a and b compares the failure behaviors and damaged states of the RC beam under two different blast loading scenarios of B2-1 and B2-4 between the results from the FE simulations and those from the experimental tests. In Figure 6.5a, it is seen that when the beam suffers a low-level concrete erosion on the shock-front surface when it is under the blast scenario of B2-1 with a relatively larger scaled distance (i.e., 0.57 m/kg$^{1/3}$). This is because of reaching the compressive blast waves on the top surface. Moreover, the global flexural damages are observed at the bottom surface of the beam. With increasing the blast intensity and reducing the scaled distance to a value of 0.4 m/kg$^{1/3}$, the beam suffers more severe compressive damage at the front surface due to the compressive stress waves. Besides, the beam experiences substantial spallation at the bottom surface with a length of 0.15 m because of reflected tensile waves from the free bottom surface as shown in Figure 6.6b. Totally, good agreements are seen between the failure behaviors of the beam simulated by the FE simulations and those from the experimental tests. Therefore, the capability of the FE simulation approach presented in this chapter using LS-DYNA to model the beam under air blast load can be captured from these validated results and observations.

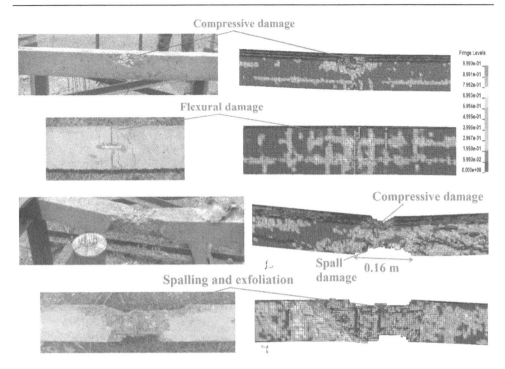

Figure 6.6 Comparing of the beam damage states between those from FE simulations and experimental tests for (a) B2-1 and (b) B2-4.

Besides, from the examination of the mesh size for the beam under impact loading, it was obtained that the results from FE simulations are converged for a mesh size of 10 mm for the concrete and the reinforcements. Accordingly, the ratio of the concrete mesh size to the beam depth (equal to 250 mm) is 0.04 which is smaller than that used for the validation study of the beam under blast loading. Totally, the beam under impact loading has 65,120 elements (i.e., 63,750 for the concrete and 1,370 for the reinforcements). Two different impact loading scenarios with velocities of 2.425 and 6.86 m/s are considered to examine the reliability of the beam FE model. According to the experimental work done by Fujikake et al. [4], the corresponding impact heights for the impact velocities of 2.425 and 6.86 m/s are 0.3 and 2.4 m, respectively. Figure 6.7a and b demonstrates the comparison of the impact force and displacement results of the RC beam S2222 between those from FE simulations and experimental tests. Although some discrepancies are observed due to the existence of some limitations in developing FE models of the beam supports and the coupling model between the concrete and reinforcements, the FE simulation results are in a good and acceptable agreement with the experimental results in terms of the peak impact forces, the impact duration, the peak displacements, and the residual displacements. Figure 6.8 shows the comparison of the damage patterns of RC beam subjected to different impact loading rates between those from FE simulations and the experimental tests. It is seen that the beam suffers a flexural-shear failure mode along with the inclined flexural-shear cracks for a low-rate impact velocity with an impact height of 0.3 m. However, appearing localized erosion of concrete on the impacted surface and the exfoliation damage on the side faces of the beam are observed during the impact scenario with a relatively higher

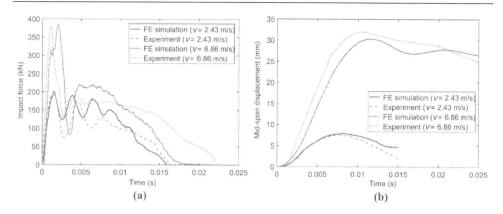

Figure 6.7 Comparing of the FE simulation and experimental results of the beam under different impact loading rates: (a) impact forces and (b) displacements at the mid-span.

impact velocity with an impact height of 2.4 m. From these observations, the capability of LS-DYNA to simulate the beam FE model with the aforementioned characteristics under different intensities of impact loading is obtained.

6.3.4 Applying the combined loads

In this chapter, a simply-supported RC beam considered by Fujikake et al. [4] with a section identification of *S*2222 is studied under the middle-tare impact loads combining with the close-in blast loads with middle-range energies. As discussed in Sections 6.3.1 and 6.3.2, a desired ratio of the mesh size to the beam depth equal or smaller than 0.05 captures the converged results in good agreement with those from the experimental tests. Therefore, it is reasonable to select a mesh size of 10 mm for the concrete and reinforcement elements for the RC beam with a depth of 250 mm (10/250 < 0.05) under the combination of impact and blast loading as shown in Figure 6.9. To clarify

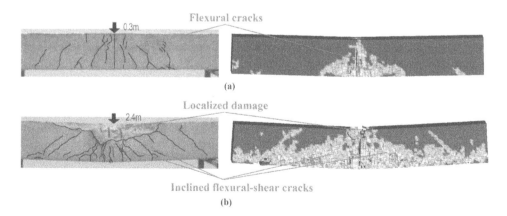

Figure 6.8 Comparing of the damaged states of the beam *S*2222 between those from FE simulations and experimental tests under (a) a low-rate impact velocity of 2.43 m/s and (b) a relatively higher impact velocity of 6.86 m/s.

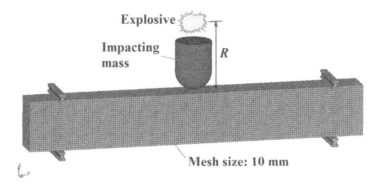

Figure 6.9 FE numerical model of the simply-supported RC beam named S2222 subjected to combined impact-blast loading developed in LS-DYNA.

the methodology of combined loading scenarios applying simultaneously and subsequently on the RC beams, several assumptions are considered as follows:

- Both impact and blast loads are applied at the mid-span of the beams.
- For combined loadings with applying of impact load before the sequent blast loading, the contact between the beam and the impacting mass is interrupted with the initiation of blast loading.
- For the parametric study and vulnerability assessment of the beam, the occurrence time of the second peak impact force owing to the rebounding of the beam which is independent of the intensity of impact loading is considered the time lag between the initiation of impact and blast loads. This time depends on the structural parameters, including the beam dimensions and the velocity of stress wave propagation in the beam [38].
- For the parametric study, a close-in and middle-energy blast load with a scaled distance of $Z = 0.25$ m/kg$^{1/3}$ (i.e., charge weight of $W_{TNT} = 0.2$ kg and a standoff distance of $R = 0.15$ m) is considered for all combined impact-blast loading scenarios.

6.4 Results and discussion

6.4.1 Damage index (vulnerability assessment)

For vulnerability assessment of RC beams under a combination of impact and blast loadings, a DI based on the residual flexural capacity of the beam cross-section is proposed. In this approach, motion-based flexural moments (rotations) are applied at the ends of simply-supported RC beams after applying impact and blast loads to capture the residual flexural capacity of the beam cross-section at the mid-span as an example shown in Figure 6.10. Therefore, the DI can be calculated as follows:

$$DI = 1 - \left(\frac{M_r}{M_N} \right)$$

(6.5)

where M_r and M_N are the residual and nominal flexural capacities, respectively, of the beam cross-section. M_N can be calculated according to ACI-318 [62] by considering

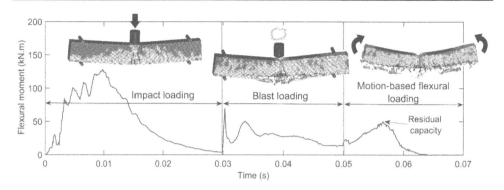

Figure 6.10 Different loading stages to achieve the residual flexural capacity of RC beams.

strain rate effects as the DIFs in the properties of the concrete and steel materials as given in Equations (6.2)–(6.4).

6.4.2 Influence of loading sequence

In this section, the influences of the loading sequence for each applied impact and blast loads are evaluated on the responses and failure behaviors of RC beams under the combination of impact and blast loadings. In this way, a reference beam $S2222$ as mentioned in Section 6.3.2 is considered under different combined loading scenarios varying in the loading sequence. A combined loading scenario in which the impact load is applied before blast loading is named "impact-blast" loading. Another combined loading is named "blast-impact" loading in which blast loading is applied on the beam before impact loading. From a modal analysis of the beam $S2222$ with a span length of 1.4 m, it was found that the first natural period of the beam is about 0.0025 s which is half of the occurrence time of the second peak impact force (i.e., 0.005 s) in this beam as shown in Figure 6.7a. According to aforementioned earlier and previous studies [38], the occurrence time of the second peak impact force indicates the time of arriving the reflected stress waves from the boundaries to the impact point (i.e., rebounding) which is dependent on the dimensions (especially, the span length) and the velocity of stress wave propagation in the structure. Therefore, a time of 0.005 s is considered for this beam for the initiation of the sequent loading in both impact-blast and blast-impact loading scenarios.

Figure 6.11a and b shows the failure behaviors of the beam under different combined loadings by varying the loading sequence, the impact velocity (V), and the scaled distance of the explosive (Z). It is observed that the beam suffers more severe spallation in the beam depth that causes the cross-sectional breach damage under a combined impact-blast loading than that under combined blast-impact loading. This is because of the discrepancies between the applying mechanism of impact and blast loads on the structure. Since the impact load (herein, dropping mass) is applied in a concentrated point on the structure, it is more likely to result in flexural-shear stresses and damages locally around the impact point as shown in Figures 6.1a and 6.12 (step 1) rather than a distributed blast loading on the structure as shown in Figure 6.2a. Therefore, occurring of more severe spallation in the beam depth is more expectable when the local stresses and damages from a concentrated impact load are associated with the tensile

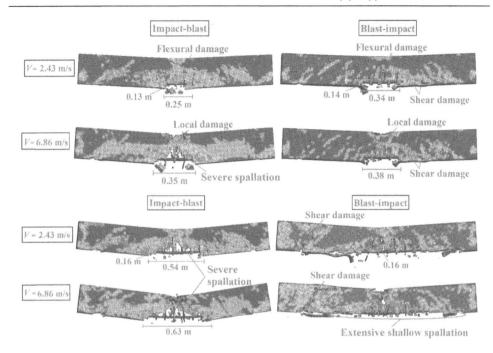

Figure 6.11 Failure behaviors of the beam to different combined loadings by varying the loading sequence and the impact velocity with scaled distances of (a) Z = 0.25 m/kg$^{1/3}$ and (b) Z = 0.15 m/kg$^{1/3}$.

stresses due to the spallation resulting from a close-in blast loading during a combined impact-blast loading (see Figure 6.12 (step 2)). However, when an unimpacted beam is subjected to a blast loading without any localized stresses before a sequent impact loading, the concrete cover in the tensile face of the beam has more strength against the spallation as shown in Figure 6.12 (steps 1 and 2). Besides, the sequential impact

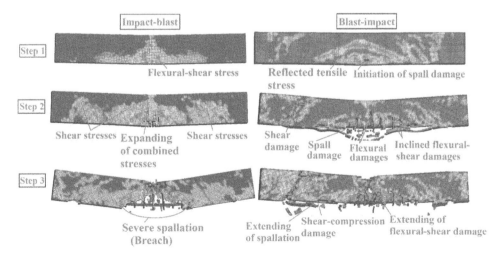

Figure 6.12 Damage states of the beam progressing under different combined loadings with V = 6.86 m/s and Z = 0.15 m/kg$^{1/3}$.

loading is not able as much as blast loading to cause the spallation in the tensile face of the beam [Figure 6.12 (steps 3)]. The damage states of the beam progressing under different steps of the combined loadings are presented in Figure 6.12.

Moreover, it is seen that the severity of spallation increases with the increase of impact velocity because of expanding the stress area around the impact zone before the initiation of the sequent blast loading. However, owing to the applying mechanism of the blast loading in a distributed manner, the length of spall damage under combined blast-impact loading is greater than those from impact-blast loading.

Figure 6.13a–c shows the comparison of the mid-span displacements, the shear forces at the critical section around the mid-span (i.e., the middle of the shear plug from the impact point propagated with an approximate angle of 45°), and the flexural moments at the mid-span of the beam under different loading scenarios by varying the loading sequence, respectively. It is seen that the beam demonstrates larger responses under only blast loading at the mid-span rather than those from only impact loading before the time of 0.005 s (i.e., the initiation time of the sequent loading). However, larger peak and residual deformations, flexural moment, and shear forces are generated in the beam under combined impact-blast loading than those from blast-impact loading. Moreover, the residual flexural capacity of the beam under different loadings is captured in Figure 6.13d according to the aforementioned procedure mentioned in

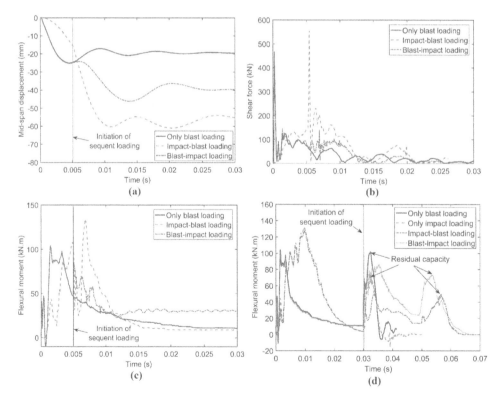

Figure 6.13 Comparing the responses of the beam to different combined loadings by varying the loading sequence with $V = 6.86$ m/s and $Z = 0.15$ m/kg$^{1/3}$: (a) mid-span displacement, (b) shear force at the critical section, (c) flexural moment at the mid-span, (d) residual flexural capacity.

Table 6.2 Damage index results for different loading scenarios

	Only impact	Only blast	Impact-blast	Blast-impact
M_r (kN m)	75.2	102	49.9	73.4
DI	0.7	0.59	0.8	0.71

Section 6.4.1 and the calculated damage indices are tabulated in Table 6.2. It is found that smaller residual flexural capacities and larger damage indices are obtained for the beam under sole impact and combined impact-blast loadings rather than those under sole blast and combined blast-impact loadings, respectively.

6.4.3 Influence of time lag between impact and blast loads

For combined impact-blast loadings, it is assumed that contact between the impacting mass and the beam is interrupted when the sequent blast loading is applied. Therefore, the time lag between the initiation of two loads will be equal to the impact duration time. In this section, the influence of the time lag (t_L) between two loads indicates the impact loading duration time is evaluated. In this way, different time stages of the responses of the reference beam S2222 under a middle-rate impact loading ($V = 6.86$ m/s) are considered as shown in Figure 6.14a–c for the initiation of the sequent blast loading with a scaled distance of 0.25 m/kg$^{1/3}$. These time stages are the occurrence times of (i) the first peak impact force at 2.1 ms (Figure 6.14a) which is simultaneous with the occurrence of the peak shear force at the critical section around the mid-span, (ii) the second peak impact force at 5 ms (Figure 6.14a) which is simultaneous with the occurrence of the peak shear force at the supports (Figure 6.14b), (iii) the unloading stage of the beam at 10 ms which is simultaneous with the occurrence of the peak flexural moment at the mid-span (Figure 6.14c), (iv) the free vibration stage at 20 ms when the contact force between the beam and the dropping mass is zero.

The failure behaviors of the beam under different combined loadings by varying of the time lag, t_L, (herein the impact loading duration) between the initiation of impact and blast loads are shown in Figure 6.15. On the left side of this figure, the trapped stress states of the beam under only impact loading (i.e., before applying the sequential blast load) and under combined impact-blast loading (i.e., after applying the sequential blast load) are demonstrated for different combined loading scenarios by varying of the time lag. Besides, the final damaged states of the beam are shown on the right side. It is observed that the severity of the spallation in the beam depth increases in proportion to the increase of t_L due to the expanding of the flexural-shear stress area around the impact zone of the beam before the initiation of the sequent blast loading. Besides, the beam suffers larger residual deformations with increasing t_L as shown in Figure 6.16a. However, Figure 6.16b and c demonstrates larger peak shear force at the critical section and the peak flexural moment at the mid-span, respectively, and more severe global shear damages are observed as shown in Figure 6.15 when the sequent blast load is applied at the same time with the occurrence of the first peak impact force at $t_L = 2.1$ ms. This is because of the accumulative effects of the brittle shear damages globally resulting from the distributed blast loading on the beam occurring almost simultaneous with the first natural period of the reference beam S2222 with a span length of 1.4 m at 2.5 ms.

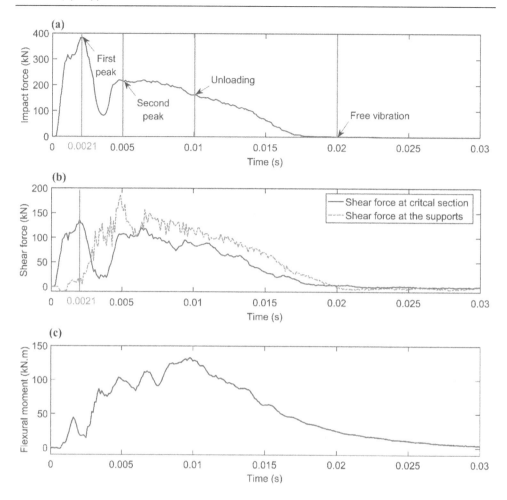

Figure 6.14 Different time stages of the responses of the reference beam S2222 a middle-rate impact loading ($V=6.86$ m/s): (a) impact force, (b) shear forces, (c) flexural moment at the mid-span.

In addition, through a vulnerability assessment using the proposed DI based on the residual flexural capacity of the beam as shown in Figure 6.16c, increasing the value of the DI is captured with the increase of t_L as given in Table 6.3.

6.4.4 Influence of beam depth

The vulnerability of the responses and failure behaviors of RC beams by varying the depth is assessed under a combined impact-blast loading with $V = 6.86$ m/s and $Z = 0.25$ m/kg$^{1/3}$. Three different beams with depths of 0.15, 0.2, and 0.25 m are considered while other structural parameters are kept constant and similar to those of the reference beam S2222. Since the changing of the beam depth is not substantially effective on the stress wave propagation velocity comparing with the beam span length, a time of 5 ms is considered the initiation of the sequent blast loading when the occurrence time of the second peak impacts force on different beams by varying the depths

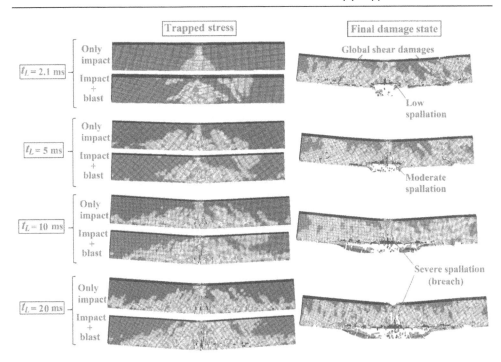

Figure 6.15 Comparing the failure behaviors of the reference beam S2222 under different combined loadings with $V = 6.86$ m/s and $Z = 0.25$ m/kg$^{1/3}$ by varying of t_L.

as shown in Figure 6.17. Also, almost the same natural periods around 2.5 ms were obtained for the modal analyses of these beams.

Figure 6.18 shows the failure behavior and damaged states of different beams by varying the depth under combined impact-blast loading comparing with those under only blast loading. It is observed that the length and depth of the spallation in the beams increase with the decrease of the depth. Besides, the beam suffers more global flexural failures under combined loading owing to the accumulation of the flexural stresses generated by impact loading and those from the sequent blast loading. Figure 6.19a demonstrates the increase of the peak and residual displacements in the beam mid-span with decreasing of the beam depth under both combined loading and only blast loading scenarios. However, decreasing the peak and residual flexural moments generated in the beam under the sole impact, sole blast, and the combination of these loads are observed with the decrease of beam depth in Figure 6.19b–d, respectively. The DI results based on the residual flexural capacity of different beams by varying the depth are given in Table 6.4. Although larger damage indices are obtained for the beams under only impact loading comparing with those from only blast loading, the combined impact-blast loading leads to the largest DI values.

Table 6.3 Damage index results for different time lags

	$t_L = 2.1$ ms	$t_L = 5$ ms	$t_L = 10$ ms	$t_L = 20$ ms
M_r (kN m)	102	85.5	68	52.3
DI	0.59	0.66	0.73	0.79

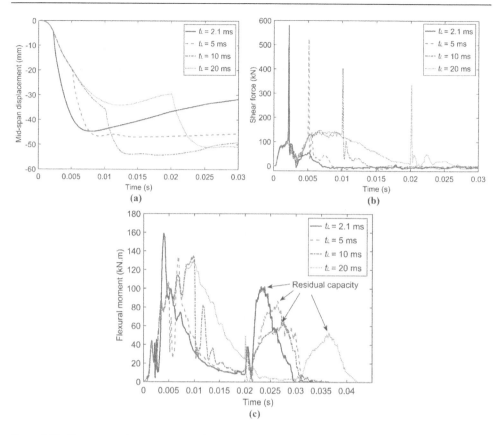

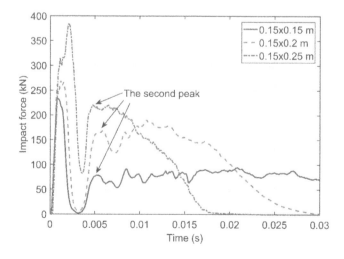

Figure 6.16 Comparing the responses of the beam to different combined loadings by varying of t_L with $V = 6.86$ m/s and $Z = 0.15$ m/kg$^{1/3}$: (a) mid-span displacement, (b) shear force at the critical section, (c) flexural moment at the mid-span.

Figure 6.17 Comparing the impact force of the different beams by varying the depth under a middle-rate impact loading with $V = 6.86$ m/s.

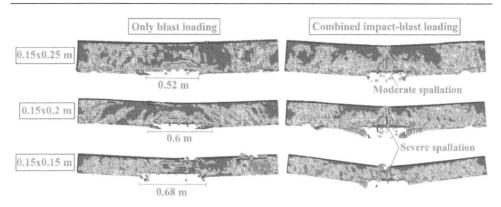

Figure 6.18 Comparing the failure behaviors of different beams by varying the depth under only blast loading and a combined impact-blast loading with $V = 6.86$ m/s and $Z = 0.25$ m/kg$^{1/3}$.

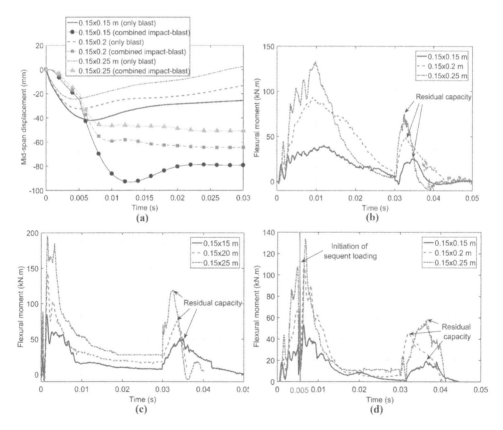

Figure 6.19 Comparing of the responses of RC beams with different depths under different loadings, (a) mid-span displacement, (b) flexural moment under only impact loading with $V = 6.86$ m/s, (c) flexural moment under only blast loading with $Z = 0.25$ m/kg$^{1/3}$, (d) flexural moment under combined impact-blast loading.

Table 6.4 Damage index results for different beam depths

	0.15 m × 0.15 m			0.15 m × 0.2 m			0.15 m × 0.25 m		
	Only impact	*Only blast*	*Impact-blast*	*Only impact*	*Only blast*	*Impact-blast*	*Only impact*	*Only blast*	*Impact-blast*
M_r (kN m)	26.9	51	22.8	51.6	75.3	45.3	75.2	119	57.9
DI	0.68	0.39	0.73	0.71	0.58	0.75	0.7	0.52	0.77

6.4.5 Influence of beam span length

The vulnerability of RC beams under combined impact-blast loading is assessed by varying the beam span length as another key structural parameter in this section. Three different RC beams with the span lengths (L_s) of 0.9, 1.4, 1.9 m are considered to study while other structural characteristics, including the cross-section, the reinforcements, and boundary conditions, are assumed similar to those of the reference beam *S2222*. Unlike the beam depth parameter, the time of both the natural period and the second peak impact force is significantly changed with changing of the beam span length. That way, the beams with the span lengths of 0.9, 1.4, and 1.9 m have the natural periods of 1.25, 2.5, and 4.1 ms, respectively, which are half of the occurrence time of the second peak impact force on these beams (i.e., 2.5, 5, and 8.2 ms, respectively) as shown in Figure 6.20. Therefore, the occurrence time of the second peak impact force is considered the initiation time of the sequent blast load in the combined impact-blast loading.

The failure behaviors of the beams with different span lengths under the only blast and combined impact-blast loadings are shown in Figure 6.21. It is observed that although the beams experience almost similar spall damages under only blast loading, the severity of spallation in the depth of the beam increases with increasing of the span length when they are subjected to a combined impact-blast load with $V = 6.86$ m/s and $Z = 0.25$ m/kg$^{1/3}$, because the beams endure more flexural moments in the mid-span with increasing the span length as shown in Figure 6.22a–c under different loadings. Besides, the accumulation of the flexural stresses and damages under impact loading

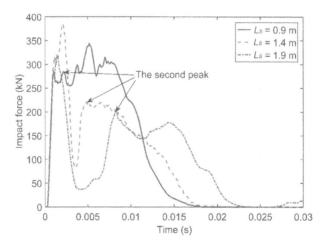

Figure 6.20 Comparing the impact force of the different beams by varying the span length under a middle-rate impact loading with $V = 6.86$ m/s.

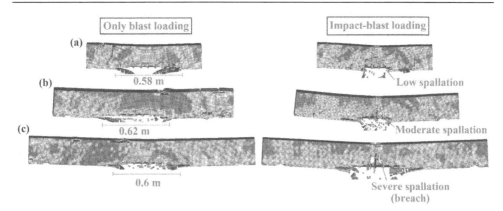

Figure 6.21 Failure behaviors of RC beams with different span lengths of (a) 0.9 m, (b) 1.4 m, (c) 1.9 m, under only blast and a combined impact-blast loading with $V = 6.86$ m/s and $Z = 0.25$ m/kg$^{1/3}$.

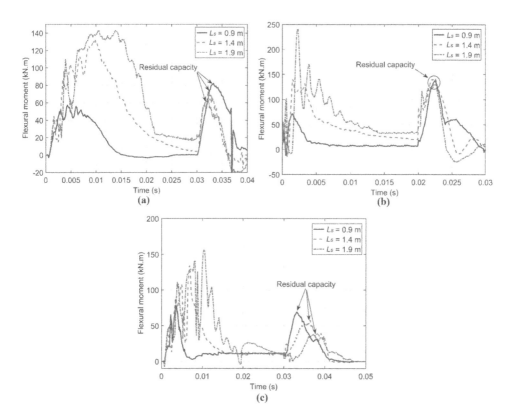

Figure 6.22 Comparing the flexural moments generated in the beams with different span lengths under (a) only impact loading, (b) only blast loading, (c) combined impact-blast loading with $V = 6.86$ m/s and $Z = 0.25$ m/kg$^{1/3}$.

Table 6.5 Damage index results for beams with different span lengths

	Ls = 0.9 m			Ls = 1.4 m			Ls = 1.9 m		
	Only impact	Only blast	Impact-blast	Only impact	Only blast	Impact-blast	Only impact	Only blast	Impact-blast
M_r (kN m)	82.3	141	68.9	73.6	130	57.9	60.9	129	37
DI	0.67	0.44	0.72	0.71	0.48	0.77	0.76	0.48	0.85

and the tensile stresses resulting from the blast loading generated in the bottom cover of beams causes more severe spallation and even breach damage in the beam depth under combined impact-blast loading.

From the vulnerability assessment of beams by varying span length under different loading scenarios, it was obtained that the beams demonstrate more damage indices under a combined impact-blast loading comparing with those from only blast and impact loadings as presented in Table 6.5. The residual flexural capacities of the beams captured according to the aforementioned procedure mentioned in Section 6.4.1 are illustrated in Figure 6.22a–c.

6.4.6 Influence of beam longitudinal reinforcement

The influence of longitudinal reinforcements on the failure behaviors of RC beams subjected to combined impact-blast loadings with $V = 6.86$ m/s and $Z = 0.25$ m/kg$^{1/3}$ is evaluated. Two different arrangements for longitudinal reinforcements are considered while other geometrical and structural parameters are kept similar to the reference beam S2222. One of these arrangements represents a relatively low-flexural resistance that includes the compressive (C-bar) and tensile (T-bar) reinforcements with diameters of 13 and 16 mm named C13T16, respectively, with a tensile reinforcement ratio (ρ_t) of 1.26%. Another arrangement indicates a higher flexural resistance which includes the compressive and tensile reinforcements with diameters of 20 and 25 mm named C20T25, respectively, with a tensile reinforcement ratio of 2.46%. Also, the longitudinal rebar arrangement of the reference beam S2222 is named C22T22 that represents a middle-range flexural strength with a tensile reinforcement ratio of 3.09%.

Figure 6.23 shows the final damaged states of the RC beams with different longitudinal rebar arrangements under only blast loading comparing and combined impact-blast

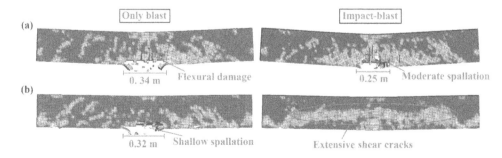

Figure 6.23 Damage patterns of RC beams with different longitudinal reinforcements of (a) C13T16, (b) C20T25, under only blast and a combined impact-blast loading with V = 6.86 m/s and Z = 0.25 m/kg$^{1/3}$.

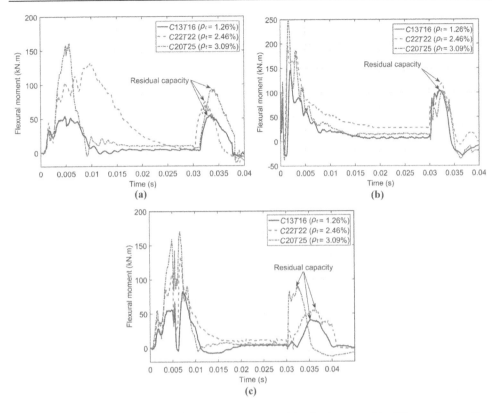

Figure 6.24 Comparing the flexural moments generated in the beams with different longitudinal rebar: (a) only impact loading, (b) only blast loading, (c) combined impact-blast loading with $V = 6.86$ m/s and $Z = 0.25$ m/kg$^{1/3}$.

loading. It is observed that both beams $C13T16$ and $C20T25$ suffer shallow spallation associated with low-level flexural damages when are under only blast loading, because the longitudinal reinforcements do not significantly affect the bond resistance in the bottom cover against the spallation. Besides, it is seen that the flexural rebar is more effective on the damage patterns of the beams when they are under combined impact-blast loading. Moderate spallation in the third beam depth of $C13T16$ and global shear damages in $C20T25$ is seen subjected to combined impact-blast loading.

By assessing the vulnerability of RC beams with different longitudinal rebar under combined impact-blast loading, larger flexural moments (including the peak and residual values) and smaller damage indices are obtained with increasing the ratio of tensile reinforcements as shown in Figure 6.24a–c and Table 6.6.

Table 6.6 Damage index results for different longitudinal reinforcement ratios

	C13T16 ($\rho_t = 1.26\%$)			C22T22 ($\rho_t = 2.46\%$)			C20T25 ($\rho_t = 3.09\%$)		
	Only impact	Only blast	Impact-blast	Only impact	Only blast	Impact-blast	Only impact	Only blast	Impact-blast
M_r (kN m)	51.6	105	45.5	116	57.9	57.9	92.9	100	89.6
DI	0.67	0.44	0.72	0.71	0.48	0.77	0.76	0.48	0.85

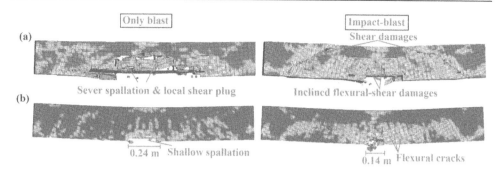

Figure 6.25 Failure behaviors of RC beams with different transverse reinforcements under sole blast and a combined impact-blast with $V = 6.86$ m/s and $Z = 0.25$ m/kg$^{1/3}$: (a) the beam with the transverse rebars of 6 mm @ 15 cm, (b) the beam with the transverse rebars of 16 mm @ 5 cm.

6.4.7 Influence of beam transverse reinforcement

Two different arrangements for transverse rebar are considered for RC beams under combined impact-blast loading. One of these arrangements represents the beam with a low-shear strength which includes the transverse reinforcements with a diameter of 6 mm spacing of 15 cm. Another arrangement represents the beam with a sufficient shear strength that includes the transverse reinforcements with a diameter of 16 mm spacing of 5 cm. It should be noted that the other geometrical and structural parameters are assumed the same as those of the reference beam $S2222$.

Figure 6.25 demonstrates a significant influence of transverse rebar on the localized spallation and global failures of beams under only blast loading. When the beam with a low-shear strength and a low bond resistance is under only blast loading, it suffers severe spallation associated with a local shear plug damage in the trapped area for the reflected tensile shock waves from the bottom free surface (i.e., free boundary condition). As the concrete spall damage has a brittle nature in RC structures, it extremely depends on the interlocking between the concrete and the reinforcements, and also on the bond resistance provided by shear rebar [16]. The beam endures only shallow spallation with sufficient shear and bond resistances. Moreover, when the beam with low-shear strength is under the combined impact-blast loading, the flexural-shear stresses generated around the impact zone prevent the expansion of the tensile stresses and brittle spallation damages in the trapped zone (i.e., the bottom cover) before the initiation of the sequent blast loading. Moreover, appearing global shear damages originating from the supports are seen owing to the global shear stresses generated in the impacted beam before the initiation of the sequent blast loading. However, when the RC beam with a sufficient shear resistance is under the combined impact-blast loading, some flexural cracks and low spallation with a smaller length rather than that generated under only blast loading are observed.

6.5 Conclusions

This chapter numerically investigated the non-linear dynamic responses and failure behaviors of simply-supported RC beams subjected to combined impact-blast loads with a middle-rate impact load and moderate-energy blast loads using LS-DYNA. This work has not been reported in the literature so far. The effects of the impact stress on the blast responses of the beams were evaluated by considering different loading sequences for each of the impact and blast loads in the combined loading scenarios. Also, a DI based on the residual flexural capacity of the beam cross-section was defined to assess the vulnerability of RC beams to different loading-related and structural-related parameters. It was found that the beam suffers more severe spallation in the depth and experiences larger deformations and internal forces when the impact load is applied before the sequent blast loading. This is because of producing the flexural-shear stresses in a trapped zone around the mid-span (i.e., the impact zone) resulting from the concentrated impact load before applying the sequent blast load. Besides, when a uniformly distributed blast load is applied prior to impact loading, the beam experiences more extensive spall damage in its cover zone associated with global shear damages originating from the supports.

In addition, the influence of impact duration that represents the time lag between the initiations of impact and blast loads was assessed with the consideration of different combined loading scenarios by varying different blast initiation times simultaneously with the occurrence of different response stages of the beam to impact loading. It was obtained that the beam suffers more severe spallation and residual plastic deformations when the blast load initiates during the free vibration stage of the response. Besides, when the blast load was applied at the same time as the occurrence of the first peak impact force, larger peak internal forces were generated in the beam.

Moreover, the influence of some structural parameters, including the beam depth, the span length, and the reinforcement configurations, was examined on the failure behaviors of RC beams subjected to combined impact-blast loadings. It was concluded that the severity of the spallation and global flexural damages increases with the decreasing of the beam depth and increasing of the span length. Accordingly, it was found that the DI results are more sensitive to these parameters when the beam is under combined loadings than those from under only blast or only impact loading. Furthermore, it was observed that the shear rebars have a more influence on the failure behaviors of the beam under both combined loading and only blast loading scenarios rather than flexural rebars. Owing to the bond resistance of the beam, the confinement and interlocking between the concrete and the reinforcements are significantly dependent on the arrangement and the strength of the shear rebar in the beam.

References

1. Carta G, Stochino F. Theoretical models to predict the flexural failure of reinforced concrete beams under blast loads. *Eng Struct* 2013;49:306–315.
2. Lu Y, Gong S. An analytical model for dynamic response of beam-column frames to impulsive ground excitations. *Int J Solids Struct* 2007;44(3):779–798.
3. Park GK, Kwak HG, Filippou FC. Blast analysis of RC beams based on moment-curvature relationship considering fixed-end rotation. *J Struct Eng* 2017;143(9):04017104.
4. Fujikake K, Li B, Soeun S. Impact response of reinforced concrete beam and its analytical evaluation. *J Struct Eng* 2009;135(8):938–950.
5. Cotsovos DM. A simplified approach for assessing the load–carrying capacity of reinforced concrete beams under concentrated load applied at high rates. *Int J Impact Eng* 2010;37(8):907–917.
6. Yan B, Liu F, Song D, Jiang Z. Numerical study on damage mechanism of RC beams under close-in blast loading. *Eng Fail Aanl* 2015;51:9–19.

7. Pham TM, Hao H. Effect of the plastic hinge and boundary conditions on the impact behavior of reinforced concrete beams. *Int J Impact Eng* 2017;102:74–85.

8. Pham TM, Hao H. Plastic hinges and inertia forces in RC beams under impact loads. *Int J Impact Eng* 2017;103:1–11.

9. Cotsovos DM, Stathopoulos ND, Zeris CA. Behavior of RC beams subjected to high rates of concentrated loading. *J Struct Eng* 2008;134(12):1839–1851.

10. Williams GD, Williamson EB. Response of reinforced concrete bridge columns subjected to blast loads. *J Struct Eng* 2011;137(9):903–913.

11. Zhang D, Yao S, Lu F, Chen X, Lin G, Wang W, Lin Y. Experimental study on scaling of RC beams under close-in blast loading. *Eng Fail Aanl* 2013;33:497–504.

12. Yao SJ, Zhang D, Lu FY, Wang W, Chen XG. Damage features and dynamic response of RC beams under blast. *Eng Fail Aanl* 2016;62:103–111.

13. Bao X, Li B. Residual strength of blast damaged reinforced concrete columns. *Int J Impact Eng* 2010;37(3):295–308.

14. Li J, Hao H, Wu C. Numerical study of precast segmental column under blast loads. *Eng Struct* 2017;134:125–137.

15. Astarlioglu S, Krauthammer T, Morency D, Tran TP. Behavior of reinforced concrete columns under combined effects of axial and blast-induced transverse loads. *Eng Struct* 2013;55:26–34.

16. Li J, Hao H. Numerical study of concrete spall damage to blast loads. *Int J Impact Eng* 2014;68:41–55.

17. Yuan S, Hao H, Zong Z, Li J. A study of RC bridge columns under contact explosion. *Int J Impact Eng* 2017;109:378–390.

18. Shi Y, Hao H, Li ZX. Numerical derivation of pressure–impulse diagrams for prediction of RC column damage to blast loads. *Int J Impact Eng* 2008;35(11):1213–1227.

19. Tai YS, Chu TL, Hu HT, Wu JY. Dynamic response of a reinforced concrete slab subjected to air blast load. *Theor Appl Fract Mec* 2011;56(3):140–147.

20. Wang W, Zhang D, Lu FY, Wang SC, Tang FJ. Experimental study on scaling the explosion resistance of a one–way square reinforced concrete slab under a close-in blast loading. *Int J Impact Eng* 2012;49:158–164.

21. Lin XS, Zhang YX, Hazell PJ. Modelling the response of reinforced concrete panels under blast loading. *Mater Des* 2014;56: 620–628.

22. Hong J, Fang Q, Chen L, Kong X. Numerical predictions of concrete slabs under contact explosion by modified K&C material model. *Constr Build Mater* 2017;155:1013–1024.

23. Jayasooriya R, Thambiratnam DP, Perera NJ, Kosse V. Blast and residual capacity analysis of reinforced concrete framed buildings. *Eng Struct* 2011;33(12):3483–3495.

24. Shi Y, Li ZX, Hao H. A new method for progressive collapse analysis of RC frames under blast loading. *Eng Struct* 2010;32(6):1691–1703.

25. Tang EK, Hao H. Numerical simulation of a cable-stayed bridge response to blast loads, Part I: Model development and response calculations. *Eng Struct* 2010;32(10):3180–3192.

26. Hao H, Tang EK. Numerical simulation of a cable-stayed bridge response to blast loads, Part II: Damage prediction and FRP strengthening. *Eng Struct* 2010;32(10):3193–3205.

27. Yi Z, Agrawal AK, Ettouney M, Alampalli S. Blast load effects on highway bridges. I: Modeling and blast load effects. *J Bridge Eng* 2014;19(4):04013023.

28. Liu H, Torres DM, Agrawal AK, Yi Z, Liu G. Simplified blast-load effects on the column and bent beam of highway bridges. *J Bridge Eng* 2015;20(10):06015001.

29. Yi Z, Agrawal AK, Ettouney M, Alampalli S. Blast load effects on highway bridges. II: Failure modes and multihazard correlations. *J Bridge Eng* 2014;19(4):04013024.

30. Williamson EB, Bayrak O, Davis C, Williams GD. Performance of bridge columns subjected to blast loads. I: Experimental program. *J Bridge Eng* 2011;16(6):693–702.

31. Williamson EB, Bayrak O, Davis C, Daniel Williams G. Performance of bridge columns subjected to blast loads. II: Results and recommendations. *J Bridge Eng* 2011;16(6):703–710.

32. Hu ZJ, Wu L, Zhang YF, Sun LZ. Dynamic responses of concrete piers under close-in blast loading. *Int J Damage Mech* 2016;25(8):1235–1254.

33. Pan Y, Ventura CE, Cheung MM. Performance of highway bridges subjected to blast loads. *Eng Struct* 2017;151:788–801.

34. Magnusson J, Hallgren M, Ansell A. Air-blast-loaded, high-strength concrete beams. Part I: Experimental investigation. *Mag Concrete Res* 2010;62(2):127–136.

35. Li J, Hao H. A two-step numerical method for efficient analysis of structural response to blast load. *Int J Protect Struct* 2011;2(1):103–126.

36. Li J, Hao H. Influence of brittle shear damage on accuracy of the two-step method in prediction of structural response to blast loads. *Int J Impact Eng* 2013;54:217–231.

37. Pham TM, Hao Y, Hao H. Sensitivity of impact behaviour of RC beams to contact stiffness. *Int J Impact Eng* 2018;112:155–164.

38. Pham TM, Hao H. Influence of global stiffness and equivalent model on prediction of impact response of RC beams. *Int J Impact Eng* 2018;113:88–97.

39. Saatci S, Vecchio FJ. Effects of shear mechanisms on impact behavior of reinforced concrete beams. *ACI Struct J* 2009;106(1):78.

40. Isaac P, Darby A, Ibell T, Evernden M. Experimental investigation into the force propagation velocity due to hard impacts on reinforced concrete members. *Int J Impact Eng* 2017;100:131–138.

41. Zhao DB, Yi WJ, Kunnath SK. Shear mechanisms in reinforced concrete beams under impact loading. *J Struct Eng* 2017;143(9):04017089.

42. Fan W, Xu X, Zhang Z, Shao X. Performance and sensitivity analysis of UHPFRC–strengthened bridge columns subjected to vehicle collisions. *Eng Struct* 2018;173:251–268.

43. Do TV, Pham TM, Hao H. Dynamic responses and failure modes of bridge columns under vehicle collision. *Eng Struct* 2018;156:243–259.

44. Fan W, Guo W, Sun Y, Chen B, Shao X. Experimental and numerical investigations of a novel steel-UHPFRC composite fender for bridge protection in vessel collisions. *Ocean Eng* 2018;165:1–21.

45. Gholipour G, Zhang C, Mousavi AA. Effects of axial load on nonlinear response of RC columns subjected to lateral impact load: ship-pier collision. *Eng Fail Anal* 2018;91:397–418.

46. Gholipour G, Zhang C, Li M. Effects of soil–pile interaction on the response of bridge pier to barge collision using energy distribution method. *Struct Infrastruct Eng* 2018;14(11):1520–1534.

47. Gholipour G, Zhang C, Mousavi AA. Analysis of girder bridge pier subjected to barge collision considering the superstructure interactions: the case study of a multiple-pier bridge system. *Struct Infrastruct Eng* 2019;15(3):392–412.

48. LS-DYNA 971. Livermore Software Technology Corporation. Livermore, CA, 2015.

49. Hao H. Predictions of structural response to dynamic loads of different loading rates. *Int J Protect Struct* 2015;6:585–605.

50. Adhikary SD, Li B, Fujikake K. Low velocity impact response of reinforced concrete beams: experimental and numerical investigation. *Int J Protect Struct* 2015;6(1):81–111.

51. Zhao DB, Yi WJ, Kunnath SK. Numerical simulation and shear resistance of reinforced concrete beams under impact. *Eng Struct* 2018;166:387–401.

52. Cotsovos DM. A simplified approach for assessing the load-carrying capacity of reinforced concrete beams under concentrated load applied at high rates. *Int J Impact Eng* 2010;37(8):907–917.

53. McVay MK. Spall damage of concrete structures. DTIC Document, 1988.

54. Murray YD, Abu-Odeh AY, Bligh RP. Evaluation of LS-DYNA concrete material model 159. Report No. FHWA-HRT-05-063, Federal Highway Administration, Washington, DC, 2007.

55. LSTC (Livermore Software Technology Corporation). LS-DYNA Keyword User's Manual Ver. 971, Livermore, CA, 2016.

56. Murray YD. Users' manual for LS-DYNA concrete material model 159. Report No. FHWA-HRT-05-062. Federal Highway Administration, Washington, DC, 2007.
57. Béton CE-Id. CEB-FIP model code 1990: design code. Thomas Telford, 1993.
58. Malvar LJ, Crawford JE. Dynamic increase factors for concrete. DTIC Document, 1998.
59. Malvar L, Crawford J. Dynamic increase factors for steel reinforcing bars. 28th DDESB Seminar, Orlando, FL, 1998.
60. Randers-Pehrson G, Bannister KA. Airblast loading model for DYNA2D and DYNA3D. DTIC Document, 1997.
61. Qu Y, Li X, Kong X, Zhang W, Wang X. Numerical simulation on dynamic behavior of reinforced concrete beam with initial cracks subjected to air blast loading. *Eng Struct* 2016;128:96–110.
62. ACI 318-14. Building code requirements for structural concrete and commentary. American Concrete Institute, Farmington Hills, MI, 2014.

Chapter 7

Loading rate effect on the responses of beam subjected to combined loads

7.1 Introduction

Concrete structures might be subjected to extreme and high-rate loadings with very short durations such as impact or blast loads. Under such loading conditions, they would show the different structural responses due to the nonlinearity and the strain rate effects of the materials compared to those under quasi-static loadings with rather long durations. A classification of loading type is based on the ratio of the loading duration (t_d) to the natural period of structure (T) proposed by Mays and Smith [1]. According to this calcification, a loading with t_d less than $0.1T$ represents an impulsive loading, a loading with t_d between the ranges $0.1T$ and $6.5T$ denotes a dynamic loading, and loading with t_d larger than $6.5T$ represents a quasi-static loading.

The collapse of concrete structures due to fatal accidents or terrorist attacks would lead to the loss of human life and economic damages. Hence, recognizing the loading mechanisms, failure behaviors, and considering the nonlinearity of materials in the design of RC structures against extreme and impulsive loads are necessary. The responses of RC structures such as RC bridge piers [2–11], framed buildings [12, 13], and RC structural members such as slabs [14–17] and columns [18–23] under impact or explosion loads have been studied by many research works using analytical [24–28], numerical [2, 29–32], and experimental [18, 27, 33, 34] techniques. In addition, the vast majority of studies focused on investigating RC beams under blast loads experimentally [33–37] and numerically [29, 38–40]. Although RC beams are mostly used in concrete structures in connection with slabs, there are some cases where isolated beams may be found without slabs such as trusses, parking ramps, and sports stadiums. Moreover, beams play a more key role in resisting the collapse of the whole structure due to their less redundancy rather than slabs. However, considering the presence of slabs attached to beams under distributed blast loads would provide a more effective loading area that leads to more critical loading conditions for the supporting beams. From this point of view, this chapter presents only a theoretical exercise to evaluate the performance of isolated RC beams under the combination of impact and explosion loads.

As an experimental study, Zhang et al. [33] evaluated the damage states of fixed-end RC beams under a series of close-range explosions. A negative effect of the scaled distance on the spalling length was concluded and the RC beams tended to flexural failure modes under far-field explosions. Similarly, from an experimental study done by Liu et al. [34], it was found that the failure modes of RC beams changed from the global flexural modes to localized spalling failure by decreasing the scaled distance

DOI: 10.1201/9781003262343-7

and increasing the explosive mass. By evaluating the influence of transverse reinforcements on the failure modes of RC beams under blast loads by Yao et al. [36], it was revealed that the increase of stirrup ratio and narrowing of the spaces between stirrups resulted in the mitigation of damage levels. Rao et al. [37] investigated the responses of RC beams under double-end-initiated close-in blast loads that generate more non-uniformly distributed blast loads than those generated by single-point detonations. Changing the beam failure mode from the flexural mode to flexural-shear mode was observed by reducing both the scaled distance and the longitudinal reinforcement ratio.

To enhance the flexural and shear strengths of concrete beams under blast loads, many research works utilized various advanced high-strength composite materials [41–46]. Most of these studies revealed that beams with a high reinforcement ratio (i.e., having high flexural strength) were suspect to fail in shear modes.

Due to the limitation of financial resources and safety considerations for conducting experimental testing of blast resistance of RC structures, high-fidelity computer software and codes such as LS-DYNA [47] can properly provide alternative approaches to accurately simulate the structural responses under extreme loads. Some numerical studies of RC beams under blast loadings are listed as follows. The mechanism of the concrete spallation in RC beams under close-range explosions owing to the overspreading of tensile stresses on the back surface (i.e., tensile zone) is described by Yan et al. [29]. Li and Hao [38] developed a numerical method including two steps for calculating the blast responses of RC beams to reduce the computation time of numerical simulations.

RC structures would demonstrate different responses and failure behaviors under impact loads compared to those under blast loads due to the discrepancies between their loading mechanisms. As such, unlike the concentrated impact loads, the reflected blast waves are applied to the target structure in distributed forms. The initial impact velocity (i.e., the loading rate) is one of the most important loading-related parameters in describing the characteristics and the type of impact loads. Despite many attempts in recognizing the influences of loading rate on the structural responses, some aspects of the dynamic behaviors of RC structures under different-rate impact loadings are still unexplored for the sake of the complexity of the behaviors of both the concrete and the reinforcement materials.

The responses of axially loaded RC columns and bridge piers are extensively assessed in the literature subjected to impact loads arising from the collision of vessels [48–54] and vehicles [55–58]. There exist many research works in investigating the impact responses of RC beams. Fujikake et al. [27] analytically categorized the impact responses of structures into local and global modes. The substantial influences of the local contact and the global stiffness of the impacted structure on the initial and the following impulses of impact loading were studied by Pham et al. [59] and [60], respectively. Besides, several works [31, 32] estimated the location of plastic hinges formed in the impacted RC beams based on the stress wave propagation theory. Cotsovos et al. [32] revealed that increasing impact rate (i.e., impact velocity) led to decreasing the lengths of plastic hinges. In-line with these studies, Gholipour et al. [48] improved the method proposed by Cotsovos et al. [32] in predicting the lengths of plastic hinges under impact loads to utilize for axially loaded RC columns under lateral impact loads.

The mechanism of stress wave propagation in RC beams under impact loads was studied by Saatci and Vecchio [61] and the influence of impact loading rate on this mechanism was investigated by Isaac et al. [62]. It was found that RC beams tended to shear failure modes with the increase of impact rate. The shear failure mechanism of RC beams under drop-weight impact loading was experimentally and numerically studied by Zhao et al. [63] by varying beam span, transverse reinforcement ratio, impact mass, and impact velocity. With increasing impact velocity, the RC beams tended to fail in shear modes through the formation of localized shear plug and diagonal cracks around the impact zone shortly after the initiation of impact loading.

The combination of impact and blast loads is possible during the collision of vehicles, vessels, or objects, which carry explosive materials with target structures. Depending on the type of impacting objects and loading scenarios, impact and blast loads would combine with different sequences and time lags. The previous work [64] presented in Chapter 6 studied the damage levels and residual capacities of RC beams with simple supports subjected to combined actions of a medium-velocity impact loading and a close-range explosion. Besides, the influences on the loading sequence, the time interval between the onsets of the applied loads, reinforcement configurations, and the beam dimensions on the responses of RC beams were evaluated by proposing a flexural-based damage index. It was found that RC beams underwent different failure modes under combined loadings compared to those under sole impact and blast loads.

From the aforementioned studies, it was found that the impact loading rate has a predominant role in the responses of RC structures. Therefore, it is worth studying the influences of this parameter on the responses and damage states of RC beams with simple supports subjected to the combined actions of impact and explosion loads. In this chapter, the sensitivity of the damage states of RC beams to impact velocity is assessed under combination modes varying in terms of the loading sequence and the time lag parameters by proposing two different damage indices based on the residual shear and flexural capacities of RC beams. The current work includes several limitations described in detail in Section 7.3. Since this chapter aims to evaluate the influences of impact loading velocity, only the variability of this parameter is assumed against the other parameters of impact loadings such as impact weight, the shape of the impactor, and the loading location of RC beams, which are kept constant for all the combined loading scenarios. Although different mass impactors striking with the same velocity on RC structures will lead to different impact loads and structural responses, considering the variability of the impact weight parameter is not in the scope of this chapter. It should be noted that the reference RC beam studied in the previous chapter is considered the case study. The detailed finite element (FE) modeling and the validation studies of the reference RC beam can be found in Chapter 6.

7.2 Theoretical background

In this section, the typical failure modes of RC beams with simple supports under different sole impact and explosion loads are presented based on the results and observations numerically or experimentally obtained by the previous works.

7.2.1 RC beams under impact loads

Impact loads on RC structures can be classified based on the ratio of loading duration to the natural period of structure [65]. According to the literature [60, 65], the enhancement of the loading rate leads to the decrease of loading duration and changing of the failure modes from global flexural modes to brittle shear failures.

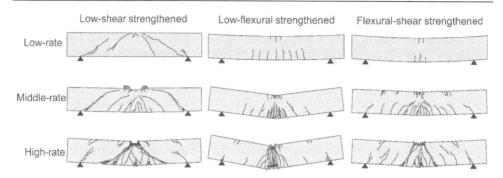

Figure 7.1 Typical failure modes of RC beams varying in shear and flexural strengths under different-rate impact loads.

In this section, it is attempted to summarize the typical failure modes of RC beams with simple supports varying in shear and flexural strengths (provided by the reinforcements) under different-rate impact loads based on the results and observations numerically or experimentally obtained by the previous works as shown in Figure 7.1. From low- and middle-rate impact loads on RC beams with low-shear and low-flexural strengths investigated by Isaac et al. [62], Adhikary et al. [66], and Kishi et al. [67], the predomination of global shear and flexural failures are observed, respectively. Moreover, the damage states of RC beams with sufficient shear-flexural strength change from minor vertical cracks to global flexural-shear damages with increasing the impact rate [27]. Besides, RC beams with sufficient flexural strengths suffer local shear plugs propagated from the impact point to the beam supports with an angle of 45° under high-rate impact loadings [62, 67]. By increasing the loading rate on these beams, the vertical cracks are propagated downward from the upper surface due to the formation of hogging moments [28].

7.2.2 RC beams under blast loads

Figure 7.2 shows a typical pressure-time-history diagram of a free-field blast loading. The shock waves from the detonation arrive at the front surface of the target

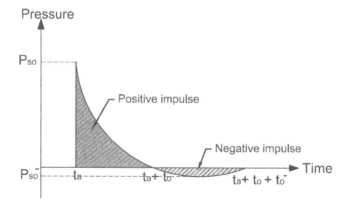

Figure 7.2 A typical plot of blast pressure-time-history.

at time t_a when the pressure instantly reaches its maximum positive value, P_{so} [68]. Then, the pressure quickly decreases to zero during the time t_o. The integrated area under the pressure-time curve during this time denotes the positive impulse of detonation, which mainly leads to localized compressive damages at the front surface of the target structure. The positive impulse is followed by a negative impulse phase with the duration of t_o^- and minimum pressure value of P_{so}^- due to the reflection of shock waves from the boundaries. This negative impulse phase mostly causes the spallation at the back surface of the structure due to the formation of suction forces and tensile stresses [29, 64, 65]. The severity of the spallation and local damages profoundly depends on the detonation energy and the standoff distance of the explosive from the target structure.

The most common criterion to classify blast loadings is Hopkinson-Cranz law that represents the intensity of blast loading based on the scaled distance (Z) of detonation given as follows:

$$Z = \frac{R}{W^{1/3}} \qquad\qquad (7.1)$$

where R is the standoff distance, and W is the equivalent weight of charge.

RC structures may demonstrate local or global responses subjected to blast loads [64]. When RC structures are exposed to close-range explosions, the shape of blast waves distributed on the target surface is more concentrated than those from far-field detonations that are uniformly distributed as shown in Figure 7.3. In such a case, the ratio of loading duration to the natural period (t_d/T) is short [65] and the occurrence of localized failures such as local concrete spallation and fragmentations is very possible as illustrated in Figure 7.3a.

Besides, when an RC beam is exposed to a far-field detonation with a relatively long duration, the structure may fail in a global shear failure mode (in a form of inclined or direct shear failure) in which severe damages are propagated from the supports to the impact zone as shown in Figure 7.3b. As the other failure mode under a far-field blast load with a large t_d/T, it is very likely to observe a global flexural failure mode with a ductile deformation and flexural cracks as shown in Figure 7.3c.

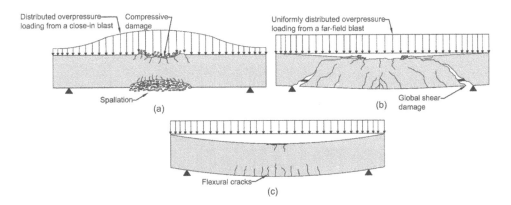

Figure 7.3 Typical failure modes of RC beams subjected to different blast loadings; (a) local spallation under a close-in blast, (b) global shear failure mode, and (c) global flexural failure mode under far-field explosions.

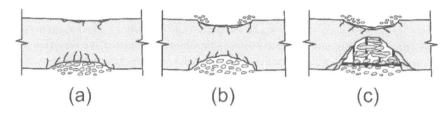

Figure 7.4 Different levels of spallation at the zone trapped blast waves on the structure; (a) low level, (b) medium level, and (c) severe level.

According to studies in the literature [21, 64, 69], which investigate the blast responses of RC beams, the severity of concrete spallation can be categorized into three different levels, including low, medium, and severe levels, as shown in Figure 7.4a–c, respectively. For low-level spallation, the structure suffers shallow spall damages with minor cracks at the bottom cover (Figure 7.4a). With increasing explosion energy, a medium level of spallation is observed in which about one-third of the structure's thickness is crushed (Figure 7.4b). Also, the severe level of spallation represents cross-sectional breach damage as illustrated in Figure 7.4c in which the structure loses its capacity.

7.3 FE modeling of RC beams and combined loading methodology

Three different RC beams with simple supports varying in terms of the reinforcement configuration under combined loadings with different-rate impact loads are studied as shown in Figure 7.5a–c. It should be noted that the other characteristics of the beams such as the boundary conditions, the material properties used for the concrete and reinforcements, and the dimensions are the same as those of the RC beam

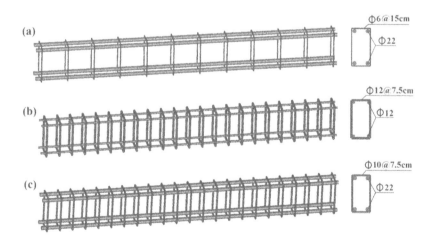

Figure 7.5 Different reinforcement configurations for RC beams; (a) L22S6, (b) L12S12, and (c) L22S10.

studied in the previous work [64] under combined actions. The various reinforcement arrangements considered in this chapter provide low shear (named $L22S6$ in which longitudinal rebars have a 22-mm diameter, and the transverse rebars have a 6-mm diameter with 15 cm spacing), low flexural (named $L12S12$ in which longitudinal rebars have a diameter of 12 m, and the transverse rebars have a diameter of 12 mm with 7.5 cm spacing), and sufficient shear-flexural (named $L22S10$ in which longitudinal rebars have a 22-mm diameter, and the transverse rebars have a 10-mm diameter with 7.5-cm spacing) strengths, respectively. The beams have a span length of 1.7 m and the cross-sectional dimensions of 0.25 m $\times$ 0.15 m in depth and width, respectively. A coupling method named CONSTRAINED_LAGRANGE_IN_SOLID was used between the reinforcements and the concrete of the beams. A contact algorithm named AUTOMATIC_SURFACE_TO_SURFACE was used between the impactor and the beam with Coulomb friction coefficients of 0.3. The nonlinear behaviors of the concrete material were modeled using MAT_CSCM_CONCRETE (MAT_159) [70] in LS-DYNA by considering the strain rate effects as described in Chapter 6. This model can simulate the concrete damages based on the concrete maximum principal strain. The material behaviors of the reinforcements were modeled using MAT_PIECEWISE_LINEAR_PLASTICITY (MAT_024) by considering strain rate effects and a rigid material model named MAT_RIGID (MAT_020) was used for the impactor and the supports. The material properties used for the various components of the beam FE model can be found in Chapter 6. Also, the blast load is simulated using the LOAD_BLAST_ENHANCED method in which shock waves are applied directly to the surface of the target structure.

Due to the lack of experimental studies in investigating the responses of RC beams under the combined actions of impact and explosion loads, the reliability of the FE models was separately examined in the previous work [64] and Chapter 6 under sole impact and explosion loads in comparison with the results of the experimental works conducted by Fujikake et al. [27] and Zhang et al. [33], respectively. To avoid the repetitive contents in this chapter, Chapter 6 is referred to find more information about the validation studies of the FE models. Based on these validation studies, a 10-mm mesh size for the concrete and rebars was captured for more reasonable and converged results in LS-DYNA.

Several assumptions limitations are considered in simulating RC beams under combined loadings as follows:

- The mid-span of RC beams is considered the location for both impact and blast loadings.
- The contact between the impactor and the RC beam is cut off simultaneously with the onset of the sequential detonation.
- The effects of blast loading on driving the impactor mass are neglected.
- All combined loadings include a close-in explosion with $Z = 0.26$ m/kg$^{1/3}$ (with a charge weight of 0.2 kg and a standoff distance of 0.15 m).

While studying the influence of impact velocity as a variable parameter, the mass of impactor is kept constant. That is, the changes in the kinetic energies of impact loadings studied in this chapter are only dependent on the variability of the impact velocity.

7.4 Results and discussion

7.4.1 RC beams under impact loads

Three different-rate (including low-, middle-, and high-rate impacts) impact loading scenarios in which an impactor with a mass of 400-kg strikes at the mid-span of the RC beams with the velocities (v) of 2, 6.86, and 12 m/s are considered in this chapter. Figure 7.6 shows the damage states of the RC beams using the fringe levels of the concrete plastic strain. As mentioned above, the damage behaviors are modeled based on the maximum principal strain of the concrete. As recommended in the previous work [64] and Chapter 6, in acquiring the converged FE simulation results, it is assumed that the concrete elements are eliminated and eroded from the simulation when the plastic strain of the concrete elements reaches 8% of the maximum plastic strain. It is observed that all RC beams suffer low-level flexural damages under a low-rate impact loading. However, the beam $L22S6$ experiences a global shear failure associated with localized damages in the beam cover with the increase of impact rate to a moderate level.

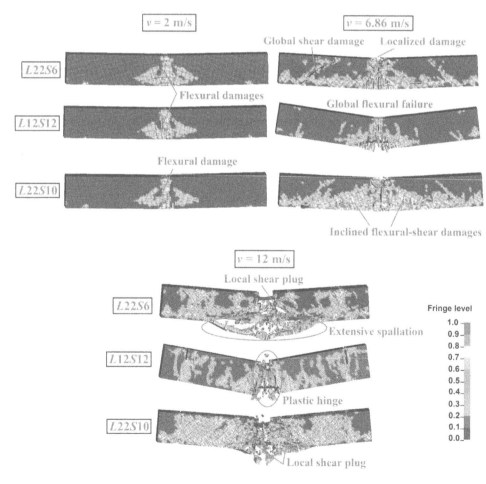

Figure 7.6 Comparing the damage states and failure behaviors of different reinforced RC beams under different-rate impact loads using the fringe levels of the concrete plastic strain.

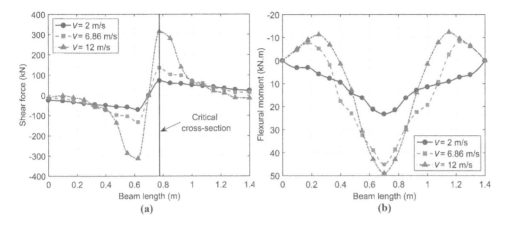

Figure 7.7 Comparing the distributions of the beam (i.e., *L22S10*) inertia and internal forces for different impact loading rates at the time of the initial peak of the impact force; (a) shear forces and (b) flexural moments.

Besides, the beam *L12S12* with a low-flexural strength endures a flexural failure, and the RC beam *L22S10* with an adequate shear-flexural strength fails in a flexural-shear mode. With the enhancement of impact rate to a high level, *L22S6* and *L22S10* tend to experience local failure modes composed of a localized shear plug and the extensive spallation, and *L12S12* fails in a flexural mode.

Figure 7.7a and b illustrates the distributions of the shear forces and bending moments generated throughout the beam *L22S10* exposed to different-rate impact loads at the time of the initial peak of impact force. The time of the initial peak of impact force denotes the time when the impact force reaches its first peak value. The location of the critical cross-section for the shear forces is at 0.1 m from the mid-span. Also, the peaks of both sagging (positive) and hogging (negative) moments increase with the increase of the impact rate.

In Figure 7.8a and b, it is revealed that the beam *L22S10* under a low-rate impact loading in which t_d/T is about 0.2 reaches its maximum displacement at 16.4 ms before the ending of the impact duration at 23 ms. However, the maximum displacement of the beam under a middle-rate impact with t_d/T about 0.14 is attained at 15 ms simultaneous with the ending of the impact duration at 17 ms. Under a high-rate impact with t_d/T about 0.06, the beam reaches the peak displacement at 21.3 ms after the ending of the impact duration at 9.2 ms. Based on a classification declared by Mays and Smith [1] on the types of impact loads and considering the aforementioned loading characteristics, the impact loads with low, middle, and high rates studied in this chapter denote the quasi-static, dynamic, and impulsive loading types, respectively.

The time-history of the shear forces generated at the location of the critical section and the supports and the mid-span bending moments of the beam under different-rate impact loads are compared in Figure 7.8c–e, respectively. The predominant failure modes on the responses of RC beams under different-rate impact loads can be clearly assessed by plotting the loading paths and moment-shear interaction diagrams as illustrated in Figure 7.8f. The moment-shear capacity curve of the beams was calculated using the software program Response-2000 [71] based on the modified compression field theory [72] by inputting the beam cross-sectional characteristics such as

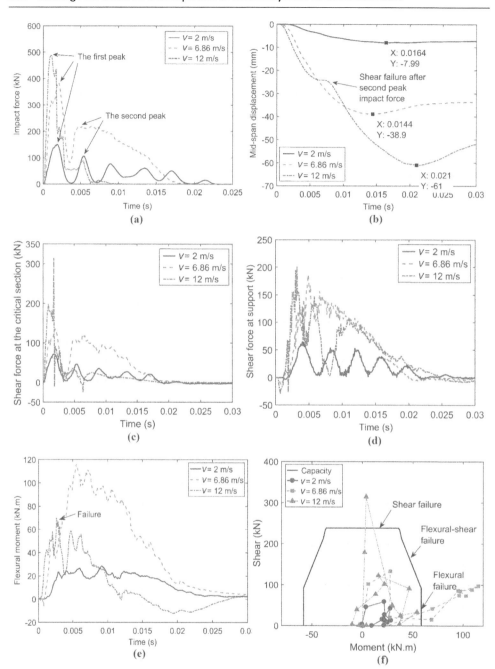

Figure 7.8 Comparison of the responses of beam *L22S10* under different-rate impact loadings. (a) Impact force-time-history, (b) mid-span displacements, (c) shear forces at 0.1 m from the mid-span, (d) shear force at the support, (e) mid-span bending moment, and (f) moment-shear interaction diagrams.

the cross-section dimensions and the material properties by considering the recommended dynamic increase factors due to the strain rate effects of the concrete and the steel reinforcements [73]. It is observed that the beam loses its flexural capacity after the occurrence of a shear failure under a high-rate impact load. However, the beam fails in a flexural failure mode subjected to a middle rate after undergoing a combined flexural-shear failure. Besides, for the beam under a low-rate impact load, the interaction diagram is bounded inside the capacity curve without any failures.

7.4.2 RC beams under combined loads

Owing to differences that exist between the loading mechanisms of impact and explosion loads [64], it is expected to capture different damage behaviors in RC structures exposed to the combined actions of impact and explosion loads compared to those under sole impact and explosion loadings. In Figure 7.8a, it is found that the occurrence time of the secondary peaks of impact forces is at 5.0 ms. According to the literature [31, 32, 60], the incidence time of the secondary peak of impact force is not dependent on the loading conditions and parameters, while it is profoundly dependent on the velocity of stress waves propagated throughout the structure. Therefore, the time stage of secondary peaks at 5.0 ms is considered the onset time of the following detonation.

Figure 7.9 demonstrates the damage behaviors of the beam $L22S10$ subjected to the combined actions of different-rate impact loads with a close-in detonation in comparison with those under sole explosion. In addition, the different phases of the trapped stresses generated in the RC beams are presented before and after the initiation of sequential explosion under impact loading alone and under combined impact-blast loadings, respectively. The beam suffers the global shear damages along with a limited-length spalling around the beam mid-span under only a close-in explosion loading. Similarly, the beam endures the overall shear damages along with low-level spallation under a combined loading with a low-velocity impact (i.e., $V = 2$ m/s). By increasing the impact velocity to a moderate level (i.e., $V = 6.86$ m/s), the global flexural-shear damages accumulated with the severe spallation lead to the breach at the beam depth

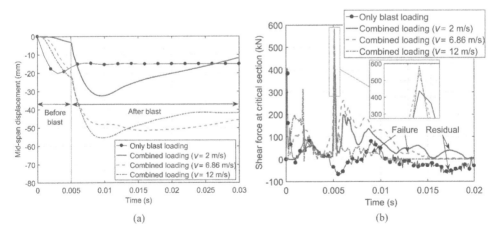

(a)   (b)

Figure 7.9 Comparing the damage behaviors of the beam *L22S10* subjected to the combined actions of different-rate impact loads with a close-in detonation using the fringe levels of the concrete plastic strain. (a) Displacement at the mid-span. (b) Shear forces at 0.1 m from the mid-span (i.e, the critical section).

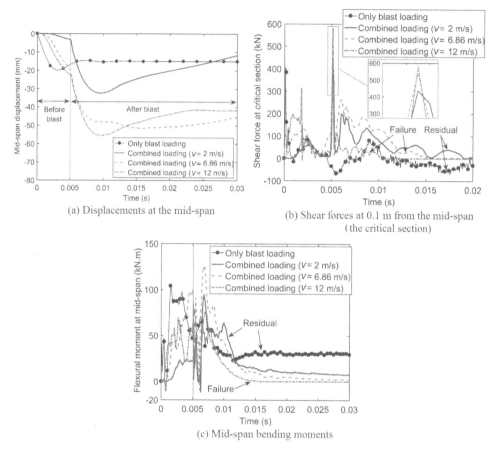

(a) Displacements at the mid-span

(b) Shear forces at 0.1 m from the mid-span (the critical section)

(c) Mid-span bending moments

Figure 7.10 Comparison of the responses of the beam *L22S10* under combined loadings with different-rate impacts: (a) displacements at the mid-span, (b) shear forces at 0.1 m from the mid-span (i.e., the critical section), and (c) mid-span bending moments.

due to the overspreading of the impact-induced stress area. In addition, when the RC beam is exposed to combined actions with a high-velocity impact (i.e., $V = 12$ m/s), the beam experiences a local shear plug along with the severe spallation and the exfoliation in the side cover.

In Figure 7.10a–c, it is observed that the beam *L22S10* undergoes larger displacements and internal forces under the combined actions of impact and blast loads compared to those under detonation alone. It is also found that although the peaks of displacements and shear forces increase with enhancing impact velocity, the residual displacement and flexural moments under high-rate impact are smaller than those under middle-rate impact loading. This is because of losing the beam cross-sectional capacity due to the occurrence of a shear plug during the impact loading phase.

7.4.3 Vulnerability assessment of RC beams under combined loadings

The vulnerability of the RC beam *L22S10* is assessed under combined loads varying in terms of the impact loading velocity, the loading sequence, and the time interval between the onsets of the applied loads. To do this, two damage indices on the basis of the residual flexural, $DI_{(M)}$, and shear, $DI_{(Q)}$, capacities of the beam are proposed as follows:

$$DI_{(M)} = 1 - \left(\frac{M_r}{M_N} \right) \tag{7.2a}$$

$$DI_{(Q)} = 1 - \left(\frac{Q_r}{Q_N} \right) \tag{7.2b}$$

where M_r and Q_r denote the residual flexural and shear capacities, respectively. M_N and Q_N are the nominal flexural and shear capacities of the RC beam, which can be obtained from the design codes of RC structures such as ACI-318 [74] by considering the dynamic increase factors due to the strain rate effects of the materials [64].

Since the dropped-weight impactor may generate both types of internal forces, including shear forces and flexural moments in the beam, considering a repetitive impact loading on the damaged beam can obtain the residual capacities of the beam through a quasi-static analysis. In this analysis, the impactor mass is applied to the beam with a constant velocity until the loss of the beam shear and flexural strengths. To acquire the residual capacities of the beam under combined loadings according to Equations (7.2a) and (7.2b), three loading phases are simulated as follows:

- First, a sole impact or explosion loading is applied to the beam according to the desired loading priority.
- Then, the following load (impact or blast) is initiated at 0.05 s when the beam reaches a stable state with a vertical velocity of less than 0.1 m/s.
- Finally, the beam is exposed to a repetitive impact loading with a constant velocity until the loss of its flexural and shear strengths as illustrated in Figure 7.11a and b.

7.4.3.1 Effects of the loading sequence

The sensitivity of the residual capacities and damage behaviors of the reference RC beam L22S10 is assessed to the loading sequence under the combined actions of a close-in explosion with different-rate impact loads. To do this, two combination scenarios of loadings with different loading priorities are considered to apply on the beam. One of these scenarios is named "impact-blast loading" scenario in which the impact loading is applied prior to the detonation. The other scenario is called "blast-impact loading" scenario that has a reverse loading sequence compared to that of impact-blast loading. In these scenarios, the onset time of the sequential loading is at 0.05 s when the structure reaches a stable state as shown in Figure 7.11a and b.

In Figure 7.12, different damage states of the RC beam are illustrated under combined loads varying in terms of the impact rate and the loading sequence. The beam undergoes localized spallation around the impact zone when exposed to impact-blast loadings. In addition, the damage severity and the spalling depth increase in proportion to the increase of impact velocity, which leads to shear plug and the breach in the beam depth. This is because of the expansion of the area of the impact-induced trapped stress with increasing impact rate before the onset of the sequential explosion as illustrated in Figure 7.9 (left side). Consequently, the beam loses its shear

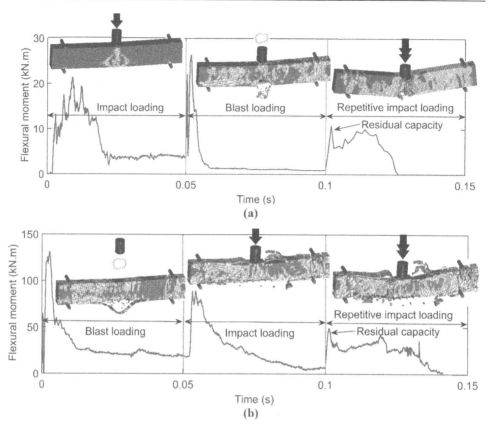

Figure 7.11 Three loading phases to acquire the residual capacities of the beam under combined loadings with the onset priority for (a) impact loading and (b) explosion loading.

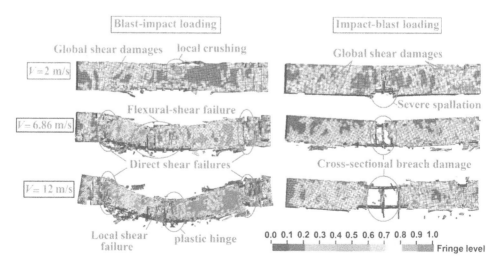

Figure 7.12 Comparing the damage behaviors of L22S10 under different-rate combined impact loadings varying in the loading sequence using the fringe levels of the concrete plastic strain.

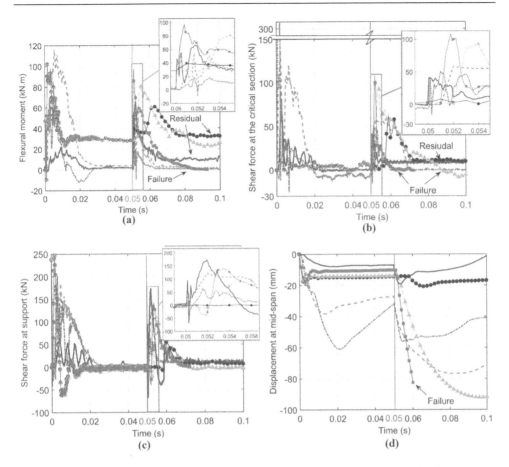

Figure 7.13 Responses of the beam L22S10 under different-rate combined impact loadings varying in the loading sequence: (a) mid-span bending moment, (b) shear forces at the critical section, (c) shear forces at the beam support, and (d) displacements at mid-span.

and flexural strengths (i.e., cross-sectional capacities) at the mid-span as shown in Figure 7.13a and b. Besides, the beam suffers more severe global shear damages when subjected to combined blast-impact loadings. The severity of the global shear damages at the supports and the mid-span flexural damages increases with the enhancement of impact velocity. Consequently, the beam endures a cumulative failure mode composed of the direct shear failures (occurred at the supports) and a plastic hinge under a blast-impact loading with a high-velocity impact as shown in Figure 7.12 (left side). The loss of shear strengths at the supports under blast-impact loadings due to the occurrence of direct shear failures can be seen in Figure 7.13c. Accordingly, larger residual displacements are produced under blast-impact loadings compared to those under impact-blast scenarios as shown in Figure 7.13d.

Figure 7.14 demonstrates the sensitivity of the damage indices to the impact velocity. To smoothly show the overall trend of the sensitivity curves, the damage indices are calculated for the seven impact velocities with the values of 1, 2, 4, 6.86, 8, 10, and 12 m/s. Accordingly, the beam damage levels can quantitatively be classified under each of the combined loading scenarios. By comparing the beam damage behaviors

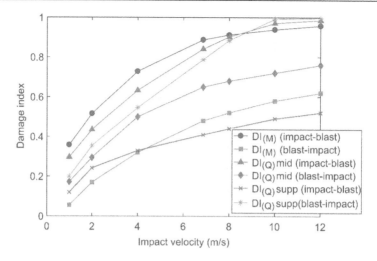

Figure 7.14 Comparing the sensitivity of the damage indices of the beam to the impact velocity under different combined loadings.

from the FE simulations and the damage index results, the beam damage states can be described as follows:

- $DI = 0$–0.2: The minor damages and cracks (flexural or shear) appear in the beam.
- $DI = 0.2$–0.4: The beam suffers moderate spallation and global damages.
- $DI = 0.4$–0.6: The severe cross-sectional damages and spallation are observed.
- $DI = 0.6$–0.8: The beam undergoes progressive damages due to the combination of flexural and shear failures and tends to the loss of cross-sectional strengths.
- $DI = 0.8$–1.0: The beam collapses through flexure (the formation of a plastic hinge) or direct shear failures due to the loss of cross-sectional strengths.

By evaluating the damage index results under impact-blast loadings, it is found that the damage indices based on the flexural moments [i.e., $DI_{(M)}$] and the shear forces at the beam mid-span [i.e., $DI_{(Q)mid}$] are more sensitive to the increase of impact velocity compared to $DI_{(Q)supp}$. Conversely, the sensitivity of $DI_{(Q)supp}$ to the increase of impact rate is more significant compared to those of other damage indices under blast-impact loading scenarios. This is due to the occurrence of direct shear failures at the supports under blast-impact loadings and the increase of their severity with the enhancement of impact velocity (also see Figure 7.12).

In addition, the damage indices under combined loadings are compared with those under sole impact and blast loadings in Figure 7.15a–c. Through the comparison of the damage index results between those captured under explosion alone and combined blast-impact loadings with a low-velocity ($V = 2$ m/s), it is revealed that a low-rate impact has a marginal influence on the blast responses of the beam, while substantial increases are observed in the damage indices under combined loadings with the increase of impact rate. Besides, the increase of the impact velocity from a low rate (i.e., $V = 2$ m/s) to a middle rate (i.e., $V = 6.86$ m/s) has more positive influences on the mid-span damage indices rather than that of high-rate velocity due to the localized failures at the mid-span under high-velocity impacts (also see Figure 7.14).

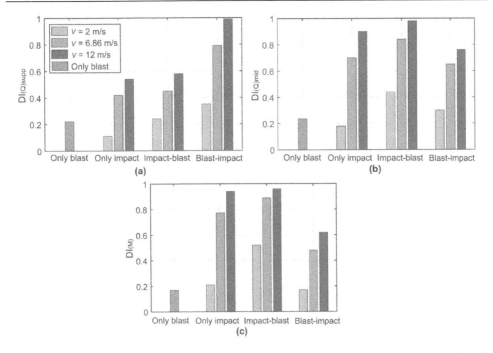

Figure 7.15 Comparison of the beam damage indices between those obtained under combined load-ings and sole impact and blast loadings; (a) $DI_{(Q)supp}$, (b) $DI_{(Q)mid}$, and (c) $DI_{(M)}$.

7.4.3.2 Effects of the time lag

The impact loading duration plays a key role in classifying the type of impact loadings. When an RC structure is exposed to a high-rate impact, the impact duration becomes very short and the structural responses are profoundly affected by the inertia and the stress wave propagation throughout the structure. In such a case, the structure tends to demonstrate local shear responses during the first impulse of impact loading rather than the global flexural responses (stiffness-related), which are activated in the follow-ing impulses [60]. However, under low-velocity impacts, the loading duration is relatively long and the overall stiffness of the structure is mobilized during the following phases.

In this section, since all combined loadings are assumed to be "impact-blast" loadings, the time lag (t_L) (i.e., the time interval) between the onsets of the applied loads represents the impact loading duration. Therefore, the impact rate can significantly affect the beam damage behaviors and structural responses under combined loads varying in t_L. To assess the influences of this parameter, the various time stages of the RC beam responses to sole impact loads, including the initial peak of impact force, the maximum shear at the sup-port, the maximum mid-span moment, and the maximum displacement, are considered the onset times of the sequential detonation. As discussed in Section 7.4.1, the occurrence time of the peak shear forces at the location of the critical section is equal to the time of the initial peaks of impact forces as shown in Figure 7.8a and c. Unlike the low-rate impact scenario, the peak of shear forces at the supports occurs at the time of the peak mid-span bending moment under middle- and high-rate impact loads.

Figure 7.16a–c illustrates the damage states of the beam $L22S10$ under combined actions varying in terms of the impact loading duration (t_L) and the impact velocity (V).

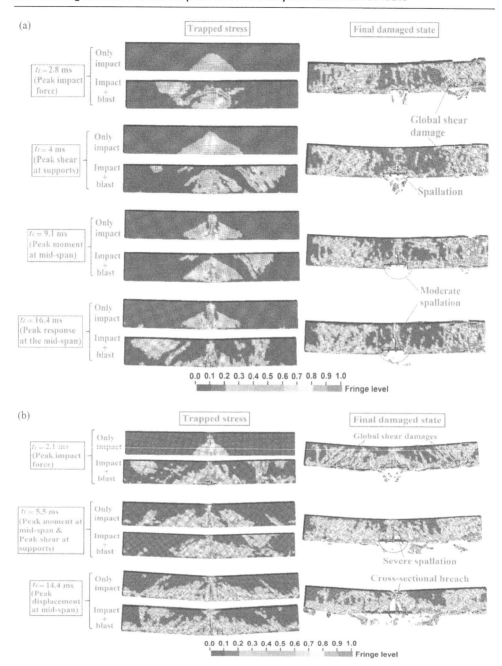

Figure 7.16 Comparing the damage behaviors of the beam L22S10 under combined loadings varying in the impact rate and t_L using the fringe levels of the concrete plastic strain. (a) The beam under combined impact-blast loadings with a low-rate impact ($V = 2$ m/s). (b) The beam under combined loadings with a medium-velocity impact ($V = 6.86$ m/s). *(Continued)*

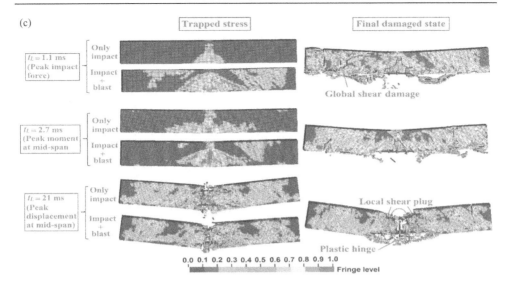

Figure 7.16 (Continued) (c) The beam under combined loadings with a high-velocity impact ($V = 12$ m/s).

Enhancing the intensity of the concrete spallation around the mid-span is observed with increasing time lag. This is because of expanding of the impact-induced stress area trapped around the mid-span by increasing t_L before applying the sequential blast loading. Moreover, the most severe damage state and the largest residual deformations occur in the beam when the sequential detonation is initiated at the time of the peak displacement as shown in Figures 7.16a–c and 7.17a–c. Also, in addition to enhancing the severity of the shear damages throughout the beam, the beam undergoes greater shear forces when the following detonation begins at the time of the initial peak of impact force as shown in Figures 7.16a–c and 7.18a–c. In addition, larger flexural moments are produced under combined loadings with low- and middle-rate impacts in which the sequential detonation is applied at the time of the peak bending moment as shown in Figure 7.19a and b. However, Figure 7.19c indicates that the beam experiences a greater peak moment under combined loading with a high-rate impact in which the explosion is initiated at the time of the initial peak of impact force.

Figure 7.20a and b illustrates the moment-shear interaction diagrams under combined actions in which the sequential detonation is initiated at the time of the peak of

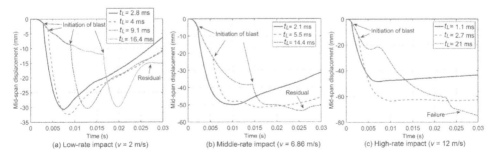

Figure 7.17 Displacements of L22S10 under combined loads with the variability of the impact rate and t_L.

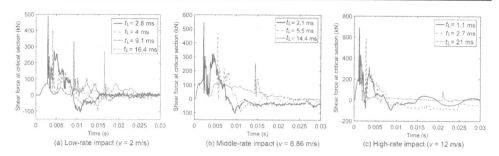

Figure 7.18 Comparing the shear forces of the beam *L22S10* at the critical cross-section under the combined loads with the variability of the impact rate and t_L.

impact force and the peak bending moment at the mid-span, respectively. It is found that the failure modes are more sensitive to the impact velocity when the sequential detonation begins at the time of the peak of impact forces compared to those under combined loadings in which the detonation is initiated at the time of the peak mid-span flexural moment. That is, the beam tends to experience the priority of the shear failures with the enhancement of impact rate when the explosion begins at the time of the peak of impact force as shown in Figure 7.20a. However, almost similar combined flexural-shear failures occur under combined loadings in which the detonation is initiated at the time of the peak mid-span bending moment as shown in Figure 7.20b.

The sensitivities of the damage indices calculated by Equations (7.2a) and (7.2b) to the impact velocity are assessed under combined loadings varying in t_L as shown in Figure 7.21a–d. Similar to the procedure proposed in Section 7.4.3.1, the damage indices are calculated for the aforementioned seven impact velocities. In Figure 7.21a and b, it is observed that the damage index based on the shear forces at the supports [i.e., $DI_{(Q)\ at\ support}$] results in larger values compared to those of other indices when the detonation is initiated at the time of the initial peak of impact force and the peak shear at the supports. A reverse approach is observed when the explosion is applied at the time of the peak mid-span displacement as plotted in Figure 7.21d.

Since the incidence time of the initial peak of impact force and the peak shear at the supports is very close (see Figure 7.8a and d), the trend for the sensitivity of the damage indices from these scenarios to the impact velocity is almost similar as shown in Figure 7.21a and b. However, different trends are observed for the sensitivity of the

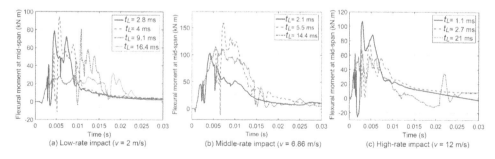

Figure 7.19 Comparing the mid-span flexural moments of the beam *L22S10* under combined loadings varying in the impact rate and t_L.

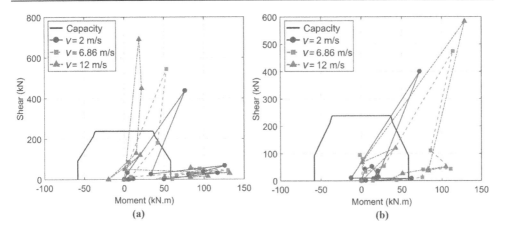

Figure 7.20 Comparing the moment-shear interaction diagrams for the beam L22S10 under combined loadings varying in the impact rate and t_L: (a) blast at the time of the first peak
impact force and (b) blast at the time of peak moment.

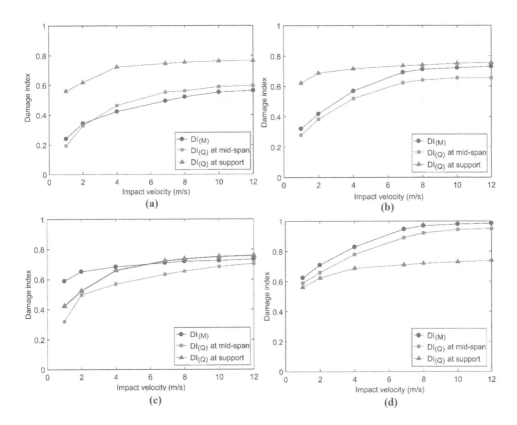

Figure 7.21 The sensitivity of the beam damage indices to the impact velocity under combined loadings varying in t_L. (a) Blast at the time of the initial peak impact force, (b) blast at the time
of peak shear force at the supports, (c) blast at the time of the peak mid-span bending
moment, and (d) blast at the time of the peak mid-span displacement.

damage indices from the scenarios in which the detonation is initiated at the time of the peak mid-span flexural moment and the peak mid-span displacement in Figure 7.21c and d. In addition, due to the predominance of the localized shear failure modes accumulated with the severe spallation by increasing the impact velocity, it is found that the mid-span flexural- and shear-based damage indices are more sensitive to the increase of impact rate compared to the damage index based on the shear at the supports, while all the damages indices have almost similar sensitivities to the enhancement of the impact velocity when the following detonation is initiated at the time of the peak mid-span bending moment as shown in Figure 7.21c.

7.4.3.3 Effects of reinforcement configuration

In this section, the RC beams with different reinforcement configurations as afore-mentioned in Section 7.3 are studied under the combined actions of different-rate impacts with a close-in explosion.

The damage behaviors of the beams L22S6 and L12S12 are illustrated in Figure 7.22 under combined loads compared to those under detonation alone. The accumulation of shear damages and the extensive shallow spallation are observed in the beam L22S6 under a close-in explosion alone due to inadequate bonding strength of the concrete cover. However, the spalling length becomes shorter and it is limited to the impact-induced zone (i.e., around the beam mid-span) under combined loads. Furthermore, the severity of spallation and the localized shear plug increases with enhancing the impact rate.

Besides, the failure behaviors of the beam L12S12 change from the flexural damages under sole explosion loading to localized spalling under combined loadings. Consequently, the beam fails in flexural failure mode through the progress of concrete

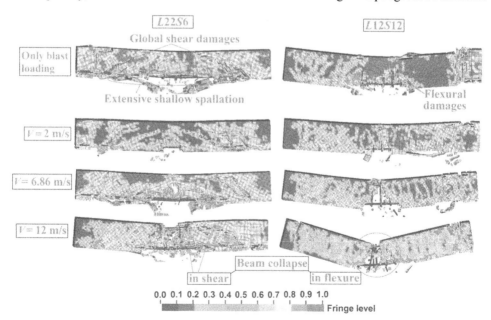

Figure 7.22 Comparing the damage states of the RC beams varying in the reinforcement configuration under the combined actions of a close-in detonation with different-rate impact loads using the fringe levels of the concrete plastic strain.

spalling and the strength loss of reinforcements (i.e., the formation of a plastic hinge) with increasing impact rate.

7.5 Conclusions

In this chapter, the influences of impact loading rate on the residual capacities and the damage behaviors of simply supported RC beams are evaluated, subjected to the combination of different-rate impacts with a close-in explosion ($Z = 0.26$ m/kg$^{1/3}$) varying in the loading sequence and the time interval between the onsets of the applied loads. In addition, two damage indices based on the beam residual shear and flexural capacities under combined loadings were proposed to assess the sensitivity of the failure modes to impact rate.

By evaluating the effects of the loading sequence, it was observed that the beam undergoes more severe spalling when the impact loads were applied prior to the detonation (i.e., impact-blast loading). Besides, a global shear failure mode was predominant on the beam responses when the detonations were applied prior to impact loads (i.e., blast-impact). Moreover, it was found that the beam flexural and shear modes-based damage indices at the mid-span were more sensitive to the increase of impact rate under impact-blast loading scenarios. However, the shear-based damage index at the supports was more sensitive to impact velocity under blast-impact loading scenarios.

By assessing the influences of the time lag parameter on the damage behaviors of the RC beam with the adequate flexural and shear strengths (i.e., $L22S10$), the following conclusions were achieved:

- The beam suffered more global shear damages and greater shear forces when the sequential detonation was initiated at the time of the peak of impact force shortly after the onset of impact loading.
- The severity of the concrete spallation increased with the enhancement of the time lag. Accordingly, the beam underwent the most severe spallation and the greater residual displacements under a combination mode in which the sequential detonation was initiated at the time of the peak mid-span displacement.
- The beam experienced greater peak moments under combined loadings with low- and middle-rate impacts in which the following detonation was initiated at the time of the initial peak of impact force.
- The internal forces exhibited greater peak values under a combined loading with a high-velocity impact in which the following detonation is applied at the time of the initial peak of impact force.

References

1. Mays G, Smith PD, editors. *Blast effects on buildings: design of buildings to optimize resistance to blast loading.* Thomas Telford, London, 1995.
2. Williams GD, Williamson EB. Response of reinforced concrete bridge columns subjected to blast loads. *J Struct Eng* 2011;137(9):903–913.
3. Tang EK, Hao H. Numerical simulation of a cable-stayed bridge response to blast loads, Part I: Model development and response calculations. *Eng Struct* 2010;32(10):3180–3192.
4. Hao H, Tang EK. Numerical simulation of a cable-stayed bridge response to blast loads, Part II: Damage prediction and FRP strengthening. *Eng Struct* 2010;32(10):3193–3205.

5. Yi Z, Agrawal AK, Ettouney M, Alampalli S. Blast load effects on highway bridges. I: Modeling and blast load effects. *J Bridge Eng* 2014;19(4):04013023.
6. Liu H, Torres DM, Agrawal AK, Yi Z, Liu G. Simplified blast-load effects on the column and bent beam of highway bridges. *J Bridge Eng* 2015;20(10):06015001.
7. Liu L, Zong ZH, Li MH. Numerical study of damage modes and assessment of circular RC pier under noncontact explosions. *J Bridge Eng* 2018;23(9):04018061.
8. Williamson EB, Bayrak O, Davis C, Williams GD. Performance of bridge columns subjected to blast loads. I: Experimental program. *J Bridge Eng* 2011;16(6):693–702.
9. Williamson EB, Bayrak O, Davis C, Daniel Williams G. Performance of bridge columns subjected to blast loads. II: Results and recommendations. *J Bridge Eng* 2011;16(6):703–710.
10. Hu ZJ, Wu L, Zhang YF, Sun LZ. Dynamic responses of concrete piers under close-in blast loading. *Int J Damage Mec* 2016;25(8):1235–1254.
11. Pan Y, Ventura CE, Cheung MM. Performance of highway bridges subjected to blast loads. *Eng Struct* 2017;151:788–801.
12. Jayasooriya R, Thambiratnam DP, Perera NJ, Kosse V. Blast and residual capacity analysis of reinforced concrete framed buildings. *Eng Struct* 2011;33(12):3483–3495.
13. Shi Y, Li ZX, Hao H. A new method for progressive collapse analysis of RC frames under blast loading. *Eng Struct* 2010;32(6):1691–1703.
14. Tai YS, Chu TL, Hu HT, Wu JY. Dynamic response of a reinforced concrete slab subjected to air blast load. *Theor Appl Fract Mec* 2011;56(3):140–147.
15. Wang W, Zhang D, Lu FY, Wang SC, Tang FJ. Experimental study on scaling the explosion resistance of a one–way square reinforced concrete slab under a close-in blast loading. *Int J Impact Eng* 2012;49:158–164.
16. Lin XS, Zhang YX, Hazell PJ. Modelling the response of reinforced concrete panels under blast loading. *Mater Des* 2014;56:620–628.
17. Hong J, Fang Q, Chen L, Kong X. Numerical predictions of concrete slabs under contact explosion by modified K&C material model. *Constr Build Mater* 2017;155:1013–1024.
18. Bao X, Li B. Residual strength of blast damaged reinforced concrete columns. *Int J Impact Eng* 2010;37(3):295–308.
19. Li J, Hao H, Wu C. Numerical study of precast segmental column under blast loads. *Eng Struct* 2017;134:125–137.
20. Astarlioglu S, Krauthammer T, Morency D, Tran TP. Behavior of reinforced concrete columns under combined effects of axial and blast-induced transverse loads. *Eng Struct* 2013;55:26–34.
21. Li J, Hao H. Numerical study of concrete spall damage to blast loads. *Int J Impact Eng* 2014;68:41–55.
22. Yuan S, Hao H, Zong Z, Li J. A study of RC bridge columns under contact explosion. *Int J Impact Eng* 2017;109:378–390.
23. Shi Y, Hao H, Li ZX. Numerical derivation of pressure–impulse diagrams for prediction of RC column damage to blast loads. *Int J Impact Eng* 2008;35(11):1213–1227.
24. Carta G, Stochino F. Theoretical models to predict the flexural failure of reinforced concrete beams under blast loads. *Eng Struct* 2013;49:306–315.
25. Lu Y, Gong S. An analytical model for dynamic response of beam-column frames to impulsive ground excitations. *Int J Solids Struct* 2007;44(3):779–798.
26. Park GK, Kwak HG, Filippou FC. Blast analysis of RC beams based on moment-curvature relationship considering fixed-end rotation. *J Struct Eng* 2017;143(9):04017104.
27. Fujikake K, Li B, Soeun S. Impact response of reinforced concrete beam and its analytical evaluation. *J Struct Eng* 2009;135(8):938–950.
28. Cotsovos DM. A simplified approach for assessing the load–carrying capacity of reinforced concrete beams under concentrated load applied at high rates. *Int J Impact Eng* 2010;37(8):907–917.

29. Yan B, Liu F, Song D, Jiang Z. Numerical study on damage mechanism of RC beams under close-in blast loading. *Eng Fail Anal* 2015;51:9–19.
30. Pham TM, Hao H. Effect of the plastic hinge and boundary conditions on the impact behavior of reinforced concrete beams. *Int J Impact Eng* 2017;102:74–85.
31. Pham TM, Hao H. Plastic hinges and inertia forces in RC beams under impact loads. *Int J Impact Eng* 2017;103:1–11.
32. Cotsovos DM, Stathopoulos ND, Zeris CA. Behavior of RC beams subjected to high rates of concentrated loading. *J Struct Eng* 2008;134(12):1839–1851.
33. Zhang D, Yao S, Lu F, Chen X, Lin G, Wang W, Lin Y. Experimental study on scaling of RC beams under close-in blast loading. *Eng Fail Anal* 2013;33:497–504.
34. Liu Y, Yan JB, Huang FL. Behavior of reinforced concrete beams and columns subjected to blast loading. *Defence Tech* 2018;14(5):550–559.
35. Nagata M, Beppu M, Ichino H, Matsuzawa R. A fundamental investigation of reinforced concrete beams subjected to close-in explosion. *Int J Protect Struct* 2018;9(2):174–198.
36. Yao SJ, Zhang D, Lu FY, Wang W, Chen XG. Damage features and dynamic response of RC beams under blast. *Eng Fail Anal* 2016;62:103–111.
37. Rao B, Chen L, Fang Q, Hong J, Liu ZX, Xiang HB. Dynamic responses of reinforced concrete beams under double-end-initiated close-in explosion. *Defence Tech* 2018;14(5):527–539.
38. Li J, Hao H. A two-step numerical method for efficient analysis of structural response to blast load. *Int J Protect Struct* 2011;2(1):103–126.
39. Li J, Hao H. Influence of brittle shear damage on accuracy of the two-step method in prediction of structural response to blast loads. *Int J Impact Eng* 2013;54:217–231.
40. Qu Y, Li X, Kong X, Zhang W, Wang X. Numerical simulation on dynamic behavior of reinforced concrete beam with initial cracks subjected to air blast loading. *Eng Struct* 2016;128:96–110.
41. Magnusson J, Hallgren M, Ansell A. Air-blast-loaded, high-strength concrete beams. Part I: Experimental investigation. *Mag Concrete Res* 2010;62(2):127–136.
42. Li Y, Algassem O, Aoude H. Response of high-strength reinforced concrete beams under shock-tube induced blast loading. *Constr Build Mater* 2018;189:420–437.
43. Magnusson J, Hallgren M. Reinforced high strength concrete beams subjected to air blast loading. *WIT Trans Built Environ* 2004;73:1–10.
44. Yusof MA, Norazman N, Ariffin A, Zain FM, Risby R, Ng CP. Normal strength steel fiber reinforced concrete subjected to explosive loading. *Int J Sustain Constr Eng Technol* 2011;1(2):127–136.
45. Astarlioglu S, Krauthammer T. Response of normal-strength and ultra-high-performance fiber-reinforced concrete columns to idealized blast loads. *Eng Struct* 2014;61:1–12.
46. Yoo DY, Banthia N. Mechanical and structural behaviors of ultra-high-performance fiber-reinforced concrete subjected to impact and blast. *Constr Build Mater* 2017;149:416–431.
47. LS-DYNA 971. Livermore Software Technology Corporation, Livermore, CA, 2015.
48. Gholipour G, Zhang C, Mousavi AA. Effects of axial load on nonlinear response of RC columns subjected to lateral impact load: ship-pier collision. *Eng Fail Anal* 2018;91:397–418.
49. Gholipour G, Zhang C, Li M. Effects of soil–pile interaction on the response of bridge pier to barge collision using energy distribution method. *Struct Infrastruct Eng* 2018;14(11):1520–1534.
50. Gholipour G, Zhang C, Mousavi AA. Analysis of girder bridge pier subjected to barge collision considering the superstructure interactions: the case study of a multiple-pier bridge system. *Struct Infrastruct Eng* 2018;15(3):392–412.
51. Gholipour G, Zhang C, Mousavi AA. Reliability analysis of girder bridge piers subjected to barge collisions. *Struct Infrastruct Eng* 2019;15(9):1200–1220.
52. Gholipour G, Zhang C, Mousavi AA. Nonlinear numerical analysis and progressive damage assessment of a cable-stayed bridge pier subjected to ship collision. *Mar Struct* 2020;69:102662.

53. Sha Y, Hao H. Nonlinear finite element analysis of barge collision with a single bridge pier. *Eng Struct* 2012;41:63–76.
54. Jiang H, Wang J, Chorzepa MG, Zhao J. Numerical investigation of progressive collapse of a multispan continuous bridge subjected to vessel collision. *J Bridge Eng ASCE* 2017;22(5):04017008.
55. AuYeung S, Alipour A. Evaluation of AASHTO suggested design values for reinforced concrete bridge piers under vehicle collisions. *Transp Res Rec* 2016;2592:1–8.
56. Abdelkarim OI, ElGawady MA. Performance of bridge piers under vehicle collision. *Eng Struct* 2017;140:337–352.
57. Abdelkarim OI, ElGawady MA. Performance of hollow–core FRP–concrete–steel bridge columns subjected to vehicle collision. *Eng Struct* 2016;123:517–531.
58. Sharma H, Gardoni P, Hurlebaus S. Performance–based probabilistic capacity models and fragility estimates for RC columns subject to vehicle collision. *Comput Aided Civ Inf* 2015;30(7):555–569.
59. Pham TM, Hao Y, Hao H. Sensitivity of impact behaviour of RC beams to contact stiffness. *Int J Impact Eng* 2018;112:155–164.
60. Pham TM, Hao H. Influence of global stiffness and equivalent model on prediction of impact response of RC beams. *Int J Impact Eng* 2018;113:88–97.
61. Saatci S, Vecchio FJ. Effects of shear mechanisms on impact behavior of reinforced concrete beams. *ACI Struct J* 2009;106(1):78.
62. Isaac P, Darby A, Ibell T, Evernden M. Experimental investigation into the force propagation velocity due to hard impacts on reinforced concrete members. *Int J Impact Eng* 2017;100:131–138.
63. Zhao DB, Yi WJ, Kunnath SK. Shear mechanisms in reinforced concrete beams under impact loading. *J Struct Eng* 2017;143(9):04017089.
64. Zhang C, Gholipour G, Mousavi AA. Nonlinear dynamic behavior of simply-supported RC beams subjected to combined impact-blast loading. *Eng Struct* 2019;181:124–142.
65. Hao H. Predictions of structural response to dynamic loads of different loading rates. *Int J Protect Struct* 2015;6:585–605.
66. Adhikary SD, Li B, Fujikake K. Low velocity impact response of reinforced concrete beams: experimental and numerical investigation. *Int J Protect Struct* 2015;6(1):81–111.
67. Kishi N, Mikami H, Matsuoka KG, Ando T. Impact behavior of shear-failure-type RC beams without shear rebar. *Int J Impact Eng* 2002;27(9):955–968.
68. TM 5-855-1. Fundamentals of protective design for conventional weapons. Technical Manual, US Department of the Army, Washington, DC, 1986.
69. McVay MK. Spall damage of concrete structures. DTIC Document, 1988.
70. LSTC (Livermore Software Technology Corporation). LS–DYNA keyword user's manual ver. 971, Livermore, CA, 2016.
71. Bentz E, Collins MP. Response-2000, V. 1.0. 5. Toronto University, Toronto, Canada, 2000.
72. Vecchio FJ, Collins MP. The modified compression-field theory for reinforced concrete elements subjected to shear. *ACI J Proc* 1986;83(2):219–231.
73. CEB-FIP. CEB-FIP model code 1990: design code. In *Comité Euro-International*, Thomas Telford, London, UK, 1993.
74. ACI 318-14. *Building code requirements for structural concrete and commentary.* American Concrete Institute, Farmington Hills, MI, 2014.

Chapter 8

RC columns subjected to the combination of impact and blast loads

8.1 Introduction

The need for realizing the dynamic behaviors of reinforced concrete (RC) structures under extreme and impulsive loads with short durations such as blast and impact loads has been increased with rising terrorist attacks on civil structures and infra-structures in recent years. Many studies can be found in the literature investigating the blast responses of RC bridge superstructures [1–5], piers [6–12], and beams [13–16]. In addition, the responses of axially loaded RC columns under sole explosions have widely been assessed in the literature numerically [17–21], and experimentally [22, 23]. A damage index (*DI*) was proposed by Shi et al. [17] based on the residual axial load-carrying capacity of RC columns under blast loads. As an extension study in devel-oping the pressure-impulse diagrams based on the SDOF models of RC columns, a numerical-based method was proposed, which captured more efficient results com-pared to those from SDOF models. As a parametric study, the residual axial resist-ances of RC columns under close-in explosions were numerically investigated by Bao and Li [19] with the variations of structural-related and loading-related parameters, including the ratio of transverse and longitudinal reinforcements, the column height, and the axial load ratio (ALR). A negative influence of the ALR was found on the lateral deformations and the residual capacity of the column. However, the residual capacity of columns increased by decreasing the column height and increasing the reinforcement ratios. Besides, Kyei and Braimah [18] revealed the substantial influ-ences of the spacing between the transverse reinforcements and the axial load on the failure behaviors of typical RC columns commonly used in concrete-framed build-ings under close-in and far-field blast loads. Moreover, significantly negative influ-ences of the axial load on the resistances of RC columns were concluded. Unlike the conclusions of the studies above, Wu et al. [22] demonstrated that the increase of the ALR may lead to the enhancement of the residual capacity of RC columns under blast loads. However, exploring the distinct sensitivity thresholds of the residual capacities of RC columns to the variation of their ALR is the research gap of previous studies. Zhang et al. [24] comprehensively reviewed the responses and failure behaviors of var-ious structural members under blast loads, and the sensitivity of the blast responses of these structures to various structural-related and loading-related parameters.

As much as the possibility of applying explosion loadings on RC columns, axi-ally loaded RC columns may be exposed to lateral impact loadings. The vulnera-bility assessments of RC columns used in large-scale infrastructures such as bridge piers during the collision of vehicles [25–27] or vessels [28–33], and rather small-scale

DOI: 10.1201/9781003262343-8

RC columns [28, 34–37] commonly used in low-rise framed buildings under impact loads, have been widely reported in the literature. Liu et al. [34] conducted a series of low-velocity impact tests on circular RC columns experimentally and numerically. A substantial positive influence of the ALR was revealed on the axial load-carrying capacity of RC columns against impact loads when the columns endured relatively small deformations. However, no sensitivity level of the column resistance was determined to the variation of the ALR. To fill this research gap, Gholipour et al. [28] found that the ALRs between the ranges of 0.3 and 0.5 increased the column impact resistance. Also, it was concluded that the peak values of impact forces increased due to the decrease of the elevation of impact loads, and the column failure mode tended to overall shear modes.

Furthermore, numerous research studies focused on exploring the impact responses and the damage states of RC beams under low-rate [38–41] and relatively high-rate [42–44] impact loadings. The majority of these studies investigated the influences of structural-related parameters such as the beam inertia [40, 45], the configuration and the ratio of reinforcements [46–48], and loading-related parameters such as impact velocity and weight [38, 39, 46].

Under more critical loading conditions, RC structures may be subjected to the combined actions of impact and blast loads during the intentional terrorist attacks or accidental events arising from the collision of vehicles, vessels, etc., carrying explosive materials. The collision of such explosive carriers with civil structures such as bridges and framed buildings can cause subsequent explosions after the onset of impact loadings. Under such combined loadings, impact and blast loadings may apply simultaneously or subsequently with a significant time lag between their onsets. Besides, the loading sequence can be considered a variable under different combined loading scenarios and conditions. In the previous study [49], the damage states and the dynamic responses of RC beams with simple supports were evaluated under the combination of middle-rate impact loadings and close-in detonations with the variation of the loading sequence, the time interval between the onsets of the applied loads, and several key structural-related parameters, including the configuration of reinforcements, the beam depth, and the span length. To do this, a DI based on the beam residual flexural capacity was proposed. It was found that applying the impact loads prior to the explosions and the initiation of the sequent detonation at the free vibration phase of the beam response provided more critical loading conditions. In-line with this study, Gholipour et al. [50] studied the effects of the impact loading rate on the blast responses of RC beams under their combined actions by proposing two damage indices based on the residual shear and flexural capacities of the beams. It was revealed that the beam damage indices based on the flexural and shear capacities at the beam mid-span were more sensitive to the enhancement of the impact velocity when it had been subjected to combined ladings with the priority of impact loading. Although the predominance of global shear failures was obtained on the beam response when the sequent explosion impact was applied simultaneously with the time of initial peak impact force, it suffered a more severe damage state for the initiation of the sequent detonation at the free vibration phase of the beam response to sole impact loading.

Owing to different applications and resistance mechanisms of axially loaded RC columns compared to other structural members such as beams or slabs, it is worthy to investigate the vulnerabilities of RC columns subjected to combined actions of impact and explosion loadings that have not been reported in the literature. This chapter presents a theoretical study on the effects of possible combined loading scenarios varying in terms of several loading-related parameters, including the loading location, the loading sequence, the time lag between the onsets of the applied loads, the ALR, and the impact velocity (V) on the dynamic responses and failure behaviors of a typical RC column commonly used in medium-rise buildings as the reference column studied by Feyerabend [51] using numerical simulations in LS-DYNA. To evaluate the damage levels and vulnerability of the RC column, a *DI* obtained from a proposed multi-step loading procedure based on the column's residual axial loading capacity is utilized. Since there exists no experimental work in the literature to investigate the RC columns under combined actions of impact and blast loads, this chapter separately examines the reliability of the FE models developed in the present study under impact and blast loads, comparing the FE numerical results with those from experimental impact and blast tests carried out by Feyerabend [51] and Siba [23], respectively. Also, a simplified model is proposed to estimate the location of the formed plastic hinges when subjected to combined loadings applied at the column mid-height.

8.2 Methodology of analysis

From the mesh convergence tests carried out for both validation studies of impact and blast loadings as presented in the following section, it was found that a 10-mm mesh for FE modeling of RC columns captured converged results. The adopted concrete martial model (i.e., MAT_159) which was similarly utilized for the RC columns in both validation studies was efficiently able to simulate the failure behaviors. Therefore, the reference RC column studied in the validation study of impact loading with similar mesh size (i.e., 10 mm) and concrete material model (i.e., MAT_159) to those of the RC columns examined in the validation study of blast loading is considered the case study subjected to the combinations of centric axial (without considering torsional effects), lateral impact, and explosion loadings. Due to different response mechanisms of axially loaded columns compared to beams, this chapter employs a different methodology from those utilized in the previous studies [49, 50] to calculate the damage indices and analyze the vulnerability of RC columns under combined loadings.

To calculate the damage indices, a multi-step (four-step) loading method composed of an axial compressive static loading and dynamic transient loadings (i.e., impact and blast loadings) on columns is utilized based on the column residual axial load capacity. This approach is different from those adopted in the previous works [49, 50] in which the damage indices were calculated based on the residual shear and flexural moment capacities of beams during a three-step loading method. Also, the effects of some new loading-related parameters, including ALR and the loading location, are assessed in addition to those studied in the previous works [49, 50] in investigating the vulnerability of beams.

To arrange more reasonable and realistic loading scenarios, RC columns are subjected to different combined loadings varying in the location of impact loading applied to the heights of 1.0 m (named *IMP*1) and 2.0 m (i.e., impact ate the column mid-height named *IMP*0) from the column base. However, the location of blast loading is kept constant at the height of 2.0 m (named *BLT*0) from the column base for all combined loading scenarios except those arranged in Section 8.4.1, which indicates the elevation of the explosive charge located in the vehicle cargo area. Considering such loading scenarios can realistically represent the collision of different types of vehicles varying in size and the relative height between the bow and rear portions.

8.2.1 Damage index

The vulnerability of the RC columns under combined loadings varying in several loading-related parameters is assessed using a *DI* based on the column residual axial loading capacity. This residual capacity is obtained through a multiple-step loading procedure as shown in Figure 8.1. Given an example of a combined loading scenario in which the impact loading is applied before the sequent detonation to the column mid-height, a multi-step loading method is sketched as follows:

- First, the axial loading is gradually applied to the top end of the column to reach its ALR at a desired service load level during a stable state with an axial velocity of less than 0.1 m/s using a dynamic relaxation analysis in LS-DYNA after 0.12 s.
- Then, the lateral impact and explosion loadings are, respectively, applied to the column while the axial load is kept constant at the desired ALR throughout the transient analyses. It is noteworthy that the procedure presented in this section has been sketched for subsequently combined loading scenarios in which the time lag between the onsets of the applied loads is significant (i.e., 0.2 s).
- Finally, to capture the residual axial load capacity of the column, the column head is exposed to a displacement-based axial loading during a gradual-incremental manner until reaching the column failure similar to the procedure proposed by Shi et al [17].

Accordingly, the *DI* is defined as follows:

$$DI = 1 - \frac{P_r - P_L}{P_N - P_L} \qquad (8.1)$$

where P_r is the residual axial load-carrying capacity of the column, P_L is the service axial load, P_N is the nominal axial loading capacity that can be computed using ACI-318 [52] by considering the dynamic increase factors (DIFs) due to the strain rate effects of the utilized materials as discussed in the previous work [49].

As recommended in the literature [53], the ramp duration of axial loading to reach a stable state should be greater than the first periodic time of the structure. This time is considered equal to 0.12 s, which is greater than 0.008 s as the periodic time of the

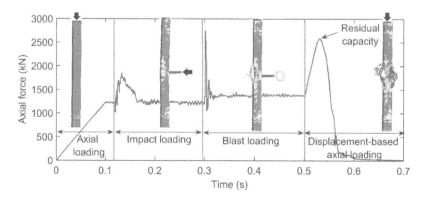

Figure 8.1 Loading steps to capture the residual axial load capacity of the column.

column studied in this chapter. Besides, the stability of the column during the multi-step loading in this study is yielded when the velocities of all nodes reach less than 0.1 m/s after 0.12 s similar to the loading rates considered in several previous research works [17, 54, 55]. Accordingly, the corresponding strain rate of the column studied in this chapter is about 4.17×10^{-3} 1/s (for a maximum axial deflection of 2.0 mm after 0.12 s). Hence, since the value of the strain rate is very low, the effects of *DIF*s of the concrete and steel materials are negligible as discussed in the previous work [49].

The methodology proposed in this chapter has several limitations in simulating the combined loading scenarios and analyzing the results listed as follows:

- The contact between the impactor and the column is cut off with the initiation of the sequent blast loading.
- The effects of the sequent detonation are neglected on driving the impactor mass.
- For all combined loading scenarios except those in Section 8.4.1, the location of the explosive is kept constant at the height of 2.0 m from the column base (named *BLT*0), which represents the elevation of the explosive located in the vehicle cargo area.
- According to a classification of blast loads by ASCE SEI 59-11 [56], blast loads with scaled distances less than 1.2 m/kg$^{1/3}$ are identified as close-in detonations, and higher than this value are far-field detonations. Therefore, based on this classification, two different blast loading scenarios with the scaled distances of $Z = 0.45$ m/kg$^{1/3}$ and $Z = 1.40$ m/kg$^{1/3}$ are considered the close-in and far-field explosions, respectively. Two blast loading scenarios with the scaled distances of $Z = 0.45$ m/kg$^{1/3}$ and $Z = 1.40$ m/kg$^{1/3}$ are considered the close-in and far-field explosions, respectively.
- For all combined loading scenarios except those in Section 8.4.2, the impact loading is applied prior to the blast loading on the column.
- For all combined loading scenarios except those in Sections 8.4.2 and 8.4.3, the time of the second peak impact force, which occurs at half of the first periodic time of the column, is considered the onset time of the sequent explosion loading.
- For all combined loading scenarios except those in Section 8.4.4, an ALR of 0.3 (i.e., ALR = 0.3) is kept constant on the top end of the column during all loading scenarios as the service load level.
- For all combined loading scenarios except those in Section 8.4.5, a medium velocity of 5.0 m/s (i.e., $V = 5.0$ m/s) is considered the impact velocity.

8.2.2 Simplified model for calculating the length of plastic hinge

A simplified analytical method is proposed to calculate the length of plastic hinges formed in the column with simple supports when subjected to the combinations of dynamic concentrated impact loading, $F(t)$ (applied to the mid-height), uniformly distributed blast loading, $w(x, t)$, and static axial loading, P_L, as shown in Figure 8.2a. In this chapter, it is assumed that the inertial force is linearly distributed along the column length with a peak of $q(t)$, and the reaction forces at the supports, R_a, are deterministic. To obtain the inertial force at the peak, the vertical force equilibrium is written between the acting loads on the column and the reaction forces as given by Equation (8.2). It is also assumed that the plastic hinges are formed in the impact-induced column immediately after the onset of the sequent blast loading. Figure 8.2b illustrates the column deformation after the formation of plastic hinges

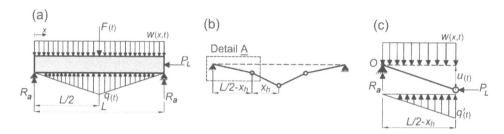

Figure 8.2 A simplified model for the length of the plastic hinge; (a) acting different loads and forces on the idealized model of the column, (b) simplified model of the column after the formation of plastic hinges, and (c) acting loads in the detail A of the simplified model.

with a length of x_h from the column mid-span and Figure 8.2c shows the acting forces in detail A of the simplified model. Based on the flexural moment equilibrium equation at the left support (i.e., point O), the relation between the moments in detail A can be written as given by Equation (8.3). Finally, the length of plastic hinges from the column mid-span can be calculated by solving Equation (8.5) obtained by substituting Equations (8.2) and (8.4) into Equation (8.3). It is noteworthy that the proposed model is limited for the case in which both impact and blast loadings are applied at the column mid-span.

$$F(t) + w(x,t) \cdot L - \frac{q(t) \cdot L}{2} - 2R_a = 0 \tag{8.2}$$

$$\sum M_O = 0 \Rightarrow \frac{w(x,t)}{2} \cdot \left(\frac{L}{2} - x_h\right)^2 + P_L \cdot u(t) - \frac{q'(t)}{3} \cdot \left(\frac{L}{2} - x_h\right)^2 = 0 \tag{8.3}$$

where the peak of inertial force at the location of the plastic hinge, $q'(t)$, can be calculated as follows:

$$q'(t) = \frac{2q(t)}{L} \cdot \left(\frac{L}{2} - x_h\right) \tag{8.4}$$

$$\frac{4}{3L^2}[F(t) + w(x,t) \cdot L - 2R_a] \cdot \left(\frac{L}{2} - x_h\right)^3 - \frac{w(x,t)}{2} \cdot \left(\frac{L}{2} - x_h\right)^2 - P_L \cdot u(t) = 0 \tag{8.5}$$

As recommended in the literature [17, 20], a triangular shape with a zero rise time gives a more accurate estimation of the positive phase of the blast loading. Therefore, the uniformly distributed blast loading along the column length is considered equal to the triangular blast pressure-time history on the front surface of the column mid-span captured from the FE simulation of the blast loading with a $Z = 0.45$ m/kg$^{1/3}$ as shown in Figure 8.3. It is observed that although the peak pressure value (w_0) is identical for both the equivalent triangular pulse and the blast pulse from the FE simulation, their durations (t_0) are different. Since the column under the blast loading with $Z = 0.45$ m/kg$^{1/3}$ reaches its maximum displacement after dropping the pressure to ambient pressure, t_0 is equal to 0.07 ms through equating the area under the blast pulse from the FE simulation [57].

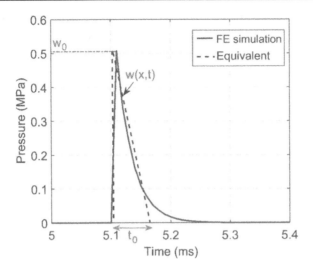

Figure 8.3 Equivalent triangular blast pulse based on FE simulation of blast loading with $Z = 0.45$ m/kg$^{1/3}$.

8.3 FE modeling and validation of RC columns under impact and blast loads

Since there exists no reported experimental work studying RC columns under the combined actions of impact and explosion loadings, the reliability of the FE models developed in LS-DYNA is separately examined under sole impact and sole explosion loadings compared to the results from the experimental works carried out by Feyerabend [51] and Siba [23], respectively.

To validate the FE model of the reference RC column under impact loads, an axially loaded RC column studied by Feyerabend [51], with a square cross-section (0.3 m × 0.3 m) and a 4.0-m span length subjected to a 1,140-kg cylindrical drop-weight hammer (with a diameter of 100 mm) with an impact velocity (V) of 4.5 m/s, is evaluated as illustrated in Figure 8.4a. The column was laterally fixed at the supports. That is, it was axially restrained at one end and the other end was attached to a mass of 20,000 kg by prestressed bars representing the service axial load. Figure 8.4b demonstrates the details of the FE model of the reference RC column in which the concrete and rebars are simulated by 8-node solid and beam elements, respectively, with a mesh size of 10 mm. Also, the contacts between the dropping mass and the column, and between the column and the rigid supports are modeled using a contact algorithm named AUTOMATIC_SURFACE_TO_SURFACE with a friction coefficient of 0.3 as recommended in Refs. [53, 58]. The axial loading is modeled by applying a uniformly distributed pressure to the column axially free end during a dynamic relaxation analysis in LS-DYNA named CONTROL_DYNAMIC_RELAXATION as a separate loading phase before the transient analysis (i.e., the impact loading phase). More detailed information about the FE modeling of the reference column can be found in the previous study [28].

Besides, an RC column with a square cross-section with 0.3 m × 0.3 m dimensions and a height of 3.2 m restrained by Y-shaped fixed supports studied by Siba [23] is selected to validate the FE model of RC columns exposed to explosions as shown in Figure 8.5a. The column contains four 25-M (with a diameter of 19.5 mm) longitudinal

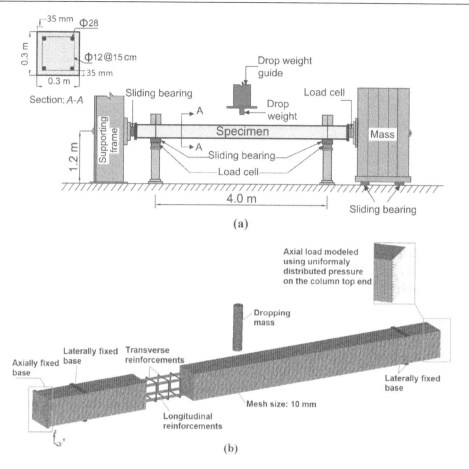

Figure 8.4 Drop-weight impact testing of the column; (a) experimental test conducted by Feyerabend [51] and (b) FE modeling by LS-DYNA in the present study.

rebars and 10-M (with a diameter of 11.3 mm) transverse reinforcements as illustrated in Figure 8.5b. The concrete of the column is modeled using 8-node solid elements, while the steel reinforcement rebars are modeled using beam elements in LS-DYNA. Moreover, the column footing with a height of 0.3 m and the cross-sectional dimensions of 0.7 m × 0.7 m contains eight *U*-shaped 25-M rebars. The column is subjected to different explosions varying in the explosive standoff distance (*R*) when the explosive charge is located at the height of 1.0 m from the column base. The responses of the column were captured using a series of pressure gauges attached to the elevations of 0.25, 1.25, and 2.2 m, and also using string potentiometers mounted on the wood studs at the elevations of 1.0, 1.5, and 2.0 m from the column base as shown in Figure 8.5a.

The nonlinear behaviors of the concrete are simulated in LS-DYNA by MAT_CSCM_CONCRETE (MAT_159) [54] considering strain rate effects, and the damage behaviors in the concrete are modeled using an erosion parameter (*ERODE*) based on the maximum principal strain of the concrete material as discussed in the previous study [59]. As such, the concrete elements are eliminated from the FE simulation when *ERODE* has a value greater than 1.0 in LS-DYNA [60, 61]. Also, the ranges between

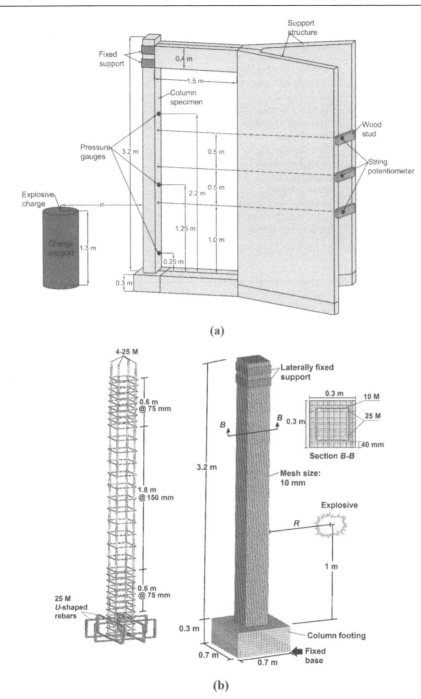

Figure 8.5 Explosion testing of the column; (a) experimented by Siba [23] and (b) FE modeling by LS-DYNA in the present study.

Table 8.1 Material properties used for the RC columns under blast and impact loadings

Model	Parameter	Case study	Value
Concrete	Mass density (kg/m³)	Impact and blast	2,400
	Unconfined compression strength (MPa)	Blast	41.3
		Impact	50
	Elements erosion (*ERODE*)	Impact and blast	1.08
	Maximum aggregate size (mm) (*Dagg*)	Impact and blast	10
	Rate effects	Impact and blast	Turn on
Rebar (longitudinal and	Mass density (kg/m³)	Impact and blast	7,800
transverse)	Young's modulus (GPa)	Impact and blast	210
	Poisson's ratio	Impact and blast	0.3
	Failure strain	Impact and blast	0.15
	Strain rate parameter C_{strain}	Impact and blast	40.4
	Strain rate parameter P_{strain}	Impact and blast	5.0
Longitudinal rebar	Yield stress (MPa)	Impact	523
		Blast	474.4
Transverse rebar	Yield stress (MPa)	Impact	507
		Blast	465.2

1.05 and 1.1 for *ERODE* (which represent that the concrete elements are eroded when the maximum principal strain of the concrete reaches the values between 0.05 and 0.1) were recommended by Murray et al. [60, 61] to capture more reasonable responses and damage patterns for RC structures under dynamic and impulsive loadings. Moreover, an elastoplastic material model MAT_PIECEWISE_LINEAR_PLASTICITY (MAT_024) considering strain rate effects is utilized to simulate the isotropic and kinematic hardening behaviors of the rebars. Also, a rigid material model MAT_RIGID (MAT_020) is used to simulate the supports and drop-weight hammer. Table 8.1 presents the values of each parameter used for the material models of RC columns, which are studied subjected to impact and explosion loads.

A series of impact and blast simulations by varying the mesh size for the concrete, including 5, 10, and 15 mm, were carried out. Figures 8.6 and 8.7 show the column

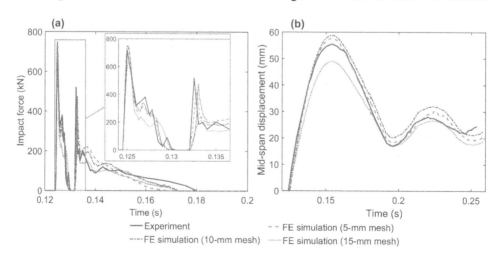

Figure 8.6 Comparing the results from FE simulation and experimental tests of the column under impact loading with V = 4.5 m/s: (a) impact force and (b) mid-span displacements.

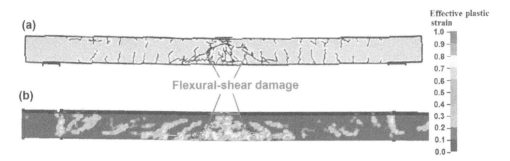

Figure 8.7 Comparing the column damage behaviors under impact loading with V = 4.5 m/s between those from: (a) experimental test done by Feyerabend [51] and (b) FE simulation with 10-mm mesh.

responses and damage states under 1,140-kg drop-weight hammer with an impact velocity (V) of 4.5 m/s resulted from the FE simulations compared to the experimental results obtained by Feyerabend [51]. It is observed that the results have good agreement with those from the experimental test for the mesh sizes smaller than 10 mm. However, the mesh size of 15 mm captures underestimated results. Hence, to reduce the computation time of the simulations, 10-mm mesh is selected for the concert elements and the length of reinforcements to utilize in the following simulations. Also, it was found that a value of 1.08 for the *ERODE* parameter (i.e., the elements are eliminated when the maximum principal strain reaches a value of 0.08) gave acceptable results, which agreed well with the experimental results.

Besides, three different explosion scenarios named *SEIS*-4, *SEIS*-8, and *SEIS*-9 similar to those considered by Siba [23] by varying the scaled distance (Z) of the detonation are selected for the validation study of the FE model of RC columns under explosion. The scaled distance for *SEIS*-4 is 0.34 m/kg$^{1/3}$ (in which the standoff distance, R, is 1.7 m and the TNT charge weight, W_{TNT}, is 123 kg), for *SEIS*-8 is 0.58 m/kg$^{1/3}$ (in which R = 2.5 m and W_{TNT} = 82 kg), and for *SEIS*-9 is 0.52 m/kg$^{1/3}$ (in which R = 2.6 m and W_{TNT} = 123 kg), respectively. An explosion function LOAD_BLAST_ENHANCED based on the empirical formulas captured from experimental blast tests as discussed in Ref. [62] is utilized to simulate the detonation in LS-DYNA. In this method, the shock waves directly interact with the front surface of the structure, which is selected using a keyword named LOAD_BLAST_SEGMENT_SET. Since the center location of detonation is not very close to the ground (the lowest blast scenario includes the blast center located at the height of 1.0 m from the base of the column), the amplification effects of the initial shock wave reflected by the ground are not considered in this chapter.

Similar to the FE mesh convergence tests of the RC column under impact loading, a 10-mm mesh gives reasonable and accurate results compared to those from experimental tests of *SEIS*-8 and *SEIS*-9 as shown in Figure 8.8a and b. Furthermore, Figure 8.9a–c shows the comparison of the column failure behaviors under *SEIS*-4, *SEIS*-8, and *SEIS*-9 between those from the FE simulations with 10-mm mesh and experimental tests [23]. It is observed that the FE model of RC columns with a mesh size of 10 mm for both concrete and reinforcement rebars can accurately simulate the damage states from the low-level flexural and scabbing damage under *SEIS*-8 and *SEIS*-9 loading scenarios to severe spallation and the exfoliations in the cover concrete under *SEIS*-4 scenario.

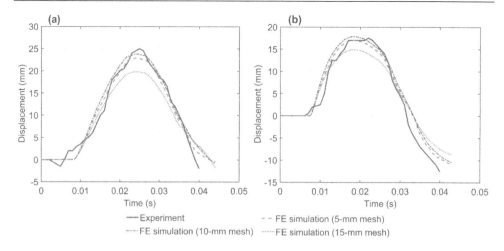

Figure 8.8 Comparing the column deflections at the height of 2.0 m from the FE simulations under explosions compared to the experimental results of (a) *SEIS-8* and (b) *SEIS-9*.

8.4 Results and discussion

The effects of several loading-related parameters, including the loading location, the loading sequence, the time lag (i.e., time interval) between the onsets of the applied loads, the ALR, the impact velocity (V) on the dynamic responses, failure behaviors, and the *DI* of the reference RC column, are evaluated. Since the reference column studied in this chapter represents a typical column used in the ground floor of low- and medium-rise RC buildings, which commonly undergoes an ALR in the ranges from 0.2 to 0.4 [22], a service load with ALR = 0.3 is assumed for all loading scenarios studied in this chapter except Section 8.4.4 in which the vulnerability of the column to the variation of ALR is evaluated under combined loadings.

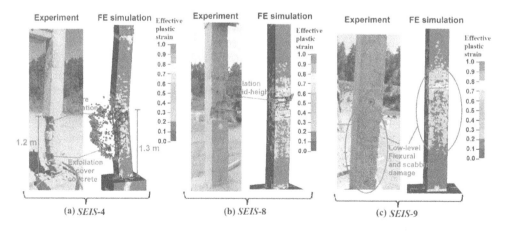

Figure 8.9 Comparing the column damage behaviors from the FE simulations under different explosions compared to the experimental results from Siba [23]: (a) *SEIS-4*, (b) *SEIS-8*, and (c) *SEIS-9*.

8.4.1 Effects of loading location

Impact and explosion loadings may be applied to different locations of RC columns during the collision of different-size vehicles carrying explosive materials in real life. Hence, it is worth investigating the effectiveness of the loading location of each impact and explosion load on the failure behaviors and the residual capacity of the RC column. In this section, three combined impact-blast loadings (i.e., with the impact loading priority than the following detonation) are applied to different locations on the reference RC column under a service axial load with ALR = 0.3 as shown in Figure 8.10a–c. These loading scenarios are described as follows:

1 **IMP0-BLT0:** this scenario represents applying both impact (*IMP*) and blast (*BLT*) loadings at the column mid-height as shown in Figure 8.10a.
2 **IMP1-BLT0:** this scenario denotes that the impact loading is applied at the height of 1.0 m from the column fixed base while the explosion is applied at the mid-height, respectively, as shown in Figure 8.10b.
3 **IMP1-BLT1:** this scenario represents applying both impact and blast loadings at the height of 1.0 m from the column fixed base as shown in Figure 8.10c.

The aforementioned combined loading scenarios include the combined actions of a middle-rate impact loading with $V = 5.0$ m/s and the explosion loadings with different scaled distances of 0.45 m/kg$^{1/3}$ (in which the standoff distance is 0.3 m and the explosive weight is 0.3 kg) and 1.40 m/kg$^{1/3}$ (in which the standoff distance is 2.0 m and the explosive weight is 3.0 kg) representing close-in and far-field detonations, respectively.

Figure 8.11a shows the comparisons of the impact force results from sole impact loadings applied at the heights of 1.0 m (*IMP1*) and 2.0 m (*IMP0*) from the column base. Although different peak values are captured for the impact forces, the first and second peaks of the impact forces simultaneously occur at 0.121 and 0.124 s, respectively.

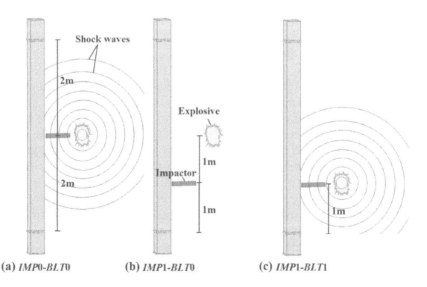

(a) *IMP0-BLT0*   (b) *IMP1-BLT0*   (c) *IMP1-BLT1*

Figure 8.10 Different loading locations for each of the applied loads on the column.

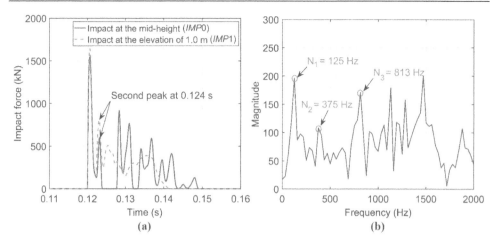

Figure 8.11 (a) Comparing the impact forces between those from sole impact loading scenarios applied at different elevations on the column and (b) natural frequencies of the column.

As reported in the literature [41], the occurrence time of the second peak impact is profoundly dependent on the propagation mechanism of stress waves throughout the impacted structure force while it is independent of the loading conditions. Besides, Figure 8.11b demonstrates that the first natural frequency (N_1) of the column is 125.0 Hz and the corresponding natural period (i.e., the first natural period) is 8.0 ms. Therefore, the second peak impact force occurs at the half of the first periodic time of the structure when is considered the onset time of the sequent detonation in this study except Section 8.4.3 in which the effectiveness of the time lag parameter between the onsets of the impact and blast loadings is investigated.

Figure 8.12a–c illustrates the corresponding failure behaviors of the column when subjected to combined loading scenarios sketched in Figure 8.10a–d in which a medium-rate impact loading with $V = 5.0$ m/s is combined with a close-in blast loading ($Z = 0.45$ m/kg$^{1/3}$). Besides, Figure 8.13a–c shows the damage states of the column when subjected to the combinations of the identical impact loading with a far-field detonation ($Z = 1.40$ m/kg$^{1/3}$) with the same sketch of loading locations as presented in Figure 8.10a–c. It is noteworthy that the damage states demonstrated in Figures 8.12a–c and 8.13a–c refer to the damaged phases of the column under their service axial loads before applying the motion-based axial loading.

In Figure 8.12a–c, it is observed that the column endures more severe and cumulative damage states composed of localized shear failures along with the formation of plastic hinges when both impact and explosion loadings are applied at the same elevation for loading scenarios *IMP0-BLT0* and *IMP1-BLT1* than *IMP1-BLT0* in which the column experienced moderate-level spallation damage. This is due to the increase of the standoff distance from the impact point in scenario *IMP1-BLT0*, which leads to smaller transmitted pressure-impulse and less detonation intensity at the impact-induced zone on the column than those in *IMP0-BLT0* and *IMP1-BLT1*. A larger residual axial load capacity of the column under *IMP1-BLT0* with a close-in detonation can be observed in Figure 8.14a, which leads to smaller *DI* than those under *IMP0-BLT0* and *IMP1-BLT1* as given in Table 8.2. Moreover, it is found that the column undergoes a larger *DI* and a more severe failure with

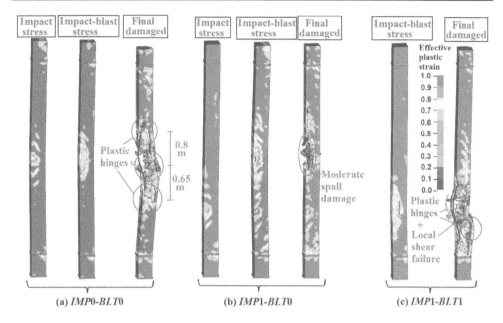

(a) *IMP0-BLT0*   (b) *IMP1-BLT0*   (c) *IMP1-BLT1*

Figure 8.12 Comparing the column failure behaviors under different combined loadings associated with a close-in detonation and varying the loading location.

more extensive lengths for the formed plastic hinges under *IMP0-BLT0* than those under *IMP1-BLT1* as illustrated in Figure 8.12. This is because of the discrepancies between the trapped overpressure areas on the column from the detonations with different locations as shown in Figure 8.10. That is, the shock wave spherically reaches the column under *IMP0-BLT0*, which leads to a more extensive overpressure area than that under *IMP1-BLT1* in which most of the shock waves are almost hemispherically applied to the column.

In contrast, when the identical impact loading is combined with a relatively far-field explosion with $Z = 1.40$ m/kg$^{1/3}$ in Figure 8.13a–c, applying both impact and blast loadings at the same elevation causes fewer damage states under *IMP0-BLT0* and *IMP1-BLT1* than *IMP1-BLT0*. This is also quantitatively can be found in Table 8.2. In Figure 8.13b, it is observed that the column suffers a global failure at the bottom support location under *IMP1-BLT0*, while it withstands the collapse under other scenarios (see Figure 8.13a and c) since the column presents larger residual capacities as shown in Figure 8.14b. This is because that when the column is subjected to an almost uniform distribution of pressure under a far-field detonation, it undergoes more severe global failure when the impact and blast loadings are applied at different elevations to the column. However, the predominance of localized failure modes is observed on the column behaviors when the impact-induced zone is exposed to the more concentrated distribution of loading under a close-in explosion. Also, owing to the more extensive pressure-induced area on the column under mid-height detonation than that under lower elevation explosion, *IMP0-BLT0* loading causes more intensive global shear damage states at the support locations and spallation at the column mid-height than *IMP1-BLT1*.

Table 8.2 Damage indices of the column under different combined loadings varying in the loading location

	IMP0-BLT0 (close-in)	IMP1-BLT0 (close-in)	IMP1-BLT1 (close-in)	IMP0-BLT0 (far-field)	IMP1-BLT0 (far-field)	IMP1-BLT1 (far-field)
P_r (kN)	1,890	2,290	1,510	2,612	1,720	3,046
DI	0.84	0.57	0.76	0.48	0.81	0.34

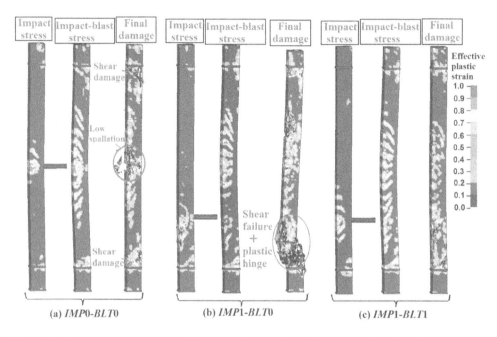

(a) IMP0-BLT0   (b) IMP1-BLT0   (c) IMP1-BLT1

Figure 8.13 Comparing the column failure behaviors under different combined loadings associated with a far-field detonation and varying in the loading location.

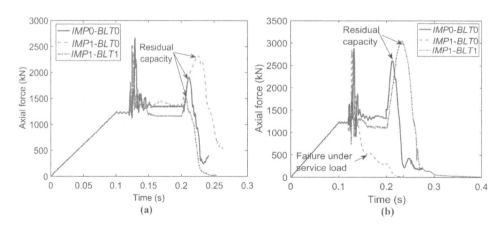

Figure 8.14 Comparing the column axial force results under different location combined loadings associated with (a) a close-in detonation and (b) a far-field detonation.

8.4.2 Effects of loading sequence

The post-impact and post-blast resistances of the RC column are evaluated by considering different combined loadings scenarios varying in the sequence of the applied loads. One of these loading scenarios is named "impact-blast" in which the impact loading is applied before the following detonation, and the other is named "blast-impact" in which the explosion loading is applied prior to the following impact loading. To prevent any interference, effects between the impact and blast loading phases, a time lag of 0.03 s is considered between the initiations of the applied loads when the column reaches a stable state with a lateral velocity of less than 0.1 m/s.

Figure 8.15a and b shows the damage states of the column under impact-blast and blast-impact loadings applied at the column mid-height (i.e., $IMP0$-$BLT0$ and $BLT0$-$IMP0$, respectively) from the combinations of a middle-rate (i.e., $V = 5$ m/s) impact loading with close-in ($Z = 0.45$ m/kg$^{1/3}$) and far-field ($Z = 1.40$ m/kg$^{1/3}$) explosion loadings, respectively. In addition, Figure 8.16a–d illustrates the column failure behaviors under the combinations of a middle-rate (i.e., $V = 5.0$ m/s) impact loading applied at the height of 1.0 m from the column base with close-in ($Z = 0.45$ m/kg$^{1/3}$) and far-field ($Z = 1.40$ m/kg$^{1/3}$) explosions applied at the column mid-height with the variation of the loading sequence (i.e., $IMP1$-$BLT0$ and $BLT0$-$IMP1$).

In Figure 8.15a, it is observed that the column experiences similar spallation under sole close-in explosion and the combined blast-impact loadings. However, more severe spallation is suffered by the column at the mid-height under an impact-blast scenario than that generated under the blast-impact loading. This is due to losing the bond strength of the concrete in the impact-induced zone under the cumulative effects of the combined flexural-shear failures and the tensile stresses generated owing to the following blast impulse. Compared to the localized spallation under a close-in

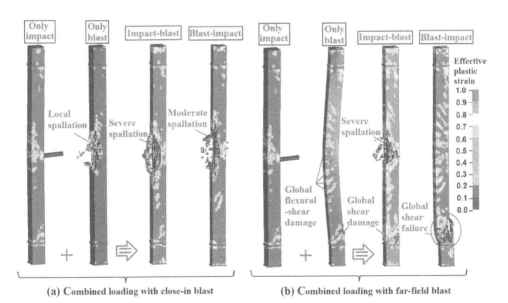

(a) Combined loading with close-in blast   (b) Combined loading with far-field blast

Figure 8.15 Damage states of the column under the combinations of a middle-rate impact loading ($V = 5.0$ m/s) with close-in ($Z = 0.45$ m/kg$^{1/3}$) and far-field ($Z = 1.40$ m/kg$^{1/3}$) explosions applied at the column mid-height varying in the loading sequence.

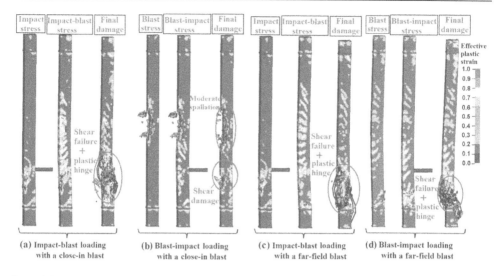

(a) Impact-blast loading with a close-in blast

(b) Blast-impact loading with a close-in blast

(c) Impact-blast loading with a far-field blast

(d) Blast-impact loading with a far-field blast

Figure 8.16 Damage states of the column under the combinations of a middle-rate impact loading ($V = 5.0$ m/s) applied at the height of 1.0 m from the column base with close-in ($Z = 0.45$ m/kg$^{1/3}$) and far-field ($Z = 1.40$ m/kg$^{1/3}$) explosions applied at the column mid-height varying in the loading sequence.

explosion in Figure 8.15a, a global flexural-shear failure is observed when the column is subjected to a rather far-field explosion as shown in Figure 8.15b. Also, the column fails in a global shear mode at the location of the bottom support under $BLT0$-$IMP0$. Therefore, the critical location for the initiation of the column progressive collapse is more sensitive to the loading sequence for the combined loadings associated with a far-field detonation than that with a close-in explosion. Moreover, by comparing the residual axial force results in Figure 8.17a and the DI values in Table 8.3, it is found that $BLT0$-$IMP0$ with a close-in explosion gives the lowest DI (equal to 0.64) and the highest residual capacity compared to other scenarios. As such, unlike the other scenarios, the column withstands the collapse under its service axial load (i.e., ALR = 0.3)

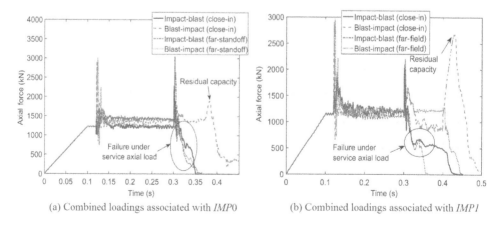

(a) Combined loadings associated with $IMP0$

(b) Combined loadings associated with $IMP1$

Figure 8.17 Axial forces of the column under combined loadings varying in the loading sequence.

Table 8.3 Damage indices of the column under different combined loadings varying in the loading sequence with mid-height impacts (IMP0)

	Impact-blast (close-in)	Blast-impact (close-in)	Impact-blast (far-field)	Blast-impact (far-field)	Only blast (close-in)	Only blast (far-field)	Only impact (IMP0)
P_r (kN)	1,480	1,900	1,360	1,380	2,960	2,130	3,490
DI	0.91	0.64	0.92	0.90	0.41	0.59	0.24

under *BLT0-IMP0* with a close-in explosion. Besides, although the impact-blast scenarios lead to more severe damage states and higher *DI* values than those under blast-impact scenarios, the far-field explosion provides more critical loading conditions than the close-in explosion when they are applied prior to the impact loading during the combined loading scenarios.

The prominence of global failure modes at the location of the column bottom support is observed due to a significant increase of shear damage generated around the impact zone, which propagated to the bottom support under lower elevation impact loading when the column is subjected to combined loading in which the impact loading is applied at the lower height of 1.0 m from the column base, Consequently, the column collapses under its service axial load under all scenarios except *BLT0-IMP1* with a close-in explosion as shown in Figure 8.16b. Although the column suffers a moderated spallation at the mid-height and severe shear damage at the impact zone, it withstands the collapse under its service axial load. This can be also quantitatively concluded from Figure 8.17b and Table 8.4 in which the column has the highest residual capacity and the lowest *DI* value of 0.44 than those from other scenarios. Accordingly, since applying the impact loading before explosions caused more critical loading conditions than blast-impact loading, the combined impact-blast loadings are assumed to study the effectiveness of the other loading parameters in the following sections of this chapter.

8.4.3 Effects of time lag

The time lag (t_L) between the onsets of the impact and explosion loadings can play a determinant role in recognizing the failure modes of structures under combined loading scenarios because this parameter specifies the contributions of the structural responses to each impact and explosion loads into their combined actions. To investigate the influences of this parameter during the combined impact-blast loading scenarios, different time stages of the column response under a medium-velocity ($V = 5.0$ m/s) impact loading applied at the column mid-height (*IMP0*) and the height of 1.0 m from the column base (*IMP1*) when axially exposed to ALR = 0.3 are considered, as shown

Table 8.4 Damage indices of the column under different combined loadings varying in the loading sequence with lower elevation impacts (IMP1)

	Impact-blast (close-in)	Blast-impact (close-in)	Impact-blast (far-field)	Blast-impact (far-field)	Only blast (close-in)	Only blast (far-field)	Only impact (IMP0)
P_r (kN)	1,480	1,900	1,360	1,380	2,960	2,130	3,490
DI	0.91	0.64	0.92	0.90	0.41	0.59	0.24

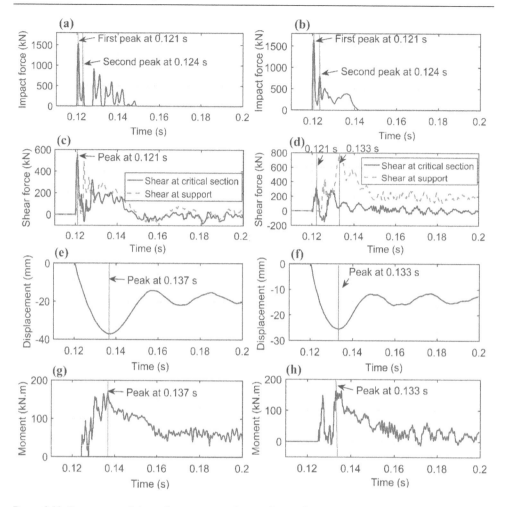

Figure 8.18 Responses of the column to a medium-velocity ($V = 5.0$ m/s) impact loading; (a) impact force under *IMP0*, (b) impact force under *IMP1*, (c) shear forces under *IMP0*, (d) shear forces under *IMP1*, (e) mid-height displacement under *IMP0*, (f) mid-height displacement under *IMP1*, (g) mid-height bending moment under *IMP0*, and (h) mid-height bending moment under *IMP1*.

in Figure 8.18a–h as the onset times of the sequent explosions in different combined loading scenarios as follows:

- The time of the initial peak impact force when it is simultaneous with the peak shear force at the critical section (around the impact zone as discussed in Ref. [47]) under *IMP0* as shown in Figure 8.18a and c, and under *IMP1* as shown in Figure 8.18b and d.
- The time of the second peak impact force at 0.124 s under *IMP0* and *IMP1* as illustrated in Figure 8.18a and b, respectively.
- The time of the peak displacement at 0.137 s when it is simultaneous with the peak bending moment at the mid-height under *IMP0* as shown in Figure 8.18e and g. This time is 0.133 s under *IMP1* when it is simultaneous with the peak bending

moment at the impact level and the peak shear at the supports as shown in Figure 8.18d, f, and h.
- The time of the free vibration phase for the column response at 0.16 s when the contact between the impactor and the column is interrupted as shown in Figure 8.18a and b.

The column failure behaviors under combined impact-blast loadings applied at the column mid-height (i.e., $IMP0$-$BLT0$) varying in the initiation time of a close-in explosion ($Z = 0.45$ m/kg$^{1/3}$) after an impact loading with $V = 5.0$ m/s are illustrated in Figure 8.19a–d. In Figure 8.19a, sufferings of more extensive damage states and more severe global failures are observed in the column under the combined loading in which the following detonation is applied at the time of the initial peak impact force (when it is simultaneous with the occurrence of the initial peak shear force at the critical section around the mid-height) at $t_L = 0.121$ s than those under other $IMP0$-$BLT0$ scenarios in which the sequent explosion is initiated at the following time stages. This can be also found from the decrease of the length of plastic hinges and tending the column to localize failure modes at the mid-height by increasing the time lag as shown in Figure 8.19c and d. Since columns are axial load-carrying structural members, the interaction diagram between the column axial load-carrying capacity and its bending moment capacity (named P-M capacity) calculated based on ACI-318 [52] is considered the assessment criterion of the plastic hinge locations from the observations in the FE simulations as shown in Figure 8.20a–d. A comparison between the calculated lengths of plastic hinges using Equation (8.5) and those observed from FE simulations in Figure 8.21 is made. It is found that the proposed simplified model for the case of $IMP0$-$BLT0$ is efficiently able to estimate the location of the plastic hinges compared to those from FE simulations. Besides, it is observed that the location of plastic hinges formed in the final damaged state of the column is extremely related to the location of the initial negative moments generated in the column under sole impact loading, which was named "effective length" by Cotsovos et al. [43].

In addition, the column dynamic responses under different combined loadings applied at the mid-height and varying in t_L are illustrated in Figure 8.22a–e. It is observed that the column experiences larger displacement when the detonation initiates simultaneously with the second peak impact force when the column has the most velocity (i.e., the steepest slope of the pier displacement) as shown in Figure 8.22a. Besides, a larger flexural moment is obtained when t_L is set at the time of the peak bending moment and the peak displacement (i.e., $t_L = 0.124$s) at the mid-height as shown in Figure 8.22b. Moreover, the column undergoes greater shear forces and peak axial forces as shown in Figure 8.22c–e when the sequent detonation initiates at the time of the initial peak impact force (i.e., $t_L = 0.121$ s). Although the column suffers more extensive global damage states and larger shear forces for the case $t_L = 0.121$ s, increasing the time lag causes more dangerous failure modes and thereby increases the DI value as shown in Figure 8.22f due to the decrease in the column residual capacity (Figure 8.22e). As such, the column failure mode changes from the global modes to localized failures owing to severe spallation in the column depth. Consequently, the column collapses through the formation of a mid-height plastic hinge under its service axial load shortly after the onset of the sequent detonation at $t_L = 0.16$ s during the free vibration phase of the column response to sole impact loading. In addition, almost the same trend for the DI values varying in t_L is observed in

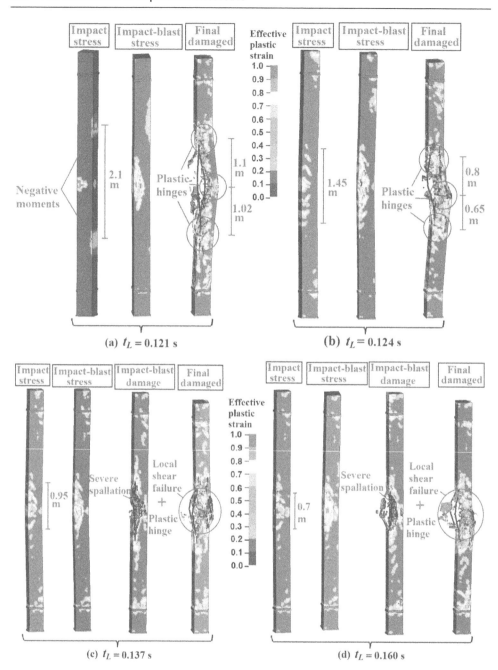

Figure 8.19 Damage states of the column under *IMP0-BLT0* loadings varying in t_L.

Figure 8.22f when the column is under *IMP0-BLT0* with a far-field explosion, while lower *DI* is obtained when $t_L = 0.124$ s.

The vulnerability of the column to the variation of t_L is also assessed when it is subjected to a lower elevation impact loading applied at the height of 1.0 m from the column base combined with blast loadings applied at the mid-height (i.e., *IMP1-BLT0*) as

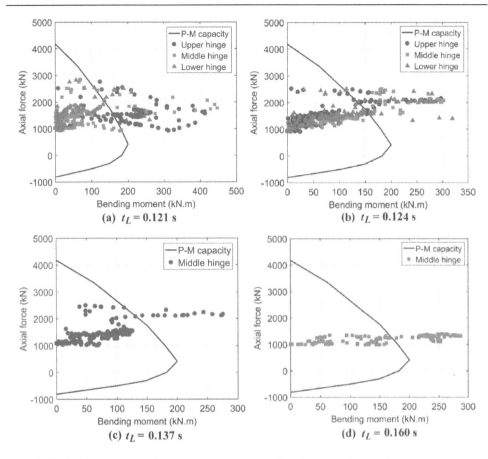

Figure 8.20 Axial force versus bending moment interaction diagrams of the column at the locations of the formed plastic hinges under *IMP0-BLT0* loadings varying in t_L.

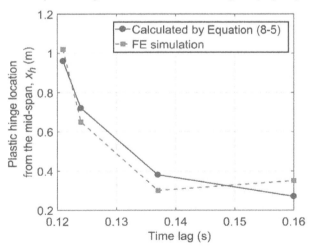

Figure 8.21 Comparing the plastic hinge locations from the column mid-height between the lengths calculated by Equation (8.5) and those observed from the FE simulations under *IMP0-BLT0* varying in t_L.

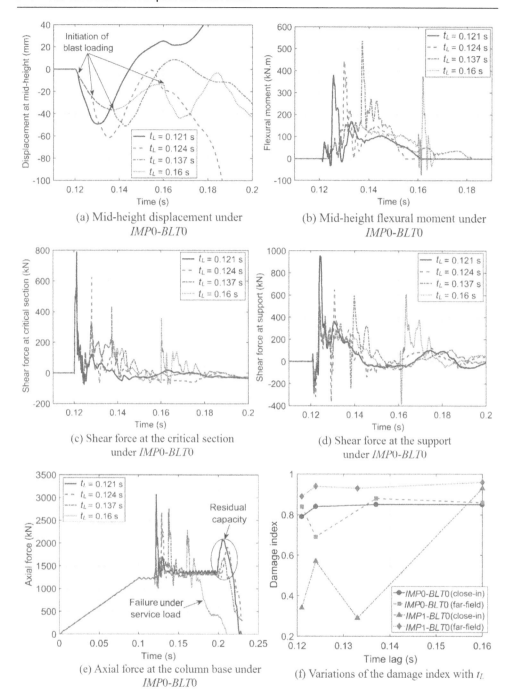

(a) Mid-height displacement under
IMP0-BLT0

(b) Mid-height flexural moment under
IMP0-BLT0

(c) Shear force at the critical section
under *IMP0-BLT0*

(d) Shear force at the support
under *IMP0-BLT0*

(e) Axial force at the column base under
IMP0-BLT0

(f) Variations of the damage index with t_L

Figure 8.22 Comparing the column responses under combined impact-blast loadings varying in t_L.

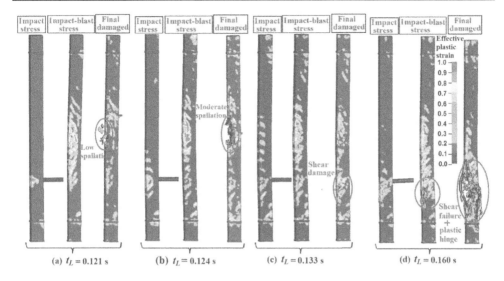

Figure 8.23 Damage states of the column under *IMP1-BLT0* with a close-in detonation and varying in t_L.

shown in Figure 8.23a–d. In Figure 8.23a and b, it is revealed that although the spallation level in the column depth increases with increasing the t_L until the incidence time of the second peak impact force at 0.124 s, a safer state with a relatively lower *DI* value is obtained when the sequent explosion is initiated simultaneously with the peaks of shear force at supports, flexural moment, and displacement at the impact zone (i.e., $t_L = 0.133$ s) as illustrated in Figures 8.22f and 8.23c. However, the most dangerous failure mode is taken place when the detonation is initiated during the free vibration phase of the column response at $t_L = 0.16$ s. Under this loading condition, a cumulative failure mode composed of localized shear failure and the formation of a plastic hinge is predominant on the column behavior as shown in Figure 8.23d. Besides, the predominance of global failure modes on the column behavior is observed in Figure 8.24a–d when it is subjected to *IMP1-BLT0* with a far-field explosion. Therefore, it can be found that although the column failure mode is not sensitive to the variation of the time lag, the column endures more severe failures with higher *DI* values (Figure 8.22f) than those under *IMP1-BLT0* with a close-in detonation.

8.4.4 Effects of axial load ratio (ALR)

As reported in the literature [28, 30], the axial load of columns and bridge piers can play a key role in the determination of failure modes. In this section, the vulnerability of the reference of RC column to various ALRs, including 0.1, 0.3, 0.5, and 0.8, is assessed subjected to combined actions of a medium-velocity ($V = 5.0$ m/s) impact loading and a close-in detonation ($Z = 0.45$ m/kg$^{1/3}$). Figure 8.25a and b demonstrates the comparison of impact force results of the column exposed to different ALRs with it is laterally under sole impact loadings ($V = 5.0$ m/s) applied at the mid-height (*IMP0*) and the height of 1.0 m from the column base (*IMP1*), respectively. It is revealed that although different peak values are obtained by the variation of ALR, the secondary peaks of the impact forces take place at the same time under both *IMP0* and *IMP1*.

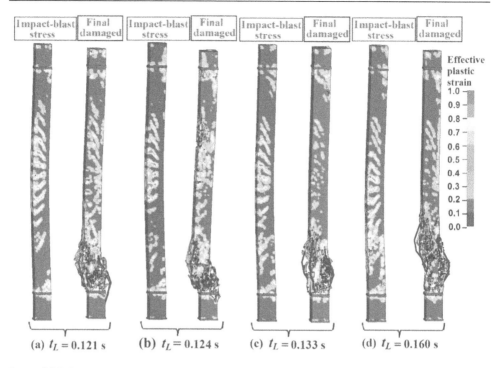

(a) $t_L = 0.121$ s   (b) $t_L = 0.124$ s   (c) $t_L = 0.133$ s   (d) $t_L = 0.160$ s

Figure 8.24 Damage states of the column under *IMP1-BLT0* with a far-field detonation and varying in t_L.

This is basically due to the independence of the secondary peak time from the loading conditions as discussed in Section 8.4.1. Hence, the time of the second peak impact forces at 0.124 s is considered the onset time of the sequent detonation.

The damage behaviors of the column exposed to different ALRs under *IMP0-BLT0* and *IMP1-BLT0* compared to those under sole impact or sole explosion loadings are

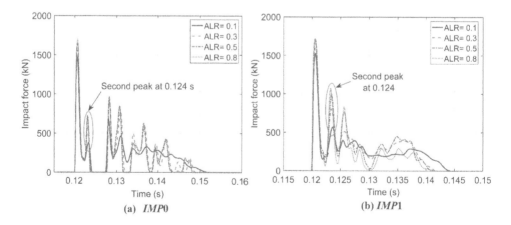

(a) *IMP0*   (b) *IMP1*

Figure 8.25 Comparing the impact forces of the column under different ALRs when laterally subjected to medium-velocity impact loadings ($V = 5.0$ m/s) at the mid-height (*IMP0*) and the height of 1.0 m from the column base (*IMP1*).

illustrated in Figures 8.26a–d and 8.27a–d, respectively. Under the application of loadings at the column mid-height, it is observed in Figure 8.26a–d that the length of the concrete spallation decreases with increasing ALR until a range of 0.5 when the column is under the only detonation due to the positive effects of ALR on the resistance capacity of the column. Similarly, increasing the column resistance under only impact loadings is observed through the decrease of impact-induced stress area until ALR = 0.5. However, the column fails in global flexural (plastic hinge forma-tion) and local shear modes under sole impact or sole explosion loadings, respectively, when ALR reaches 0.8. Therefore, the case ALR = 0.5 is recognized as a sensitivity level of the column resistance against sole impact and explosion loadings. Besides, when the column is subjected to combined *IMP0-BLT0* loadings, more severe dam-age states are observed through the formation of plastic hinges on both sides of the column mid-height than those under sole impact and explosion loadings. The length of plastic hinges decreases with the increasing ALR. Insofar, as ALR increases to more than 0.3, it leads to a localized failure when ALR = 0.5 as shown in Figure 8.26c, and an overall flexural failure through a mid-height plastic hinge when ALR = 0.8 as shown in Figure 8.26d. In Figure 8.28, it can be obtained that the calculated lengths of plastic hinges using Equation (8.5) for different ranges of ALR are in good agree-ment with those observed from the FE simulations given in Figure 8.26. Moreover, by comparing the damage indices based on the residual axial loading capacities as shown in Figure 8.29a and c, it is found that the column with ALR = 0.3 has the low-est *DI* value under combined *IMP0-BLT0* loadings. However, this sensitivity level is ALR = 0.5 under sole impact or sole blast loading as discussed above and as shown in Figure 8.29c. Therefore, the sensitivity level of the column resistance to ALR decreases from 0.5 to 0.3 when the column is subjected to combined actions of impact and explo-sion loadings. It is noteworthy that the damage states demonstrated in Figure 8.26a and b for ALRs of 0.1 and 0.3, respectively, refer to the damaged phase after applying the motion-based axial loading beyond their service axial loads. The purpose of such a magnified presentation was to illustrate the location of the formed plastic hinges in a more obvious manner.

Similar to the trend of the column *DI* under *IMP0-BLT0*, the sensitivity level of the column *DI* to ALR is reduced from 0.5 to 0.3 when it is exposed to *IMP1-BLT0* as shown in Figure 8.29b and c. However, it undergoes totally different failure modes under *IMP1* and *IMP1-BLT0* compared to those under *IMP0* and *IMP0-BLT0*. As such, the level of the shear damage at the impact level (i.e., at 1.0 m from the column base) under sole impact loading (i.e., *IMP1*) decreases until ALR = 0.5, while a global flexural failure mode is yielded through the formation of a plastic hinge at the impact level under ALR = 0.8. Besides, the extent of the damaged zone in the column under *IMP1-BLT0* is significantly reduced when the range of ALR increases from 0.1 to 0.3 (Figures 8.27a and b). Beyond this level (i.e., ALR = 0.3), the column fails in global modes through the formation of plastic hinges particularly around the column mid-height as illustrated in Figure 8.27c and d.

8.4.5 Effects of impact velocity

The vulnerability of the reference RC column with ALR = 0.3 is assessed under the combinations of different-rate impact loadings with the velocities (V) of 1, 3, 5, and 10 m/s with a close-in ($Z = 0.45$ m/kg$^{1/3}$) explosion. Since the independence of the

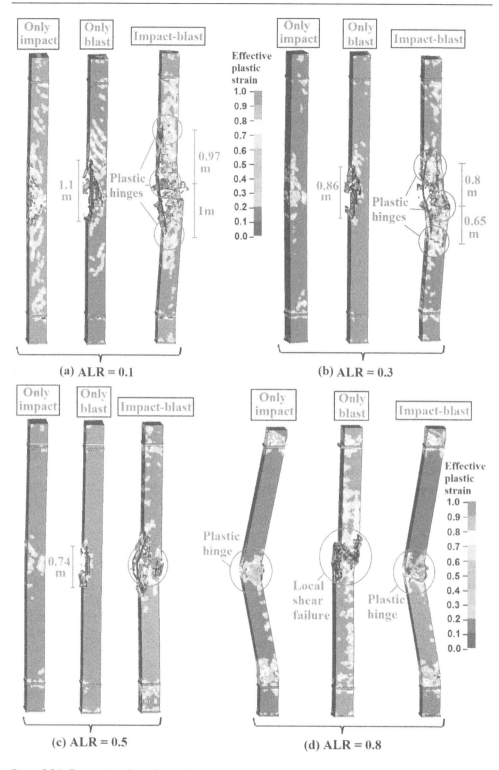

Figure 8.26 Comparing the column damage states exposed to different ALRs under *IMP0-BLT0*.

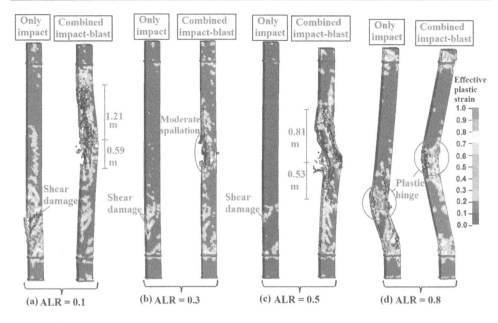

Figure 8.27 Comparing the column damage states exposed to different ALRs under *IMP1-BLT0*.

incidence time of the second peak impact forces at 0.124 s from the impact velocity under *IMP0* and *IMP1* is found from Figure 8.30a and b, respectively, this time stage is considered the onset time of the sequent explosion in this section. However, substantial positive influences of the impact velocity on the peak values are observed. The damage states of the column under *IMP0-BLT0* and *IMP1-BLT0* varying in *V*

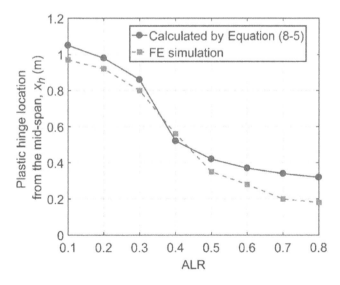

Figure 8.28 Comparing the plastic hinge locations from the column mid-height between the lengths calculated by Equation (8.5) and those observed from the FE simulations exposed to different ALRs under *IMP0-BLT0*.

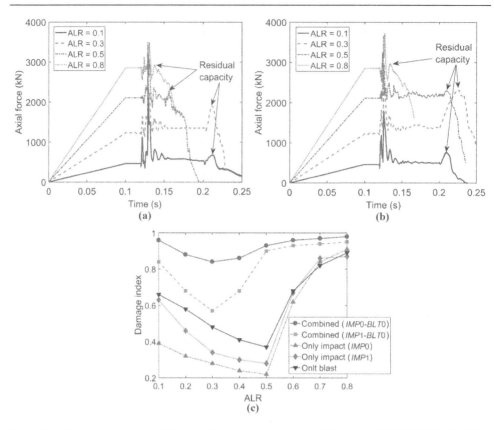

Figure 8.29 Comparing the axial forces of the column varying in ALR under; (a) IMP0-BLT0, (b) IMP1-BLT0, and (c) the damage index results varying in ALR under various loading scenarios.

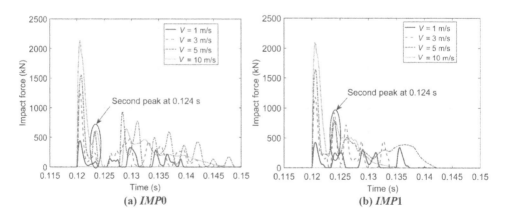

Figure 8.30 Comparing the impact forces of the column under different-velocity impact loadings applied at the mid-height (IMP0) and the height of 1.0 m from the column base (IMP1).

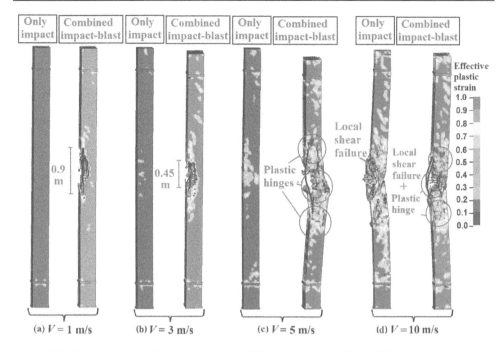

| Only impact | Combined impact-blast | Only impact | Combined impact-blast | Only impact | Combined impact-blast | | Only impact | Combined impact-blast |

Figure 8.31 Damage states of the column under different-velocity *IMP0-BLT0* loadings compared to those under sole impact loadings.

are illustrated in Figures 8.31a–d and 8.32a–d, respectively, compared to those under sole impact loadings. Under sole impact loadings, it is observed that the column tends to localized shear failure mode at the impact level. Besides, when the column is subjected to *IMP0-BLT0*, the spallation length decreases until $V = 3.0$ m/s, while the column withstands the collapse. Beyond this velocity range, the column suffers cumulative failure modes composed of localized shear failures and plastic hinges formed around the mid-height, which leads to the progressive collapse of the column as shown in Figure 8.31c and d. Similar to the manner of presentation of failure behaviors in Figure 8.26, it is noteworthy that the damage states illustrated in Figure 8.31c refers to the failure phase of the column after applying the motion-based axial load beyond its service axial load.

In Figure 8.33, the capability of the proposed simplified model with good accuracy in estimating the location of plastic hinges formed in the column under *IMP0-BLT0* is demonstrated in comparison with those observed from the FE simulations. In addition, the residual axial load capacity and corresponding *DI* results of the column under *IMP0-BLT0* are presented in Figure 8.34a and c. It is found that $V = 3.0$ m/s represents a sensitivity level of the column failure to the impact loading velocity under these combined loadings.

The sensitivity level of the column axial resistance to the impact velocity increases from 3.0 to 5.0 m/s, decreasing the impact loading elevation from the mid-height to the height of 1.0 m from the column base in the combined loadings (i.e., *IMP1-BLT0*) as shown Figures 8.32 and 8.34b and c. That is, the column fails under *IMP1-BLT0* with the impact velocities higher than 5.0 m/s. Consequently, unlike the localized failure

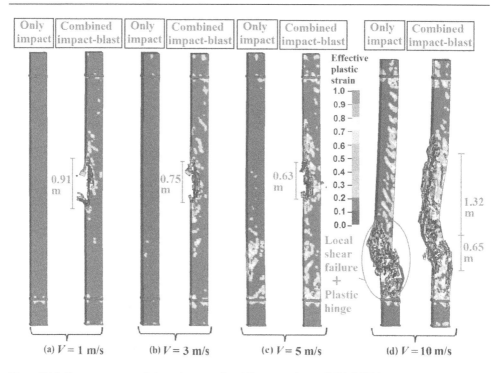

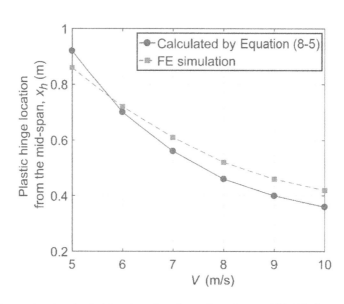

Figure 8.32 Damage states of the column under different-velocity *IMP1-BLT0* loadings with a close-in explosion compared to those under sole impact loadings.

Figure 8.33 Comparing the plastic hinge locations from the column mid-height between the lengths calculated by Equation (8.5) and those observed from the FE simulations under *IMP0-BLT0* varying in *V*.

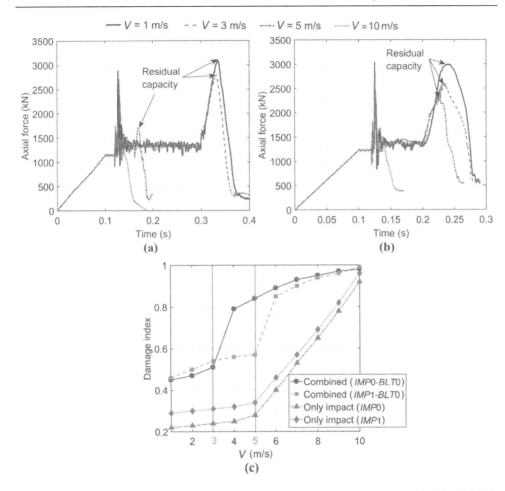

Figure 8.34 Comparing the axial forces of the column varying in V under (a) *IMP0-BLT0*, (b) *IMP1-BLT0*, and (c) the damage index results varying in V under various loading scenarios.

mode suffered by the column under *IMP0-BLT0* with $V = 10$ m/s, the column endures a global flexural failure mode with more extensive lengths for the formed plastic hinges under *IMP0-BLT0* with $V = 10$ m/s as shown in Figure 8.32d.

8.5 Conclusions

This chapter presents a theoretical study on the effects of combined impact and blast loadings on the dynamic responses and failure behaviors of a typical RC column commonly used in medium-rise buildings using numerical simulations in LS-DYNA. Since there exists no experimental work in the literature to investigate RC columns under combined actions of impact and blast loads, this chapter attempts to ideally assess the vulnerability of an RC column subjected to possible combined loading scenarios varying in terms of several loading-related parameters, including the loading sequence, the time interval between the onsets of the applied loads, the loading location, the ALR, and the impact velocity (V) through a *DI* defined based on the residual axial load-carrying capacity of the column. To do this, a multi-step loading method was utilized to calculate the damage indices that were different from those adopted in the previous works. In addition, a simplified model is proposed to estimate the location of the plastic hinge formed in the column when subjected to combined loadings applied at the mid-height. It was concluded that the proposed model is efficiently able to estimate the lengths of plastic hinges in comparison with those observed from the FE simulations.

Based on the results from the numerical simulations, the following conclusions can be drawn:

- The column suffered more severe damage states and a higher *DI* when both impact and explosion loadings were applied at the same elevation on the column. In addition, when a lower elevation impact loading was applied at the height of 1.0 m from the column base, *IMP1* was combined with a far-field detonation (i.e., *IMP1-BLT0*), the column failed in a global shear mode. Therefore, more dangerous loading conditions were provided by *IMP1-BLT0* than scenarios in which both impact and blast loadings are applied at the same elevation (i.e., *IMP0-BLT0* and *IMP1-BLT1*).
- Applying impact loadings prior to explosions resulted in more severe localized spallation failure modes in the column. In addition, it was found that the failure mode of the column was not significantly dependent on the loading sequence when the column was subjected to combined loadings associated with a close-in explosion applied at the mid-height (i.e., *IMP0-BLT0*). However, the failure mode of the column was more sensitive to the loading sequence when the column was subjected to combined loadings associated with a far-field detonation. Besides, the predominance of global shear failure modes on the column response was mainly concluded when the column was exposed to the combined actions of a low-elevation impact loading (i.e., *IMP1-BLT0*) and a far-field detonation.
- The column experienced larger shear forces and more severe global failures when the following detonation was initiated at the time of the initial peak impact force under *IMP0-BLT0*. By increasing the time lag in scenarios *IMP0-BLT0* and *IMP1-BLT0*, the column failure mode tended to a localized failure at the impact level leading to an immediate collapse under its service axial load. Besides, the column failure mode was almost independent of the time lag parameter when a far-field detonation was combined with the low-elevation impact loading since the column underwent global shear failures at the bottom support under all combined loading scenarios.
- ALR = 0.5 was accepted as the sensitivity threshold of the column resistance against the collapse under sole blast or impact loading. However, the change of the column failure mode from an overall flexural mode to a localized failure mode under combined loadings was sensitive to ALR = 0.3 and beyond.
- The sensitivity level of the column failure to the impact velocity decreased from a value of 5.0 m/s under only impact loading to $V = 3.0$ m/s under combined loadings associated with a close-in explosion applied at the column mid-height (i.e., *IMP0-BLT0*). However, the sensitivity threshold of the *DI* to the impact velocity remained constant for the loadings scenarios of *IMP1* and *IMP1-BLT0* (i.e., a sensitivity level of $V = 5.0$ m/s). Therefore, the *IMP0-BLT0* scenario captured a lower sensitivity threshold in the *DI* of the column than that under *IMP1-BLT0*.

References

1. Foglar M, Kovar M. Conclusions from experimental testing of blast resistance of FRC and RC bridge decks. *Int J Impact Eng* 2013;59:18–28.
2. Islam AA, Yazdani N. Performance of AASHTO girder bridges under blast loading. *Eng Struct* 2008;30(7):1922–1937.
3. Son J, Lee HJ. Performance of cable-stayed bridge pylons subjected to blast loading. *Eng Struct* 2011;33(4):1133–1148.
4. Foglar M, Hajek R, Fladr J, Pachman J, Stoller J. Full-scale experimental testing of the blast resistance of HPFRC and UHPFRC bridge decks. *Constr Build Mater* 2017;145:588–601.
5. Pan Y, Ventura CE, Cheung MM. Performance of highway bridges subjected to blast loads. *Eng Struct* 2017;151:788–801.
6. Winget DG, Marchand KA, Williamson EB. Analysis and design of critical bridges subjected to blast loads. *J Struct Eng* 2005;131(8):1243–1255.
7. Hu ZJ, Wu L, Zhang YF, Sun LZ. Dynamic responses of concrete piers under close-in blast loading. *Int J Damage Mech* 2016;25(8):1235–1254.
8. Fujikura S, Bruneau M. Experimental investigation of seismically resistant bridge piers under blast loading. *J Bridge Eng* 2010;16(1):63–71.
9. Yi Z, Agrawal AK, Ettouney M, Alampalli S. Blast load effects on highway bridges. I: modeling and blast load effects. *J Bridge Eng* 2013;19(4):04013023.
10. Yi Z, Agrawal AK, Ettouney M, Alampalli S. Blast load effects on highway bridges. II: failure modes and multihazard correlations. *J Bridge Eng* 2013;19(4):04013024.
11. Williams GD, Williamson EB. Response of reinforced concrete bridge columns subjected to blast loads. *J Struct Eng* 2011;137(9):903–913.
12. Williamson EB, Bayrak O, Davis C, Daniel Williams G. Performance of bridge columns subjected to blast loads. II: results and recommendations. *J Bridge Eng* 2011;16(6):703–710.
13. Yan B, Liu F, Song D, Jiang Z. Numerical study on damage mechanism of RC beams under close-in blast loading. *Eng Fail Anal* 2015;51:9–19.
14. Zhang D, Yao S, Lu F, Chen X, Lin G, Wang W, Lin Y. Experimental study on scaling of RC beams under close-in blast loading. *Eng Fail Anal* 2013;33:497–504.
15. Yao SJ, Zhang D, Lu FY, Wang W, Chen XG. Damage features and dynamic response of RC beams under blast. *Eng Fail Anal* 2016;62:103–111.
16. Li J, Hao H. Influence of brittle shear damage on accuracy of the two-step method in prediction of structural response to blast loads. *Int J Impact Eng* 2013;54:217–231.
17. Shi Y, Hao H, Li ZX. Numerical derivation of pressure–impulse diagrams for prediction of RC column damage to blast loads. *Int J Impact Eng* 2008;35(11):1213–1227.
18. Kyei C, Braimah A. Effects of transverse reinforcement spacing on the response of reinforced concrete columns subjected to blast loading. *Eng Struct* 2017;142:148–164.
19. Bao X, Li B. Residual strength of blast damaged reinforced concrete columns. *Int J Impact Eng* 2010;37(3):295–308.
20. Astarlioglu S, Krauthammer T, Morency D, Tran TP. Behavior of reinforced concrete columns under combined effects of axial and blast-induced transverse loads. *Eng Struct* 2013;55:26–34.
21. Yan QS. Damage assessment of subway station columns subjected to blast loadings. *Int J Struct Stab Dy* 2018;18(03):1850034.
22. Wu KC, Li B, Tsai KC. Residual axial compression capacity of localized blast-damaged RC columns. *Int J Impact Eng* 2011;38(1):29–40.
23. Siba F. Near–field explosion effects on reinforced concrete columns: an experimental investigation. Master's Thesis, Department of Civil and Environmental Engineering, Carleton University, Ottawa, Canada, 2014.
24. Zhang C, Gholipour G, Mousavi AA. Blast loads induced responses of RC structural members: state-of-the-art review. *Compos Part B Eng* 2020;195:108066.

25. Fan W, Xu X, Zhang Z, Shao X. Performance and sensitivity analysis of UHPFRC-strengthened bridge columns subjected to vehicle collisions. *Eng Struct* 2018;173: 251–268.

26. Chen L, Xiao Y, El-Tawil S. Impact tests of model RC columns by an equivalent truck frame. *J Struct Eng* 2016;142(5):04016002.

27. Do TV, Pham TM, Hao H. Dynamic responses and failure modes of bridge columns under vehicle collision. *Eng Struct* 2018;156:243–259.

28. Gholipour G, Zhang C, Mousavi AA. Effects of axial load on nonlinear response of RC columns subjected to lateral impact load: ship-pier collision. *Eng Fail Anal* 2018;91:397–418.

29. Gholipour G, Zhang C, Li M. Effects of soil–pile interaction on the response of bridge pier to barge collision using energy distribution method. *Struct Infrastruct Eng* 2018;14(11):1520–1534.

30. Gholipour G, Zhang C, Mousavi AA. Analysis of girder bridge pier subjected to barge collision considering the superstructure interactions: the case study of a multiple-pier bridge system. *Struct Infrastruct Eng* 2018;15(3):392–412.

31. Gholipour G, Zhang C, Mousavi AA. Reliability analysis of girder bridge piers subjected to barge collisions. *Struct Infrastruct Eng* 2019;15(9):1200–1220.

32. Gholipour G, Zhang C, Mousavi AA. Nonlinear numerical analysis and progressive damage assessment of a cable-stayed bridge pier subjected to ship collision. *Mar Struct* 2020;69:102662.

33. Fan W, Guo W, Sun Y, Chen B, Shao X. Experimental and numerical investigations of a novel steel-UHPFRC composite fender for bridge protection in vessel collisions. *Ocean Eng* 2018;165:1–21.

34. Liu B, Fan W, Guo W, Chen B, Liu R. Experimental investigation and improved FE modeling of axially-loaded circular RC columns under lateral impact loading. *Eng Struct* 2017;152:619–642.

35. Cai J, Ye JB, Chen QJ, Liu X, Wang YQ. Dynamic behaviour of axially-loaded RC columns under horizontal impact loading. *Eng Struct* 2018;168:684–697.

36. Alam MI, Fawzia S, Zhao XL. Numerical investigation of CFRP strengthened full scale CFST columns subjected to vehicular impact. *Eng Struct* 2016;126:292–310.

37. Demartino C, Wu JG, Xiao Y. Response of shear-deficient reinforced circular RC columns under lateral impact loading. *Int J Impact Eng* 2017;109:196–213.

38. Adhikary SD, Li B, Fujikake K. Low velocity impact response of reinforced concrete beams: experimental and numerical investigation. *Int J Protect Struct* 2015;6(1):81–111.

39. Fujikake K, Li B, Soeun S. Impact response of reinforced concrete beam and its analytical evaluation. *J Struct Eng* 2009;135(8):938–950.

40. Pham TM, Hao H. Effect of the plastic hinge and boundary conditions on the impact behavior of reinforced concrete beams. *Int J Impact Eng* 2017;102:74–85.

41. Pham TM, Hao H Influence of global stiffness and equivalent model on prediction of impact response of RC beams. *Int J Impact Eng* 2018;113:88–97.

42. Cotsovos DM. A simplified approach for assessing the load-carrying capacity of reinforced concrete beams under concentrated load applied at high rates. *Int J Impact Eng* 2010;37(8):907–917.

43. Cotsovos DM, Stathopoulos ND, Zeris CA. Behavior of RC beams subjected to high rates of concentrated loading. *J Struct Eng* 2008;134(12):1839–1851.

44. Zhan T, Wang Z, Ning J. Failure behaviors of reinforced concrete beams subjected to high impact loading. *Eng Fail Anal* 2015;56:233–243.

45. Pham TM, Hao H. Plastic hinges and inertia forces in RC beams under impact loads. *Int J Impact Eng* 2017;103:1–11.

46. Zhao DB, Yi WJ, Kunnath SK. Shear mechanisms in reinforced concrete beams under impact loading. *J Struct Eng* 2017;143(9):04017089.

47. Ožbolt J, Sharma A. Numerical simulation of reinforced concrete beams with different shear reinforcements under dynamic impact loads. *Int J Impact Eng* 2011;38(12):940–950.
48. Saatci S, Vecchio FJ. Effects of shear mechanisms on impact behavior of reinforced concrete beams. *ACI Struct J* 2009;106(1):78.
49. Zhang C, Gholipour G, Mousavi AA. Nonlinear dynamic behavior of simply-supported RC beams subjected to combined impact-blast loading. *Eng Struct* 2019;181:124–142.
50. Gholipour G, Zhang C, Mousavi AA. Loading rate effects on the responses of simply supported RC beams subjected to the combination of impact and blast loads. *Eng Struct* 2019;201:109837.
51. Feyerabend M. Hard transverse impacts on steel beams and reinforced concrete beams. PhD Thesis, University of Karlsruhe (TH), Karlsruhe, Germany, 1988 (in German).
52. ACI 318-14. Building Code Requirements for Structural Concrete and Commentary, American Concrete Institute, Farmington Hills, MI, 2014.
53. Thilakarathna HMI, Thambiratnam DP, Dhanasekar M, Perera N. Numerical simulation of axially loaded concrete columns under transverse impact and vulnerability assessment. *Int J Impact Eng* 2010;37:1100–1112.
54. Shi Y, Stewart MG. Spatial reliability analysis of explosive blast load damage to reinforced concrete columns. *Struct Saf* 2015;53:13–25.
55. Mutalib AA, Hao H. Development of PI diagrams for FRP strengthened RC columns. *Int J Impact Eng* 2011;38(5):290–304.
56. ASCE. Blast protection of buildings. ASCE SEI 59-11, American Society of Civil Engineers Reston, VA, 2011.
57. Beshara FBA. Modeling of blast loading on aboveground structures–I. General phenomenology and external blast. *Comput Struct* 1994;51(5):597–606.
58. McCormick J, Nagae T, Ikenaga M, Zhang P, Katsuo M, Nakashima M. Investigation of the sliding behaviour between steel and mortar for seismic applications in structures. *Earthq Eng Struct D* 2009;38(12):1401–1419.
59. LSTC (Livermore Software Technology Corporation). LS–DYNA keyword user's manual ver. 971, Livermore, CA, 2016.
60. Murray YD, Abu-Odeh AY, Bligh RP. Evaluation of LS-DYNA concrete material model 159. Report No. FHWA-HRT-05-063, Federal Highway Administration, Washington, DC, 2007.
61. Murray YD. Users' manual for LS-DYNA concrete material model 159. Report No. FHWA-HRT-05-062. Federal Highway Administration, Washington, DC, 2007.
62. Randers-Pehrson G, Bannister KA. Airblast loading model for DYNA2D and DYNA3D. DTIC Document, 1997.

Chapter 9

Bridge pier subjected to vessel impact combined with blast loads

9.1 Introduction

Bridges as the strategic structures may be exposed to dynamic and impulsive loads such as impact and blast loads during their service life. Bridge failures have been reported in many cases all over the world due to the collision of vehicles or vessels and explosion loadings [1–3]. Harik et al. [1] classified the bridge failures that occurred in the United States during 1951–1988 according to various causes. It was reported that 42 of 79 (i.e., about 53%) bridges collapsed due to collisions in which 19 cases (24%) were caused by ships, 11 (14%) by trucks, and 6 (8%) by trains. Besides, in four cases (5%), the bridges collapsed due to the exploding or burning of fuel-tanker trucks. Based on a study done by Wardhana and Hadipriono [2], from 1989 to 2000 in the United States, 12% of the total bridge failures occurred due to lateral impact forces (arising from the collision of trucks, barges, ships, and trains), and 3% due to fire and explosions from 2000 to 2008. According to a study by Lee et al. [3] from 1980 to 2012 in the United States, 15.3% and 2.8% of failures occurred due to collisions, and fire and explosions, respectively.

Since the bridge piers play vital roles in the reliability and fail-safe of the whole bridge structure because of their low redundancy compared to bridge decks and superstructures, recognizing the failure behaviors and dynamic responses of bridge piers under extreme loads is very important to safely design and adopt efficient solutions. For the vessel collision studies, the impact of force-deformation relationships between low-velocity [4] and high-velocity [5] barge collisions with bridge piers were studied. Compared to the marginal effects of the pier size on the force-deformation results, Consolazio and Cowan [4] found that the pier shape and geometry have significant influences on the impact forces. As the high-energy barge collision studies with bridge piers, Kantrales et al. [5] concluded that the collision with flat-faced piers resulted in greater impact forces than those of round-faced piers.

Consolazio et al. [6] carried out a series of full-scale experimental tests of barge collisions on the old St. George Island Bridge to capture a total insight of the behaviors of an impacting vessel with the structure in real life. Afterward, some analytical methods were proposed by several previous research works [7–9] to achieve the impact responses of both the vessel and impacted structure in simplified manners. Consolazio and Cowan [7] proposed a coupled vessel impact analysis (CVIA) method in which a single degree of freedom (SDOF) vessel model collides with a multi degree of freedom (MDOF) pier model (i.e., medium-resolution technique). Thereafter, the proposed CVIA is used to analyze the dynamic responses of an equivalent one-pier,

DOI: 10.1201/9781003262343-9

two-span (OPTS) simplified bridge model by Consolazio and Davidson [8] in which the effects and characteristics of adjacent piers and spans were considered using a series of equivalent translational and rotational springs that are attached to a lumped mass of adjusted piers and spans. From an analytical study on ship-pier collisions done by Fan et al. [9], it was obtained that the strain rate effects of the steel materials used for the vessel bow had significant positive influences on the impact forces. However, the impact duration was negatively affected by the strain rate effects. The amplification effects of structural dynamic parameters of the superstructure such as inertia, on the impact responses of girder bridge piers, were numerically investigated by [10, 11]. Some of the previous works considered different approaches such as energy-based [12] and reliability [13] methods to analyze the responses of impacted structures. Gholipour et al. [12] revealed the significant effects of soil-pile interactions on the dynamic responses of girder bridge piers using an energy distribution method.

Figure 9.1 shows the typical failure modes of reinforced concrete (RC) bridge piers under different collision loads based on observed real collision events as reported by Buth et al. [13]. Different modes of shear failure, including a punching shear failure (or shear plug), a shear hinge (i.e., a shear failure mode that commonly occurs at the mid-height of the pier), and a diagonal shear failure originating from the impact point and extending to the column base, are illustrated in Figure 9.1a–c, respectively. According to the literature [13–18], shear failure modes commonly take place when bridge piers are subjected to high-velocity collision loads. Under such loading conditions, bridge columns suffer brittle failures that would increase the probability of the whole collapse of the bridge. Besides, RC bridge piers may experience combined flexural-shear failure mode and minor flexural damages as illustrated in Figure 9.1d and e, respectively, when they are exposed to collision loads with relatively lower impact velocities. Although the progression of combined flexural-shear damages may lead to the formation of a plastic hinge during the following stages of impact responses, more time would be provided until the whole collapse (i.e., a higher chance of survival) compared to shear failure modes.

The damaged states and failure behaviors of RC bridge piers with the nonlinearity under vessel collisions were numerically evaluated in several research works [17–20]. However, the dynamic strain rate effects of both concrete and steel materials on the dynamic responses were not taken into account in these studies. Gholipour et al. [17]

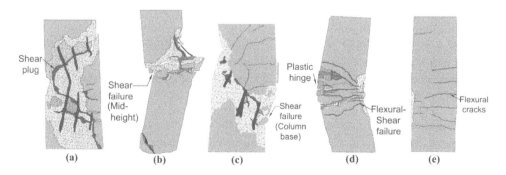

Figure 9.1 Typical failure modes of RC bridge piers under different collision loads: (a) punching shear, (b) shear hinge, (c) diagonal shear failure, (d) combined flexural-shear failure, and (e) flexural failure.

obtained the dynamic amplification and significant positive influences of the super-structure inertia on the peak impact forces. Moreover, the sensitivity thresholds of the lateral resistance of the columns to the axial load ratio were explored. It was found that the axial load positively affects the lateral resistance of RC columns when its ratio was between the ranges from 0.3 to 0.5. Recently, Wan et al. [19] experimentally studied the scaled ship collisions with RC piers with the variability of the impact velocity and the dynamic parameters related to the steel materials used in the ship bow. It was found that the nonlinearity of the concrete material used for the pier was substantially effective on the impact responses when under relatively high-energy collisions. The nonlinear dynamic responses and progressive damage process of a cable-stayed RC bridge pier were numerically studied by Gholipour et al. [20] in LS-DYNA. Furthermore, the concrete strain rate effects are analytically formulated by proposing a simplified two-degree-of-freedom system. The capability of various proposed damage indices was assessed to explore a more efficient approach in describing the damage states of the impacted pier.

The failure behaviors of bridge piers under close-in and far-field explosions have also been investigated experimentally [21] and numerically [22] in the literature. The typical failure modes of RC bridge piers under different blast loads are illustrated in Figure 9.2 based on large-scale experimental blast tests as documented in Williamson et al. [23]. Figure 9.2a–c shows three types of severe damage states of RC columns under close-in blast loads, including brittle shear failure mode at the column base, extensive spallation of concrete cover (medium spall) along a shear failure, and severe spallation leading to the breach of the cross-section, respectively. According to a classification on spall damage of RC structures under different blast loads presented by McVay [24] and reviewed by Zhang et al. [25], threshold spall represents a few cracks and exfoliation of concrete cover, medium spall denotes a damaged state from a shallow spall to a third of the column thickness, severe spall represents from over a third of the column thickness to almost breach, and the beach denotes the formation of light through to large hole in the cross-section of columns. Besides, RC columns may endure two typical failure modes, including local spall and exfoliation of concrete cover (threshold spall), and flexural cracks as shown in Figure 9.2d and e, respectively, under far-field detonations with relatively large-scaled distances.

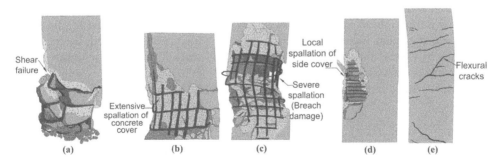

Figure 9.2 Typical failure modes of RC bridge piers under different blast loads: (a) brittle shear failure, (b) medium spall damage of the concrete cover, (c) severe spall damage (a breach in the cross-section), (d) local spall damage of cover concrete (threshold spall), and (e) flexural failure.

The influences of various structural parameters, including the cross-sectional shape, height, reinforcement ratio, type of transverse reinforcement, and loading parameters such as the scaled standoff, location of the explosion the blast responses of RC piers, were assessed using different parametric studies by [21, 22]. Williamson et al. [21] revealed that although the square columns have larger areas exposed to the overpressure of close-in blast loads than circular columns, greater resistances against shear failure were provided by square-shaped piers than round-shaped columns. Also, more severe spallation was observed in RC columns with increasing the height of blast loading location due to expanding the area exposed to striking blast waves [22]. The dynamic responses and failure behaviors of various types of RC structures under impact and blast loads have been comprehensively reviewed by Zhang et al. [18, 25].

The responses of simply-supported RC beams under the combination of close-in blast loads with the middle-rate impacts were studied [26] by varying several structural-related and loading-related parameters. It was found that the application of the impact load before the detonation resulted in more severe spall damages in the beam cross-section. Afterward, by evaluating the impact loading rate effects on the blast responses of simply-supported RC beams by Gholipour et al. [27], it was concluded that the damage index based on the shear forces at the supports was more sensitive to the increase of the impact velocity when the sequent detonation was initiated at the time of the initial peak impact force. Moreover, the vulnerability of an axially loaded RC column to several loading parameters was assessed by Gholipour et al. [28] subjected to synergetic effects of impact and blast loads. It was revealed that the sensitivity level of the column failure to impact velocity and axial load ratio parameters decreased under combined loadings compared to those under sole impact and blast loads.

Despite the aforementioned studies, an investigation is made to detect the responses and behaviors of RC bridge piers under the combination of impact and blast loadings to rectify the shortcomings of previous research works. The combined actions of impact and blast loadings may be applied to bridges during accidental or intentional (i.e., terrorist attacks) collision of vessels, vehicles, etc., which can carry explosive materials. Although the occurrence of such loading combinations has not been reported yet, it is worthy to investigate the loading mechanisms and the structural behaviors to capture some insights into the protective design of bridges against such combined loadings. Based on a review study on the security recommendations for bridges and tunnels [29], it was declared that although blast loads arising from car bombs have the highest probability to happen, the application of blast loadings by vessels on bridge piers located at major waterways is also very likelihood. Besides, the explosions by vessels may provide more intensive loading conditions due to the larger cargo-carrying capacity of vessels than vehicles.

Therefore, this chapter aims to ideally exercise the basic theory and the possible scenarios of the combination of impact and blast loadings on a typical RC girder bridge pier. To this end, the dynamic responses, internal forces, and failure behaviors of a typical RC girder bridge pier with two columns are numerically investigated with high- and moderate-resolutions in LS-DYNA [30] under vessel collisions combined with blast loadings. Moreover, the vulnerability of the bridge pier is quantitatively assessed varying in terms of several loading-related parameters, including the vessel type, impact weight, impact velocity, the location of loadings, and the time lag between the initiations of the applied loads by proposing three damage indices based on the residual axial load, shear, and flexural moment capacities of the pier column. The residual capacities of the pier are obtained through a series of multistep loading procedures applied to the simplified finite element (FE)-based model of the pier with a moderate resolution. In organizing the methodology for the loading scenarios in this study, the collisions of only two typical vessels, including 2,000-DWT (deadweight tonnage) Jumbo Hopper barge and 5,000-DWT container ship that commonly navigate in inland waterways of the United States and China with moderate design velocities (4 m/s for ship and 1.65 m/s for barge), are considered for the high-resolution FE simulations that carry 200-kg and 500-kg TNT as the explosive materials on their deck level of rear portions, respectively.

9.2 FE modeling of vessels and bridge pier

In this section, the FE models of a typical bridge pier and striking vessels are described in detail. Then, the calibration of FE models is carried out by comparing the simulation results from the present study and those from the previous experimental and numerical studies existing in the literature.

9.2.1 FE modeling of vessels

Two typical vessels varying in terms of the bow configuration, DWT, and velocity are considered in this study. One of these vessels is a 5,000-DWT container ship with a total length of 121 m and a scantling draught of 5.0 m as shown in Figure 9.1a. The details of FE models of the ship bow, including the internal stiffeners and the outer plates, are illustrated in Figure 9.3a based on the design characteristics available in the literature [31, 32]. Another vessel is a 2,000-DWT jumbo Hopper barge with a total length of 59.4 m, a depth of 4.26 m in the bow portion, and a loaded draft of 2.56 m as shown in Figure 9.1b according to design specifications available in AASHTO [33]. Based on the design drawings presented by Kantrales et al. [5], the details of FE models of the internal trusses and frames used in the bow portion of the barge are also illustrated in Figure 9.3b. All internal structures and outer plates used in the bow portions of the vessels are modeled using Hughes-Liu shell elements with five integration points through thickness [34]. The steel plates modeled using shell elements in the bow portions have thicknesses ranging from 6 to 18 mm with mesh sizes between 50 and 150 mm to accurately predict the impact forces, localized buckling, and crushing behaviors during head-on collisions. To avoid any penetration between the structural components and to take into account the possible secondary contacts during the collision, a contact algorithm named AUTOMATIC_SINGLE_SURFACE is used in LS-DYNA between the internal and outer steel plates in the bow portions of the vessels with static and dynamic friction coefficients of 0.21 as recommended by previous studies [19, 35]. Moreover, to simplify and reduce the computation time of the FE simulations, the non-bow (stern) portions of the vessels are modeled using 8-node solid elements with coarse meshes by adopting a rigid material model named MAT_RIGID (MAT_020) [34] since it is not expected to observe significant deformations in these portions. The cargo weights of the vessels are adjusted using a mass distribution method in the stern portions. In this method, the cargo weights of the vessels (herein, vessels with fully loaded cargo capacities) are divided into the total volume of the stern portions and applied to the mass density parameter of the material model defined in

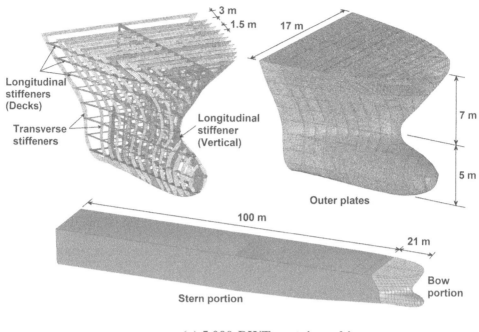

(a) 5,000-DWT container ship

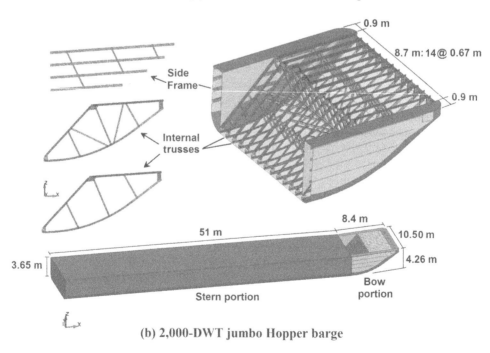

(b) 2,000-DWT jumbo Hopper barge

Figure 9.3 Detailed FE models of the vessels used in the present study. (a) 5000-DWT container ship. (b) 2000-DWT jumbo Hopper barge.

Table 9.1 Material parameters used for the vessels

Model	Parameter	Magnitude
Steel plates used for bow portions of the vessels	Mass density	7,850 kg/m³
	Young's modulus	206 GPa
	Poisson's ratio	0.3
	Yield stress	248 MPa
	Tangent modulus	885 MPa
	Failure strain	0.25
	Strain rate parameter C_{strain}	40.4
	Strain rate parameter P_{strain}	5.0
Stern portion of 2,000-DWT barge	Mass density	880 kg/m³
Stern portion of 5,000-DWT ship	Mass density	192.3 kg/m³

LS-DYNA for the rigid elements of the stern portions as given in Table 9.1. It is worth noting that a 10% added mass is considered for the stern portions of the vessels to take into account the hydrodynamic effects during FE simulations.

An elastic-plastic material model named MAT_PLASTIC_KINEMATIC (MAT_003) in LS-DYNA [34] represents the isotropic and kinematic hardening behaviors considering the strain rate effects of the steel material is used for the steel plates of the vessels' bow portions. The stress-strain characteristics of a structural steel *A*36 with a failure strain of 0.25 are employed based on the standard specification for carbon structural steel published by American Society for Testing and Materials (ASTM) [36] to input in the material model of the steel plates used in the bow portions as given in Table 9.1. Also, Getter et al. [37] recommended adopting the average of the failure strain range from 0.2 to 0.4 for mild steel *A*36 since no significant sensitivity was observed to the use of failure strain over this range. In this study, the strain rate effects of the steel materials are modeled using Cowper-Symonds coefficients of $C_{strain} = 40.4$ s⁻¹ and $P_{strain} = 5.0$ in the properties of MAT_003 in LS-DYNA. These values have been widely used by many research works in the literature [5, 20, 32, 38, 39] investigating the collision behaviors of vessels modeled using the dynamic characteristics of the materials of mild steel (especially A36 steel). According to Symonds [40], the basis for utilizing these values for Cowper-Symonds coefficients is the dynamic stresses of various steel materials resulted from extensive experimental tests conducted by Manjoine [41]. The strain rate parameters can be calculated using Cowper and Symonds's formulas as follows [39]:

$$\frac{\sigma'_d}{\sigma_s} = 1 + \left(\frac{\dot{\varepsilon}}{C_{strain}} \right)^{1/P_{strain}} \tag{9.1}$$

where σ'_d is the dynamic flow stress at uniaxial plastic strain rate $\dot{\varepsilon}$, and σ_s is the static flow stress. C_{strain} and P_{strain} are the constants of steel material.

9.2.2 FE modeling of bridge

A typical RC girder bridge pier with two hollow-section columns with the height of 38.8 m, a piles' cap with a thickness of 3.5 m, 14 concrete piles with 2.0 m in diameter and 98.0 m in height is considered to study under vessel collisions combined with blast

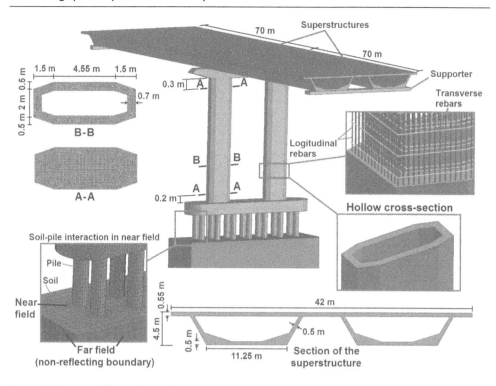

Figure 9.4 Detailed FE models of the pier components, superstructure, and the substructure of the bridge.

loads as shown in Figure 9.4. The detailed dimensions and elevations of the bridge pier and superstructure are illustrated in Figures 9.4 and 9.5a. All the concrete material of the pier, superstructures are modeled using constant stress solid elements in LS-DYNA. However, the reinforcements are modeled using 3-node beam elements in LS-DYNA as shown in Figure 9.2a considering minimum design requirements defined by design codes such as ACI-318 [42] in which the longitudinal rebars have ratios between 2% and 4%. Also, a coupling algorithm named CONSTRAINED_LAGRANGE_IN_SOLID is utilized to model the interaction between the concrete and the reinforcement bars. Figure 9.3b demonstrates the reinforcement configurations used in different parts of the bridge.

From the previous studies [8], it was found that the consideration of a part of the bridge with one pier and two adjacent spans can accurately capture the dynamic responses of the multi-span bridge. Therefore, only two spans of twin-cell box-girder superstructures with lengths of 70.0 m and weight of 25.5 tons per unit length placed on top of the pier's cap beam are considered as shown in Figure 9.4. Accordingly, the weight of the superstructure (with a partial length of 70 m) is 1,785 tons (= 70 × 25.5) which is equal to 10% of the axial load-carrying capacity of the pier. Other ends of the superstructures are freely sitting on the rigid and fixed supporters. It should be noted that an open joint was considered between the adjoining decks to avoid the significant contact forces developed during the closure of the deck joint due to the substantial displacements of the pier. From the previous studies [15, 43], it was concluded that utilizing bearing pads between the pier and superstructures had marginal influences on the

Table 9.2 Contact algorithms used in the LS-DYNA

Contact algorithm	Contact components	Contact materials	Static and dynamic friction coefficients	References
AUTOMATIC_SURFACE_ TO_SURFACE	Ship bow and pier	Steel to concrete	0.3	Consolazio and Cowan [4], El-Tawil et al. [43], Sha and Hao [44]
	Superstructure and cap beam of the pier	Concrete to concrete	0.6	Do et al. [15], ACI [42]
AUTOMATIC_SINGLE_ SURFACE	Inner and outer plates of vessels' bow	Steel to steel	0.21	Wan et al. [19], Yuan [35]
AUTOMATIC_NODES_ TO_SURFACE	Soil and piles	Clay/Sand to concrete	0.26	Brumund and Leonards [45]

impact loads and the responses of bridge piers. Therefore, a contact algorithm named AUTOMATIC_SURFACE_TO_SURFACE is used to model the stiffness between the concrete to the concrete surface of the superstructures and the cap beam of the pier with a friction coefficient of 0.6 as recommended in the literature [15, 42] instead of modeling the bearing pads to reduce the computational time of FE simulations. Also, like the previous studies [4, 43, 44], a similar contact algorithm with static and dynamic friction coefficients of 0.3 was used between the steel and concrete surface of the ship bow and the bridge pier. Table 9.2 summarizes the contact algorithms utilized between various components of the bridge with assumed friction coefficients based on the previous studies.

Moreover, it is assumed that the piles are surrounded by the soil with a mudline elevation of 8.0 m beneath the lower surface of the piles' cap as shown in Figure 9.5a. Also, the interaction between the concrete piles and the soil layer in the near-field is modeled using a contact algorithm named AUTOMATIC_NODES_TO_SURFACE with static and dynamic friction coefficients of 0.26 in LS-DYNA as recommended by Brumund and Leonards [45] between the contact surface of silty (i.e., Clay/Sand) soil and concrete materials as presented in Table 9.2. To prohibit the reflection of stress waves from the boundaries, the soil behaviors in the far-field are modeled using non-reflecting boundary conditions in LS-DYNA similar to the procedure employed in the literature [32]. In LS-DYNA, the infinite domain of soil layers is modeled using an artificial boundary condition named BOUNDARY_NON_REFLECTING. In modeling this boundary condition, to prevent any exaggerated deformations occurred in the exterior boundary, the FE model of surrounding soil contains two zones including (i) the near-field soil modeled using nonlinear material, and (ii) the far-field soil modeled using a linear elastic material. The nonlinear behavior of the soil surrounding the piles in the near-field is modeled using an elastic-plastic model MAT_MOHR_COULUMB (MAT_173) assuming the reasonable characteristics as given in the literature [12, 32]. The material properties assumed for the various components of the bridge are summarized in Table 9.3.

The nonlinear damage behaviors of the concrete material are modeled using a continuous surface cap model (CSCM) named MAT_CSCM_CONCRETE (MAT_159)

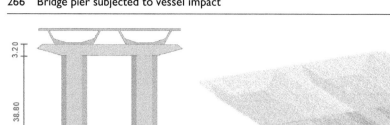

Figure 9.5 (a) Elevations and dimensions of the pier (Unit: m) and (b) reinforcement configurations in the pier and superstructure.

[46] that captures some key mechanical characteristics of the concrete such as the pressure and strain hardening. This model is easily available to use in LS-DYNA by inputting several parameters of concrete materials, including unconfined compression strength of concrete (f_c'), the element erosion (i.e., the elimination of FE elements) parameter (*ERODE*), and the maximum aggregate size (*Dagg*), which profoundly

Table 9.3 Material properties used for the FE model of the bridge

Member	Material Model	Parameters	Values
Concrete of pier and superstructures	MAT_CSCM_CONCRETE	Mass density	2,400 kg/m³
		Unconfined compression strength	50.0 MPa
		Elements erosion (*ERODE*)	1.08
		Maximum aggregate size (*Dagg*)	19.0 mm
		Rate effects	Turn on
Steel reinforcements	MAT_PIECEWISE_LINEAR_PLASTICITY	Mass density	7,865 kg/m³
		Young's modulus	210 GPa
		Poisson's ratio	0.27
		Yield stress	420 MPa
		Failure strain	0.35
		Strain rate parameter C_{strain}	40.4
		Strain rate parameter P_{strain}	5.0
Soil in the near field	MAT_MOHR_COULUMB	Mass density	1,890 kg/m³
		Shear modulus	6.32 MPa
		Poisson's ratio	0.25
		Angle of friction	32°
		Cohesion value	11.0 kPa

govern the fracture energy [46]. The damage of the concrete is modeled by using the erosion parameter *ERODE* defined in this model based on the maximum principal strain when it is set more than 1.0 in LS-DYNA. According to the previous studies [17, 26, 47], the selection of the value of *ERODE* parameter between 1.05 and 1.1 was recommended to capture accurate and reasonable failure behaviors of the concrete materials under extreme loads in comparison with those from the experiments. These ranges represent that the damage occurs when the strain of the concrete elements reaches between the ranges of 5% and 10% of the maximum principal strain, respectively. Based on the FE simulations of dropping mass impact and blast loadings on RC beams and columns as given in the previous works [26–28], selecting a value of 1.08 for *ERODE* parameter resulted in acceptable results and failure behaviors compared to those from experimental tests. More detailed validation studies on the failure behaviors of a concrete material model under impact and blast loads by adopting a value of 1.08 for the erosion parameter *ERODE* are presented in Section 9.2.3 of this chapter. Besides, the material behaviors of structural steel of *A*615-Grade 60 [48] are used for the steel reinforcements of the bridge pier with a failure strain of 0.35 according to the recommendations made by ACI-318 [42]. In addition, an elastic-plastic material model named MAT_PIECEWISE_LINEAR_PLASTICITY (MAT_024) is employed to model the behaviors of steel rebars considering the strain rate effects. The strengths of RC structures under impulsive loads may increase due to the strain rate effects of the concrete and steel materials. These enhancements in the relationship with the strain rates of the concrete and steel materials have been discussed in the previous study [26].

9.2.3 Validation of FE models

According to previous studies on ship collisions [31, 49], to accurately model the crushing behaviors of steel plates, the ratio of element length to thickness should be adopted between the ranges 8 and 10. To meet this meshing criterion in this study, the steel plates modeled using shell elements in the bow portions have thicknesses ranging from 6 to 18 mm with mesh sizes between 50 and 150 mm. Therefore, the ratio of the element to the thickness of the steel plates employed for the vessels' bow in this study is 8.33 that meets the meshing criterion proposed by Alsos and Amdahl [49]. Therefore, the internal structures and outer plates of the ship bow have 39,292 and 48,816 shell elements, respectively.

To validate the FE model of the ship bow, the force-deformation results from a series of quasi-static analyses of the ship bow collision with a rigid wall with a mesh size of 200 mm as shown in Figure 9.6a are compared with those computed by the current design codes and previous studies. Since there exists very limited experimental test data on the ship bow collisions in the literature, the force-deformation results from the ship impact simulations are compared with the empirical formulas given by AASHTO [33], Pedersen et al. [50], and China's bridge design codes (CMR [China Ministry of Railways]) [51] as shown in Figure 9.6b. A summary of key information from China's bridge design codes (CMR) [51] in estimating the equivalent collision loading is provided in Appendix A. Besides, the FE model of the ship bow developed by Fan and Yuan [31], which was verified using the aforementioned empirical formulas and met the meshing criterion proposed by Alsos and Amdahl [49] based on FE benchmark models, is additionally utilized to verify the FE model developed in the present study. It should be noted that the characteristics of the ship model studied in the present

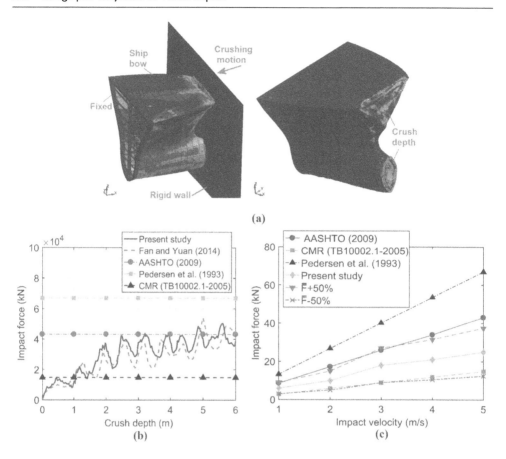

Figure 9.6 Validation of the FE model of the ship bow: (a) ship-rigid wall collision with an impact velocity of 5.0 m/s, (b) comparing the force-deformation results for an impact velocity of 5.0 m/s, and (c) comparing the trend of peak impact force versus impact velocity.

work such as the ship dimensions, the structural configuration of the bow portion, the mesh sizes adopted for the ship bow and the rigid wall, and the collision velocity (i.e., 5.0 m/s) are the same as those considered in the study done by Fan and Yuan [31]. It is found that the force-deformation results from the FE simulations are agreed well with those presented by Fan and Yuan [31] (see Figure 9.6b). Compared to the empirical formulas proposed by China's bridge design codes and Pedersen et al. [50], which capture underestimate and overestimate (i.e., conservative) results, respectively, AASHTO formulas can more accurately predict the force-deformation of the ship collision. As recommended by Fan and Yuan [31], the mean impact forces obtained from the FE simulations and calculated by Equation (9.2) are compared with those from the empirical formulas in the design codes as presented in Figure 9.6c and Table 9.4. A ±50% scatter in the mean impact forces (i.e., $\bar{F}(a)$ ±50%) from the FE simulations is considered similar to the Woisin's ship collision study given by AASHTO [33]. By comparing the results in Figure 9.6c, it is observed that the trend of results from the FE simulations by the present study reasonably agreed well with those from empirical formulas. As such, the trends of results calculated by AASHTO [33] and China's code [51] are realized as the lower and upper bounds of the mean impact force results that

Table 9.4 Comparisons between the mean impact forces from the FE simulations and empirical formulas

Impact velocity (m/s)	AASHTO [33] (kN)	CMR [51] (kN)	Pedersen et al. [50] (kN)	$\bar{F}(a)$ (Present study) (kN)	$\bar{F}(a)$+50% (kN)	$\bar{F}(a)$–50% (kN)
1.0	8.63	2.71	13.29	6.20	9.30	3.10
2.0	17.25	5.94	26.78	10.31	15.47	5.16
3.0	25.88	8.91	40.18	18.42	27.63	9.21
4.0	33.94	11.89	53.57	21.11	31.67	10.56
5.0	43.13	14.86	66.96	25.23	37.85	12.62

are very close to ±50% scatter in the mean impact forces from the FE simulations by the present study.

$$\bar{F}(a) = \frac{1}{a} E(a) = \frac{1}{a} \int_0^a F(a)\, da \qquad (9.2)$$

where $\bar{F}(a)$ is the mean of impact force $F(a)$, a denotes the crush depth, and $E(a)$ presents the crush energy of the ship bow.

Similar to the FE modeling procedure employed for the ship bow, thicknesses range from 6 to 18 mm with the mesh sizes between 50 and 150 mm used for all the steel plates of the barge bow portion that meets the meshing criterion proposed by Alsos and Amdahl [49] based on FE benchmark models in which the length-to-thickness ratio of the shell elements should be between 8 and 10. Accordingly, the internal trusses and outer plates of the barge bow have 43,185 and 40,266 shell elements, respectively. To verify the FE models of the barge bow portion, two quasi-static collision tests of the reduced-scale (0.4-scale) barge bow were simulated in LS-DYNA with the rigid blocks with flat and round noses same as those adopted in the experimental tests conducted by Kantrales et al. [5] as shown in Figure 9.7a and b. Since this chapter investigates the barge impact scenarios on the round-shaped piles' cap and the flat-shaped pier columns, the reliability of the barge bow FE model is examined subjected to both round- and flat-nose rigid blocks. The flat- and round-nose rigid blocks with the weights of 4,182 and 4,400 kg and the initial velocities of 9.16 and 9.34 m/s collide with the fixed barge bow that represents the collision scenarios named FLT1 and RND1 conducted

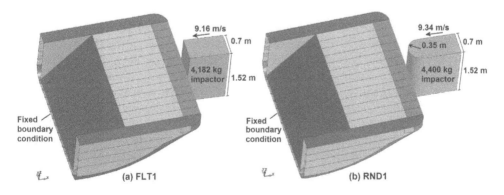

Figure 9.7 FE collision simulations of different rigid impactors with a 0.4-scale barge bow.

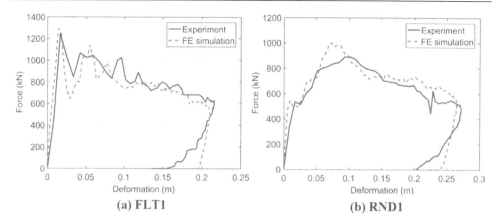

(a) FLT1   (b) RND1

Figure 9.8 Comparing the force-deformation results between those from FE simulations and experimental tests conducted by Kantrales et al. [5].

by Kantrales et al. [5], respectively. The dimensions of the rigid blocks against 0.4-scale barge bows are illustrated in Figure 9.7a and b. The impact force-deformation results and crushing behaviors of the barge bow from the FE simulations of the present study and those from the experimental tests done by Kantrales et al. [5] are compared in Figures 9.8 and 9.9, respectively. The values of the peak impact forces and the maximum deformations of barge bow are depicted in Table 9.5 by presenting their percentage of differences (i.e., Error) between those from FE simulations and the experiment [5]. It is found that the results are very close since their percentage of differences less than 5% for FLT1. For RND1, although the difference of the peak forces is higher than 10%, the difference between the maximum deformations is very low and less than 5%. Also, the trends of force-deformation curves from FE simulations are very similar to those from the experiments. Therefore, good agreements between the FE simulation and the experimental results are obtained when a mesh size of 100 mm for the solid elements of the impactor and the mesh sizes from 50 to 150 mm for the shell elements of the barge bow are selected.

The reliability of the FE model of the bridge pier is examined by investigating the sensitivity of the impact force results to the mesh size of the pier and also studying the validity of the concrete material adopted in modeling the failure behaviors of the concrete pier. Two validation studies are separately carried out to evaluate the capability of the nonlinear concrete material model (i.e., MAT_CSCM_CONCRETE) as described in Section 9.2.1 in modeling the failure behaviors of the concrete under impact and blast loads. Because there exists no available experimental test in the literature to investigate concrete structures subjected to synergetic effects of impact and blast loads, simultaneously. The first study includes the investigation of the impact responses and damage states of a simply-supported RC beam previously studied by Fujikake et al. [52] with a cross-sectional dimension of 0.25×0.15 m in depth and width, respectively, and a span length of 1.7 m subjected to a 400-kg dropping mass with different impact velocities as shown in Figure 9.10a. A schematic of this beam, including its dimensions and cross-sectional details, is illustrated on the left side of Figure 9.10a. Also, the corresponding FE model of the beam developed in LS-DYNA by the present study with a mesh size of 10 mm and a value of 1.08 for the erosion

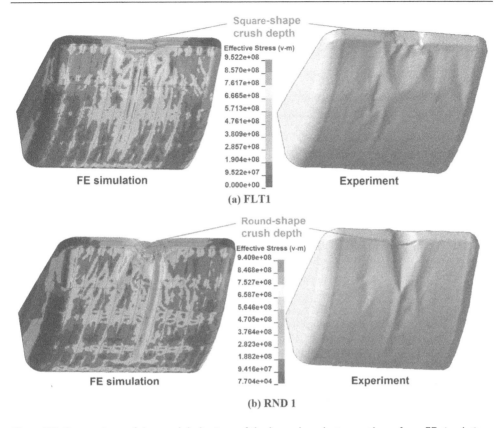

Figure 9.9 Comparison of the crush behaviors of the barge bow between those from FE simulations and experimental tests conducted by Kantrales et al. [5] using the fringe levels of effective (Von-Mises) stress. (a) FLT1 (b) RND1.

parameter *ERODE* in the concrete material model (MAT_CSCM_CONCRETE) is shown on the right side of Figure 9.10a. It should be noted that the material models used for the concrete and the reinforcements of the beam are similar to those adopted for the bridge pier as described in Section 9.2.2. More detailed information about the FE modeling and material properties of the beam considering their strain rate effects can be found in the previous work.

The crack patterns and damage states of the beam resulted from the experimental tests done by Fujikake et al. [52] under different impact loadings with velocities of 2.43 m/s (with a corresponding impact height of 0.3 m) and 6.86 m/s (with a corresponding

Table 9.5 Comparisons between the FE simulation results and those from experimental tests done by Kantrales et al. [5]

	FLT1			RND1		
		Experiment			Experiment	
Parameter	FE simulation	[10]	Error (%)	FE simulation	[10]	Error (%)
Peak force (kN)	1,295	1,243	4.2	1,000	896	11.6
Maximum deformation (m)	0.21	0.22	4.5	0.26	0.27	3.7

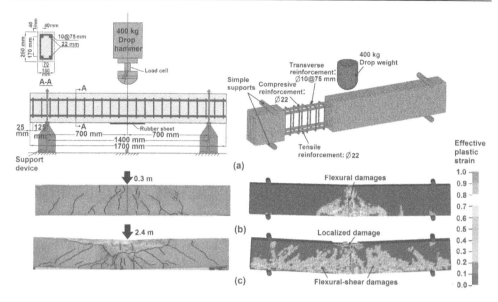

Figure 9.10 (a) A schematic of the RC beam studied by Fujikake et al. [52] (left side) and corresponding FE model of the beam developed in LS-DYNA by the present study (right side), (b) validation of the crack patterns of the beam under impact loading with a velocity of 2.43 m/s, (c) validation of the failure behavior of the beam under impact loading with a velocity of 6.86 m/s.

impact height of 2.4 m) are illustrated on the left sides of Figure 9.10b and c. Besides, the corresponding FE simulation results of the failure behaviors of the beam are presented on the right sides of Figure 9.10b and c. The damage criteria for the concrete material MAT_CSCM_CONCRETE are based on the maximum principal strain of the concrete that is controlled by using *ERODE* parameter. The fringe levels presented for the FE model of concrete denote the levels of the effective plastic strain of the concrete that is directly related to the maximum principal strain and *ERODE* parameter. The scalar ranges of this fringe level are between 0 and 1.0. In this study, the value of 1.08 is selected for the *ERODE* parameter that means that the concrete FEs will be eliminated from the analysis when the effective plastic strain of the concrete exceeds 8% of the maximum principal strain of the concrete. Therefore, the scalar value of 1.0 (the range is shown with red color) for the fringe level of the effective plastic strain of the concrete represents 8% of the maximum principal strain of the concrete material. Totally, two major damage levels are notable from the validation studies. One of these levels denotes the spallation of concrete that is modeled through the elimination of concrete elements from the FE models. For this damage level, the scalar value of the fringe level (i.e., the effective plastic strain) exceeds 1.0. Another damage level represents the major and minor cracks (with flexural or shear modes) in the concrete. The fringe levels of these crack patterns in the FE models (where the concrete elements have the potential to be eliminated from the analysis) have scalar values between 0.7 and 1.0. By comparing the damage levels of the concrete between the FE simulation and experimental results in Figure 9.10b and c, it is seen that the FE simulations are efficiently able to model the different damage levels of the RC beam from overall failure modes, including flexural cracks and flexural-shear damages to localized failures,

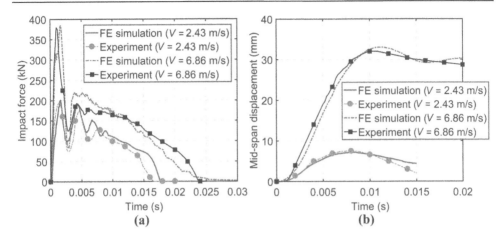

Figure 9.11 Comparison of the impact responses of the beam between those resulted from the FE simulations and the experiments: (a) impact forces and (b) mid-span displacements.

and concrete spallation under different-rate impact loadings by selecting a mesh size of 10 mm and the value of 1.08 for the erosion parameter (i.e., *ERODE*) of the concrete material model.

Furthermore, the impact responses of the RC beam resulted from FE simulations under different-rate impact loadings are compared with those from the experiments in Figure 9.11a and b and Table 9.6. It is seen that the initial peak of impact forces and maximum mid-span displacements resulted from the FE simulations are very close to those from experiments with errors less than 2% and 3%, respectively. Although the impact loading durations with velocities of 2.43 m/s and 6.86 m/s from the FE simulations have errors of 7.4% and 11.3% compared to those from the experiments, respectively, the results are generally in good agreements.

To validate the failure behaviors of the concrete material model with a value of 1.08 adopted for the erosion parameter *ERODE* under blast loads, the FE model of a fixed-base RC column with a square cross-section with 0.3 × 0.3 m dimensions and a height of 3.2 m studied by Siba [53] is developed in this study with material models similar to those adopted for the bridge pier as described in Section 9.2.2 and with a mesh size of 10 mm for the concrete and reinforcements as illustrated in Figure 9.12a. Also, the column was reinforced using four 25-M longitudinal rebars with a diameter of 19.5 mm and 10-M transverse reinforcements with a diameter of 11.3 mm. The RC column is subjected to

Table 9.6 Comparison of the impact responses of the RC beam obtained from the FE simulations with those the experiments under different-rate impact loads

Parameter	Impact with a velocity of 2.43 m/s			Impact with a velocity of 6.86 m/s		
	FE simulation	Experiment [52]	Error (%)	FE simulation	Experiment [52]	Error (%)
Initial peak force (kN)	201	197	2.0	380	376	1.1
Impact duration (ms)	17.5	16.2	7.4	27.3	24.2	11.3
Maximum displacement at mid-span (mm)	7.2	7.4	2.7	32.4	33.1	2.1

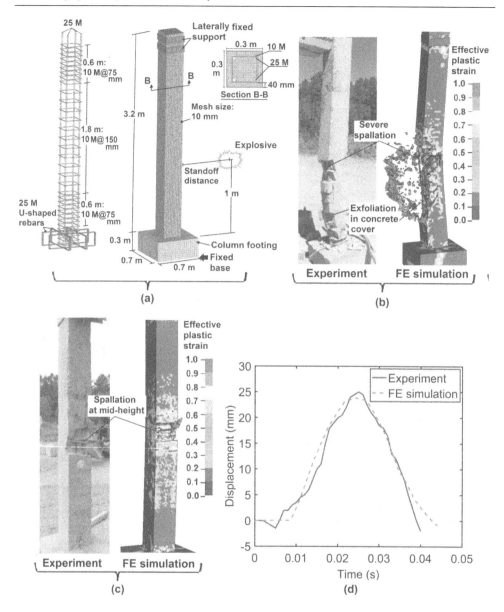

Figure 9.12 (a) FE model of the RC column studied by Siba [53] and developed in the present study, (b) validation of the failure behaviors of the column under blast loadings of *SEIS*-4, (c) *SEIS*-8, and (d) comparing the displacement of the column at the height of 2.0 m resulted from the FE simulation with that form experiment when subjected to blast loading of *SEIS*-8.

different blast loadings with the scaled distances of 0.34 m/kg$^{1/3}$ (*SEIS*-4) and 0.58 m/kg$^{1/3}$ (*SEIS*-8) varying in the explosive standoff distance (R) when the explosive charge is located at the height of 1.0 m from the column base as shown in Figure 9.12a. More detailed information about the structural characteristics and the FE modeling of the column and blast loading scenarios can be found in the literature [28, 53].

In Figure 9.12b and c, it is seen that the FE simulations can efficiently model the failure behaviors of the RC column, including the exfoliation in the concrete cover and concrete spallation. Moreover, Figure 9.12d demonstrates that the peak value and the amplitude of the column displacement at the height of 2.0 m from the column base from the FE simulation under *SEIS*-8 are very close to those from the experiment with errors less than 5%, 8%, respectively. Therefore, the nonlinear material models adopted in this study are efficiently able to model the dynamic responses and failure behaviors of RC structures under blast loads.

To study the sensitivity of the impact results to the mesh size of the pier, three collision scenarios of a 2,000-DWT barge with three pier columns with different mesh sizes of 50, 100, 200 mm were carried out with an impact velocity of 1.65 m/s and their impact forces are compared in Figure 9.13a. It is worth noting that the behaviors of the material of the barge and the pier concrete are modeled using the properties described in Sections 9.2.1 and 9.2.2. In Figure 9.13a, it is seen that a 200-mm mesh obtains an underestimated impact force compared to those from 50- and 100-mm meshes. That is, the results are not converged for a 200-mm mesh since the differences between results from 200-mm mesh and those from 50- and 100-mm are rather high. However, the results from 50- and 100-mm meshes are very close due to the trend of convergence. Therefore, to reduce the computational time of the simulations, a 100-mm mesh was selected for meshing the pier concrete rather than a 50-mm mesh. Besides, to improve the computational efficiency of the FE simulations, the superstructure and the soil layers are modeled using coarse meshes. Moreover, an hourglass control with a coefficient of 0.05 was employed in LS-DYNA to prevent excessive deformations (which causes uncontrolled energy absorption) in the mesh elements and to provide the energy balance between the vessel and the pier components in the barge-pier collision system. As shown in Figure 9.13b, the hourglass energy is about 5% of the pier internal energy that indicates the robustness of the FE modeling of the pier with a mesh size of 100 mm.

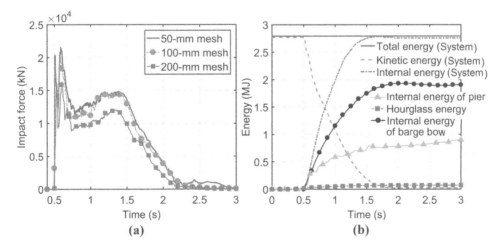

Figure 9.13 Validation of the FE model of the pier: (a) mesh convergence tests and (b) energy distributions in the barge-pier collision system.

9.3 Methodology and limitations

The combined loading scenarios studied using high-resolution FE-based simulations in this chapter are designed based on the variability of the vessel collision and explosion parameters in a reasonable manner considering those limitations existing in the current design codes. All the loading scenarios applied to the high-resolution FE model of the bridge pier are summarized in Table 9.7.

Due to the existing limitations in terms of software facilities and the complexity of FE simulations, only a limited number of FE simulations are carried out using a high-resolution approach for the bridge system as introduced in Section 9.2. This high-resolution FE-based modeling of combined loading scenarios includes some limitations listed as follows:

1. Tow typical vessels, including a 5,000-DWT container ship and a 2,000-DWT jumbo hopper barge navigating in inland waterways of the United States and China, are considered the impacting vessels with their design velocities (i.e., 4.0 m/s (≈2.06 knots) for the ship, and 1.65 m/s (≈3.21 knots) for the barge).

2. Compared to the impact elevations of container ships that are commonly around the piles' cap levels in the real world, the probability of barge collision on the pier column is considered in the implementation of high-resolution FE simulations that is more likely to happen in the real world than ship collisions due to their lower loaded drafts than those of ships. Hence, the water level is assumed to be 2.1 m above the column base (i.e., the top surface of the piles' cap) for the collision of a 5,000-DWT ship with a scantling draught of 5 m on the pier cap level. Besides, for the collision of a 2,000-DWT barge with a loaded draft of 2.65 m when it is applied to the piles' cap, and to the pier column with heights 4.5 and 7.4 m, the water levels are assumed to be 3.5 m below the column base (i.e., at the bottom surface of the piles' cap), and 2.89 and 5.79 m above the column base, respectively.

3. To study the effects of impact velocity using high-resolution FE simulations, the failure behaviors of bridge pier under a design velocity of 1.65 m/s for barge collisions (as recommended in AASHTO [33]) are compared with those under a reasonable high-velocity range of 5.0 m/s (≈10.0 knots).

4. For high-resolution simulations, it is assumed that the blast loads caused by explosive materials (herein, TNT) are placed on the deck level and the boundary line between the bow and stern portions of the vessels. Also, the distances of the explosive charge (i.e., TNT) from the pier at the onset time of detonation are assumed equal to the lengths of bow portions of the vessels.

5. Since there exists no clear and practical information about explosive charge weight for the bridge piers in the current design codes existing in the literature due to security considerations, it is assumed that the charge weight carried by 2,000-DWT barge and 5,000-DWT container ship vessels are 200 and 500 kg, respectively.

To assess the structural design permissions of the studied bridge pier under blast loadings assumed in this chapter and explained in the previous paragraph, the human tolerance and survival probabilities are checked at the farthest distance (i.e., 70-m distance) on the deck span from the center of explosive charge under the blast effects of assumed explosions by comparing the peak pressure and impulse results with those

Table 9.7 Different loading scenarios applied to the high-resolution FE model of the bridge pier varying in different impact and blast loading-related parameters

Loading case ID	Loading type	Vessel type	Impact weight (ton)	Impact velocity (m/s)	Charge weight (kg)	Scaled distance $(m/kg^{1/3})$	Loading height from column base (m)	Time lag (s)
S5V4H0	Impact	Ship	5,000	4.0	—	—	Impact at 0	—
B2V1H0	Impact	Barge	2,000	1.65	—	—	Impact at 0	—
B2V1H4	Impact	Barge	2,000	1.65	—	—	Impact at 4.5	—
B2V1H7	Impact	Barge	2,000	1.65	—	—	Impact at 7.4	—
B2V5H0	Impact	Barge	2,000	1.65	—	—	Impact at 0	—
B2V5H7	Impact	Barge	2,000	5.0	—	—	Impact at 7.4	—
S5C5H8	Blast	—	—	—	500	2.61	Blast at 8.9	—
B2C2H0	Blast	—	—	—	200	2.62	Blast at 0	—
B2C2H4	Blast	—	—	—	200	1.28	Blast at 4.5	—
B2C2H7	Blast	—	—	—	200	1.28	Blast at 7.4	—
S5C5V4TN	Impact-blast	Ship	5,000	4.0	500	2.61	Impact at 0 Blast at 8.9	0.8 (First natural period)
B2V1H0TN	Impact-blast	Barge	2,000	1.65	200	2.62	Impact at 0 Blast at 0	0.8 (First natural period)
B2V1H4TN	Impact-blast	Barge	2,000	1.65	200	1.28	Impact at 4.5 Blast at 4.5	0.8 (First natural period)
B2V1H7TN	Impact-blast	Barge	2,000	1.65	200	1.28	Impact at 7.4 Blast at 7.4	0.8 (First natural period)
B2V5H7TN	Impact-blast	Barge	2,000	1.65	200	1.28	Impact at 7.4 Blast at 7.4	0.8 (First natural period)
B2V1H7TI	Impact-blast	Barge	2,000	1.65	200	1.28	Impact at 7.4 Blast at 7.4	0.09 (Peak impact force)
B2V5H0TN	Impact-blast	Barge	2,000	5.0	200	2.62	Impact at 0 Blast at 0	0.8 (First natural period)
B2V1H7TD	Impact-blast	Barge	2,000	1.65	200	1.28	Impact at 7.4 Blast at 7.4	1.62 (Peak displacement)

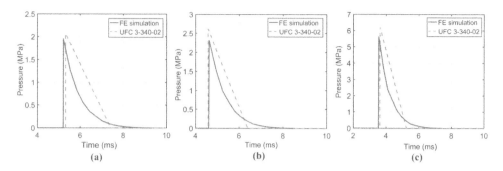

Figure 9.14 Comparison of reflected pressure and impulse results at 70-m distance on the deck span from the center of charge obtained from FE simulations with those calculated from UFC 3-340-02 [54] for blast loading scenarios of: (a) B2C2H0, (b) B2C2H7, and (c) S5C5H8.

from UFC 3-340-02 [54]. According to this manual, the structural analysis and design procedure are recommended for cases in which the survival probabilities of humans are notable when they do not suffer high lung damage and eardrum injuries by the primary effects of the blast pressure. Figure 9.14a–c shows the comparison of reflected pressure and impulse results at 70-m distance on the deck span from the center of charge obtained from FE simulations with those calculated by UFC 3-340-02 [54] for blast-loading scenarios with IDs of B2C2H0, B2C2H7, and S5C5H8, respectively. Also, the scalar magnitudes of the peak pressures (P_r), impulses (i_r), arrival times to peak pressures (t_a), and amplitudes (i.e., the time duration of impulse) (t_o) are quantitatively compared in Table 9.8. It is seen that although the peak pressures obtained from the FE simulations are about 10% higher than those calculated using UFC 3-340-02 [54] manual, the amplitudes and arrival times to peak pressures are almost in good agreement. Also, by checking the survival probabilities (S_p) of a 70-kg adult human subjected to the blast effects of three explosions at 70-m distance from the center of charge on the deck span using UFC 3-340-02 [54] manual, it is found that under a blast loading scenario named S5C5H8, S_p has a value less than 1%, which means that almost no one will survive in the radius of 70 m under this detonation. Therefore, this loading scenario (i.e., S5C5H8) and its combination with vessel impact are not worthy to be considered for design analysis and further vulnerability assessments bridge pier with the variability of loading parameters, including the impact velocity, the time lag in Sections 9.5.4 and 9.5.5, respectively.

Table 9.8 Comparison of the blast loading effects resulted from the FE simulations with those calculated UFC 3-340-02 [54] at 70-m distance (on the deck span) from the center of the explosive charge

Blast-loading case ID	P_r (UFC) (MPa)	P_r (FE) (MPa)	i_r (UFC) (MPa ms)	i_r (FE) (MPa ms)	t_a (UFC) (ms)	t_a (FE) (ms)	t_o (UFC) (ms)	t_o (FE) (ms)	Survival probability (S_p) of a 70-kg human
B2C2H0	2.07	1.94	2.15	1.41	5.33	5.23	2.08	3.07	$90\% < S_p < 99\%$
B2C2H7	2.62	2.46	2.36	1.70	4.57	4.62	1.82	3.21	$50\% < S_p < 90\%$
S5C5H8	6.21	5.82	4.99	3.81	3.62	3.55	1.61	2.65	$S_p < 1\%$

Besides, it is seen that S_p is ranged between 90% and 99% under B2C2H0, and between 50% and 90% under B2C2H7 as given in Table 9.8. Therefore, it is worthy to investigate the vulnerability of the bridge pier and its progressive failure behaviors under these detonations and their combinations with vessel impact loads. That is, the additional collapse of the pier and the bridge deck (two single spans) will cause more fatality. Hence, the failure analysis and structural design of the bridge pier are permitted as recommended by UFC 3-340-02 [54] manual when it is subjected to sole blast loadings of B2C2H0, B2C2H7, and their combinations with vessel impacts.

9.4 Failure modes of bridge pier under combined loads

In this section, the failure behaviors and dynamic responses of the pier under combined impact-blast loadings are evaluated compared to those under only vessel impact and only blast loads. To do this, a collision of a 2,000-DWT barge with a design impact velocity of 1.65 m/s ($\approx$3.21 knots) impacted on the pier column at the height of 7.4 m from the column base named B2V1H7 is considered in combination with a blast loading with a scaled distance of 1.28 m/kg$^{1/3}$ arising from a spherical explosive with a weight of 200 kg standing at the distance of 7.5 m named B2C2H7 as shown in Figure 9.15. The combination of these loadings is named B2V1H7TN as identified in Table 9.7. Before the initiation of transient loading stages (including the vessel collision and blast load), the whole bridge (including pier and superstructure) is exposed to its gravity loads under self-weight using LOAD_BODY keyword in LS-DYNA through an implicit dynamic relaxation analysis before transient impact- and blast-loading phases during explicit analysis. Then, the results obtained from the implicit dynamic relaxation analysis are transferred to explicit analysis to consider the permanent gravity loads during the transient loading phases. Therefore, the bridge undergoes the gravity loads during both stress initialization (i.e., dynamic relaxation analysis) and transient loading phases. It is also worth noting that the gravity loads should be gradually applied to the FE models for a sufficiently long time to avoid undesirable dynamic vibration in the structural responses of the bridge. In this study, the gravity

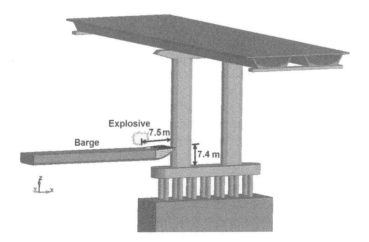

Figure 9.15 Details of the barge collision and the explosion locations on the bridge pier modeled in LS-DYNA.

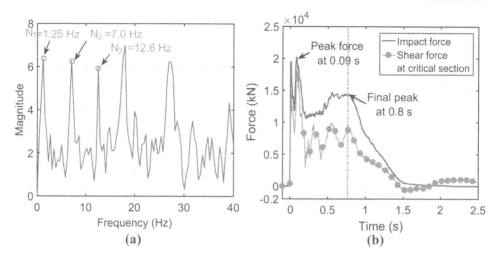

Figure 9.16 Impact responses of the bridge pier under 2,000-DWT barge collision with the impact velocity of 1.65 m/s: (a) FFT of the pier acceleration in collision direction and (b) impact and shear forces.

loads are gradually applied to the models until the bridge pier and superstructure reach equilibrium and stable state with a vertical velocity of less than 0.1 m/s after 0.5 s (during an implicit analysis). Also, global damping of 5% was used to achieve the equilibrium state of the whole bridge during a dynamic relaxation phase as recommended by bridge design codes [55]. From a modal analysis and also using the operation of fast Fourier transform (FFT) function on the acceleration response of the bridge pier in the collision direction, the first three natural frequencies of the pier are $N_1 = 1.25$ Hz, $N_2 = 7.0$ Hz, and $N_3 = 12.6$ Hz as shown in Figure 9.16a. Therefore, the first natural periodic time of the pier at $T_1 = 0.8$ s after barge collision is assumed as the initiation time of the following blast load that is almost simultaneous with the happening times of the final peaks of the impact force and the shear force at the critical section of the impacted column (i.e., the middle of the shear plug) under barge impact load as shown in Figure 9.16b.

Figure 9.17a–e demonstrates that the pier experiences greater deflections and the internal forces, including the flexural moment, shear force, and axial force under combined loadings compared to those under sole impact and blast loads. In addition, the pier loses its shear, flexural, and axial load-carrying capacities under combined actions of impact and blast loads. The failure behaviors of the pier under only barge impact, only explosion loading, and the combined impact-blast loadings (with the priority of the impact loading) at different time stages of the pier response are illustrated in Figure 9.18a–c, respectively. Furthermore, the corresponding flexural moments and shear forces distributed in the height of the front column of the pier are presented in Figures 9.19a–c and 9.20a–c, respectively.

In Figure 9.18a, the plastic strain of the concrete around the impact zone and the flexural cracks originating from the tensile surface (i.e., back surface) of the impacted column at the impact level is extended when the inertial resistance of the superstructure is mobilized at $t = 0.17$ s simultaneously with the occurrence of the peak flexural moment in the impacted column as shown in Figures 9.17d and 9.19a. Afterward, the

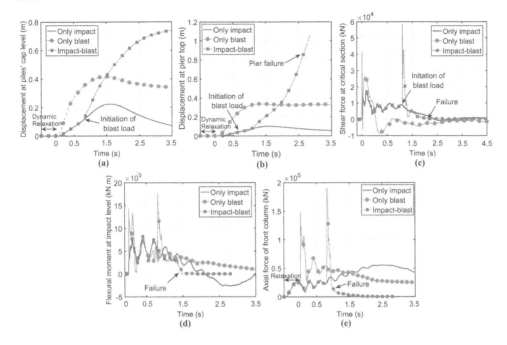

Figure 9.17 Comparing the responses of the pier under different loadings: (a) displacements at the piles' cap level, (b) displacements at the pier top, (c) shear forces at the critical section, (d) flexural moments at the impact level, and (e) axial forces in the impact column.

formation of substantial shear cracks propagated from the column base to the impact zone with an approximate angle of 45° is observed at $t = 0.8$ s. Figure 9.20a shows the shear force distribution in the height of the impacted column. It is found that the critical section for shear forces is at the height of 5.0 m from the column base around the middle of the shear damage. Although the pier withstands against the collapse under only vessel impact load as shown in Figure 9.18a, it suffers substantial shear and flexural damages at the impact and top elevations through the change of the directions of flexural moments during the free vibration phase of the pier response at $t = 1.95$ s as shown in Figure 9.19a.

Besides, the failure behaviors of the pier under the sole explosion with a scaled distance of 1.28 m/kg$^{1/3}$ at identical time stages are shown in Figure 9.18b. The predominance of global shear and flexural-shear failures in combinations with localized failures composed of a localized shear plug, spallation, and the compressive damages is observed around the explosion elevation of the blast-front column. Accordingly, an additional shear failure occurs around the height of 20 m from the column base at $t = 1.95$ s. Around this elevation, the strength losses occur in the cross-sectional capacities as shown in Figures 9.19b and 9.20b.

Figure 9.18c illustrates the failure behaviors of the pier under combined actions of impact and blast loads after the onset of the detonation at $t = 0.8$ s. It should be noted that the failure behaviors before $t = 0.8$ s are identical with those under sole impact loading as shown in Figure 9.18a. The accumulative effects of the impact-induced shear damage at $t = 0.8$ s (see Figure 9.18a) and the stresses arising from blast-loading impulses (mainly pressure) result in the formation of a localized shear failure

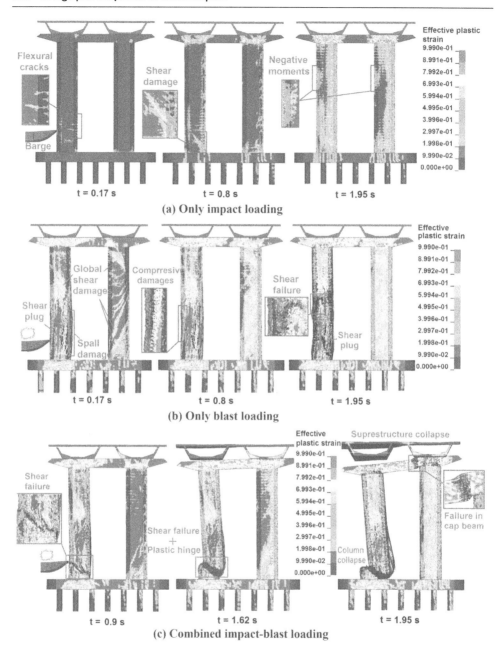

Figure 9.18 Failure behaviors of the pier under: (a) only impact loading named B2VIH7, (b) only blast-loading named B2C2H7, and (c) combined impact-blast loading named B2VIH7TN.

propagated from the column base to the impact zone shortly once the onset of the detonation at $t = 0.9$ s. Thereafter, the blast-front column loses its shear resistance around $t = 1.3$ s as shown in Figures 9.18c and 9.21c. Then, the column undergoes a flexural failure around $t = 1.5$ s as shown in Figures 9.18d and 9.20c. Finally, the column collapses under self-weight load after $t = 1.62$ s, which accordingly causes

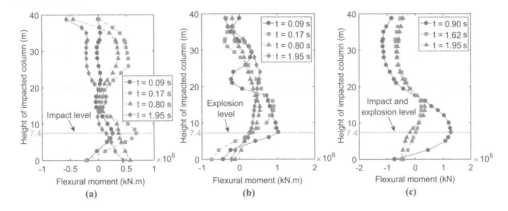

Figure 9.19 Distribution of flexural moment along with the height of impacted column under: (a) only impact loading named B2VIH7, (b) only blast loading named B2C2H7, and (c) combined impact-blast loading named B2VIH7TN.

progressive damages in the other components of the pier and the collapse of bridge superstructures at $t = 1.95$ s. Table 9.9 summarizes the peak values of the displacement and internal forces of the pier, and the failure behaviors observed from the high-resolution FE simulations in Figures 9.17 and 9.18, respectively.

9.5 Vulnerability assessment of bridge pier under combined loads

The vulnerability of the bridge pier under impact and blast loads and their combinations is assessed using high- and moderate-resolution FE simulations in LS-DYNA varying in different loading-related parameters, including the location of applied loads, vessel type (varying in terms of the weight, size, and bow configurations), vessel

Table 9.9 Comparing the dynamic responses and failure modes of the pier under sole impact and blast loads with those obtained under combined loading

Loading case ID	Loading type	Maximum displacement at piles' cap (m)	Peak shear force in front column (MN)	Peak flexural moment at column base (MN m)	Failure	Remarks
B2VIH7	Impact	0.22	18	800	Flexural-shear, diagonal shear	Minor flexural cracks along with diagonal shear damage
B2C2H7	Blast	0.42	42	1,400	Shear plug	Significant shear damages at loading-front column
B2VIH7TN	Impact-blast	0.73	58	1,750	Diagonal shear	Collapse of pier due to diagonal shear failure at loading-front column

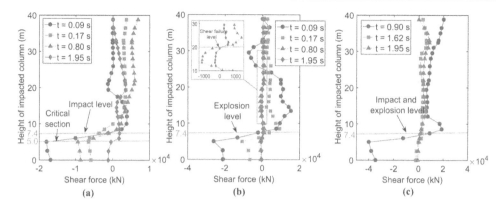

Figure 9.20 Distribution of shear force along with the height of impacted column under: (a) only impact loading named B2VIH7, (b) only blast loading named B2C2H7, and (c) combined impact-blast loading named B2VIH7TN.

impact velocity, and the time lag between the initiations of the loadings. In this section, high-resolution FE simulations are carried out to visually and qualitatively study the failure behaviors of the bridge pier under different loading conditions. Besides, a simplified FE model of the bridge pier (in which only the impact-front column and piles' cap components are developed using high-resolution FEs) is developed in LS-DYNA to quantitatively investigate the damage indices as defined in Section 9.5.2 for the parametric studies based on the moderate-resolution FE simulations.

9.5.1 Simplified FE model of the pier

To assess the vulnerability of the bridge pier to various ranges of the loading parameters through a parametric study, a simplified model of the bridge pier with a moderate resolution is developed to reduce the computation time of the FE simulations. In this model, the mass of the bridge superstructure is modeled using a lumped mass element, and its lateral stiffness and damping behaviors are modeled using a spring element in parallel with a dashpot, respectively, as shown in Figure 9.21. The equivalent stiffness and damping of the superstructure are obtained using the method proposed by Consolazio et al. [56] based on the maximum displacement of the superstructure under the partial impact loading during static analysis. More detailed descriptions of the procedure of this method can be found in the previous work [11]. According to this method, the equivalent lumped mass of the superstructure is 1.785 Mkg, and the equivalent stiffness and damping of the superstructure are 1.0×10^8 N/m and 2.7×10^7 N s/m, respectively, by assuming critical damping of 0.05 as recommended by bridge design codes [55]. Besides, the back column and the piles are modeled using 3-node elastic beam elements (MAT_001) with equivalent cross-sectional characteristics. These beam elements are attached to the lumped mass element of the superstructure and the piles' cap using rigid links. However, the front column and the piles' cap of the pier are modeled with a high-resolution approach to accurately capture the failure behaviors of the pier column that is exposed to impact and blast loads.

The soil-pile interaction between lateral and axial directions is modeled using nonlinear elastic-plastic spring elements attached to the piles' nodes. The lateral and

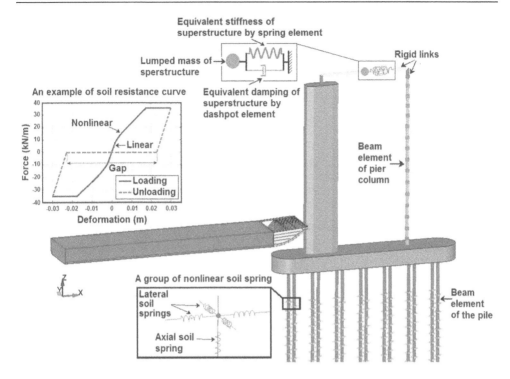

Figure 9.21 Simplified FE model of the bridge pier with moderate resolution.

axial resistances of the soil are modeled using the force-deformation curves generated according to the method proposed by Reese et al. [57] and Matlock [58], respectively, based on the strength characteristics of the soil, including unit weight, internal friction angle, and subgrade modulus. Then, these curves are defined in the material model of the nonlinear soil springs that are modeled using MAT_SPRING_GENERAL_ NONLINEAR (MAT_S06) in LS-DYNA. An example of the force-deformation curve, including both loading and unloading curves with considering the soil gap model [12], is presented in Figure 9.21. It is also noteworthy that the simplified model is only used to quantitatively investigate the damage indices defined in Section 9.5.2 for the parametric studies based on the moderate-resolution FE simulations.

To validate the simplified FE model proposed in this section, the impact forces and displacements of the pier resulted from the simplified FE model are compared with those from high-resolution simulations when the pier models are subjected to barge collisions with velocities of 1.65 and 5 m/s as shown in Figure 9.22a and b. It is seen that the initial peaks and residual phases of impact forces and the pier displacements from the simplified FE-based simulations are in good agreement with those from high-resolution simulations results. Therefore, the simplified FE model is sufficiently reliable to be used for parametric study and sensitivity assessments in the following sections.

9.5.2 Damage indices

To quantitatively assess the damage states and failure modes of the pier under combined loadings, three damage indices based on the residual axial load, shear force, and

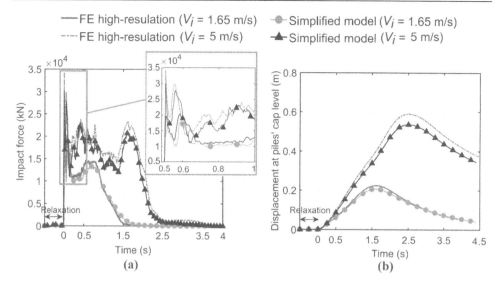

Figure 9.22 Comparing the impact responses of the simplified FE-based simulations with those from high-resolution simulations: (a) impact forces and (b) displacements at piles' cap level.

flexural moment capacities of the pier column named $DI_{(P)}$, $DI_{(Q)}$, $DI_{(M)}$, respectively, are defined by Equations (9.3)–(9.5), respectively.

$$DI_{(P)} = 1 - \frac{P_r - P_L}{P_N - P_L} \qquad (9.3)$$

$$DI_{(Q)} = 1 - \frac{Q_r}{Q_N} \qquad (9.4)$$

$$DI_{(M)} = 1 - \frac{M_r}{M_N} \qquad (9.5)$$

where P_r, Q_r, and M_r represent the residual axial load, shear force, and flexural moment, respectively. P_N, Q_N, and M_N denote the nominal axial, shear, and flexural moment capacities of the RC column that can be calculated using ACI-318 [42] considering the dynamic increase factors due to the strain rate effects of the materials. P_L denotes the service axial load due to the weight of superstructure (as described in Section 9.2.2) that is equal to 10% of the axial load-carrying capacity of the pier columns (i.e., $0.1P_N$).

The residual capacities of the pier column are computed using a series of multistep loading procedures as shown in Figure 9.23a and b. These procedures have four loading steps that are sketched as follows:

- First, the simplified model of the pier is gradually exposed to the gravity loads due to the mass of the bridge superstructure and the pier column using LOAD_BODY keyword in LS-DYNA through an implicit dynamic relaxation analysis before transient impact- and blast-loading phases during explicit analysis. To avoid

undesirable dynamic vibrations, the gravity loads are gradually applied to the FE models for a sufficiently long time until the bridge pier and superstructure reach equilibrium (i.e., stable) state with a vertical velocity less than 0.1 m/s after 0.5 s as shown in Figure 9.23a.
- Then, the pier is subjected to barge collision during transient dynamic analysis.
- The sequent blast loading is applied to the impact-induced pier at a specified time stage after the collision phase.
- Finally, the pier is exposed to a motion-based axial loading at the front column end to obtain the residual axial load-carrying capacity of the column as shown in Figure 9.23a. Also, through a separate analysis, the pier model is subjected to a repetitive motion-based barge collision with a constant velocity during a quasi-static analysis to compute the residual shear and flexural capacities of the pier column as shown in Figure 9.14b. It should be noted that the pier is fixed at the piles' cap level with the initiation of this step using a time-controlled boundary condition keyword in LS-DYNA.

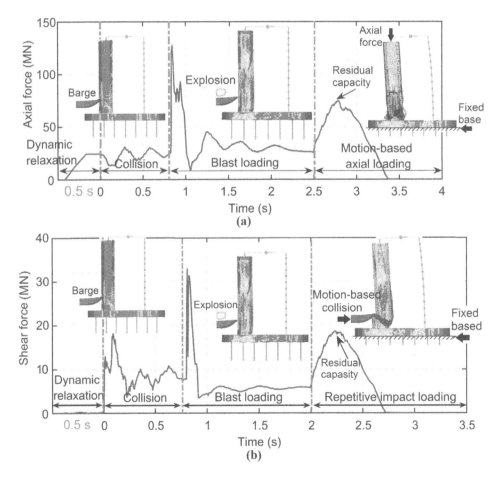

Figure 9.23 Multiple-step loading procedures to compute: (a) the residual axial load-carrying capacity and (b) the residual shear and flexural capacities of the pier column.

9.5.3 Influence of the loading location

Two different types of vessels, including a 2,000-DWT Jumbo Hopper barge and a 5,000-DTW container ship that commonly navigate in inland waterways of the United States and China, are considered for high-resolution simulations of the bridge pier under combined impact-blast loadings. To simulate the blast loading in combination with the vessel collisions, it is assumed that the explosive materials (herein, TNT) are placed on the deck level of the stern portions of the vessels. Owing to different bow configurations of barges compared to those of container ships, their bows strike on different elevations on the bridge piers. As such, the collision of a 5,000-DWT container ship with the piles' cap level with an impact velocity of 4.0 m/s, which carries 500-kg TNT placed at the elevation of 8.9 m (i.e., the deck level of ship rear) from the column base (i.e., the piles' cap level), represents a combined loading scenario named S5C5V4TN in which the vessel impact loading elevation is different from that of the explosion as shown in Figure 9.24a. In this scenario, the scaled distance of the detonation is $Z = 2.61$ kg/m$^{1/3}$. Besides, during the collision of a 2,000-DWT barge with an impact velocity of 1.65 m/s, which carries 200-kg TNT that is applied to different locations on the pier, including the piles' cap level, and the heights of 4.5 and 7.4 m from the column base, it represents combined loading scenarios named B2V1H0TN, B2V1H4TN, and B2V1H7TN, respectively. For these loading scenarios, both impact and blast loads are applied to the pier at the same elevation. One of these scenarios named B2V1H0TN represents the applying of combined loading to the piles' cap level that causes a scaled distance of 2.62 kg/m$^{1/3}$ as shown in Figure 9.24b. Two other scenarios, including B2V1H4TN and B2V1H7TN, denote the applying of the combined loading to the heights (H_i) of 4.5 and 7.4 m from the column base that causes the scaled distance of 1.28 kg/m$^{1/3}$ as shown in Figure 9.24c and d, respectively.

   Figures 9.25 and 9.26 show the local and global behaviors of the pier and vessels during only vessel collisions, respectively. Figures 9.25a and 9.26a demonstrate that the pier endures more severe localized damages at the piles' cap (i.e., impact zone) and global flexural cracks under ship collision compared to those from the barge impact on the piles' cap as shown in Figures 9.25b and 9.26b. Although the first peak impact forces from ship and barge impacts on the piles' cap have almost similar magnitudes, the vessel weight has significantly positive influences on the following peaks and the duration of impact loading as shown in Figure 9.27a. Accordingly, the pier experiences larger shear forces at piles' cap level, displacements, and the flexural moment at the base of columns compared to those from barge collision as shown in Figure 9.27b–e, respectively.

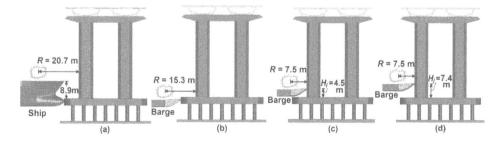

Figure 9.24 Different combined loading scenarios by varying the loading location: (a) S5C5V4TN, (b) B2V1H0TN, (c) B2V1H4TN, and (d) B2V1H7TN.

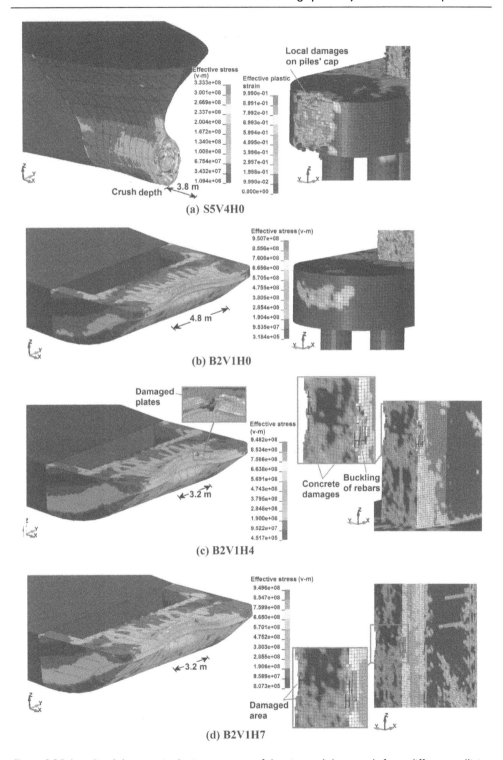

Figure 9.25 Localized damages in the impact zone of the pier and the vessels from different collision scenarios. (a) S5V4H0 (b) B2V1H0 (c) B2V1H4 (d) B2V1H7.

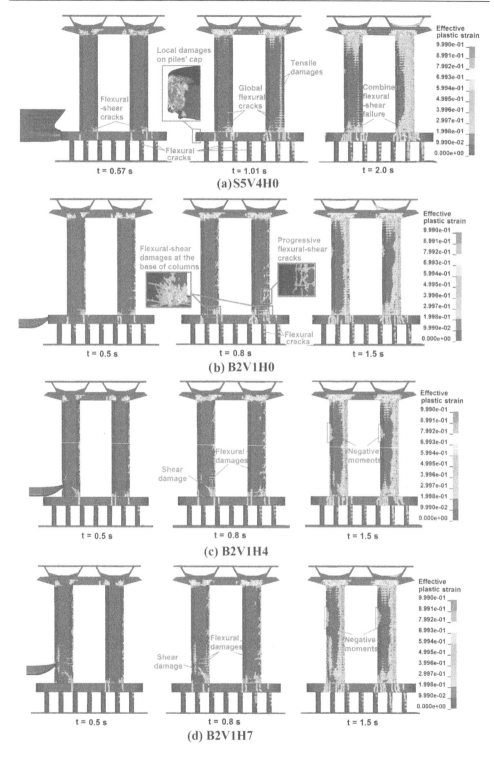

Figure 9.26 Global damages and failure behaviors of the pier from different collision scenarios. (a) S5V4H0 (b) B2VIH0 (c) B2VIH4 (d) B2VIH7.

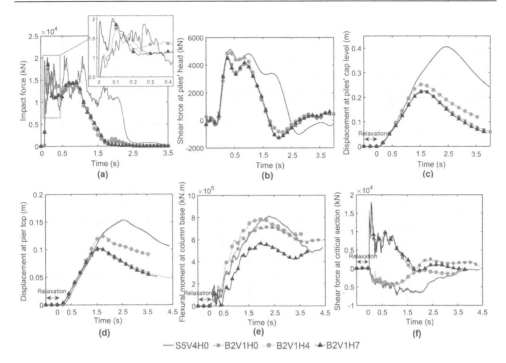

Figure 9.27 Comparing the impact responses of the pier under different vessel collision scenarios: (a) impact forces, (b) shear force at piles' head, (c) displacement at piles' cap level, (d) displacement at the pier top, (e) flexural moments at the base of the impacted column, and (f) shear forces at the critical section of the impacted column.

Besides, when the barge collides with the pier column at the upper elevations, both the barge bow and the column suffer more severe localized damages and greater peak impact forces as illustrated in Figures 9.25c, 9.25d, and 9.27a, respectively. However, the secondary peak impact force resulted from the barge impact on the piles' cap level is greater than those from the barge collision with the pier column at upper elevations due to the mobilization of more internal trusses during the following impulses. Moreover, the impacted column endures a more intense shear plug propagated from the column base to the impact point and larger shear forces at the critical section as shown in Figures 9.26c, 9.26d, and 9.27f. In Figures 9.25b and 9.27e, it is observed that the column suffers more localized damages and flexural moments at the base level in proportion to the decrease of the barge impact height on the column. A summary of the dynamic responses and failure behaviors endured by the bridge pier under different loading conditions varying in the loading location, vessel type, and impact weight is presented in Table 9.10.

The first periodic time of the pier (i.e., $t = 0.8$ s) is simultaneous with the incidence time of the final peak impact force from the barge collision as shown in Figure 9.16a and b, and the second peak impact forces from the ship collision at 0.57 s as shown in Figure 9.27a are considered the initiation times of the sequent blast loadings after barge and ship collisions, respectively.

Figure 9.28a–d illustrates the failure behaviors of the pier under different combined loadings varying in the vessel type and the location of applied loads in comparison

Table 9.10 Comparison of the dynamic responses and failure modes of the pier under different loadings varying in the loading location, vessel type, and impact weight

Loading case ID	Loading type	Maximum displacement at piles' cap (m)	Peak shear force at front column (MN)	Peak flexural moment at column base (MN m)	Failure	Remarks
S5V4H0	Impact	0.41	7.5	804	Overall flexure	Pier globally suffers flexural damages in the columns and piles
B2V1H0	Impact	0.25	4.8	800	Flexural-shear	Pier globally suffers combined flexural-shear damages
B2V1H4	Impact	0.22	14	720	Flexural-shear, and diagonal shear	Minor flexural cracks along with diagonal shear damage
B2V1H7	Impact	0.22	18	600	Flexural-shear, and diagonal shear	Minor flexural cracks along with diagonal shear damage
S5C5H8	Blast	0.52	25	1,220	Threshold spall	Extensive spallation in cover concrete, and global shear cracks
B2C2H0	Blast	0.48	18	1,015	Moderate spall, and shear plug	Front column suffers shear plug, and back column endures global shear damages
B2C2H4	Blast	0.46	36	1,375	Moderate spall, and shear plug	Front column suffers shear plug, and back column endures global shear damages
B2C2H7	Blast	0.42	42	1,400	Threshold spall	Pier suffers global shear damages
S5C5V4TN	Impact-blast	0.72	47	1,520	Moderate spall	Extensive exfoliation and spallation in cover concrete, and global flexural-shear damages
B2V1H0TN	Impact-blast	0.6	41	1,325	Moderate spall, and punching shear	Front column suffers a punching shear, and back column endures global flexural-shear damages
B2V1H4TN	Impact-blast	0.64	56	1,570	Moderate spall, and diagonal shear	Front column suffers diagonal shear, and back column endures global flexural-shear damages
B2V1H7TN	Impact-blast	0.73	58	1,750	Diagonal shear	Collapse of pier due to diagonal shear failure at loading-front column

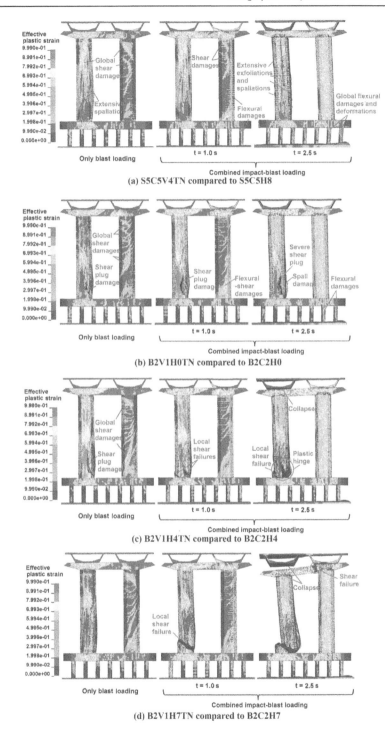

Figure 9.28 Failure behaviors of the pier under different combined loadings in comparison with those under sole blast loadings. (a) S5C5V4TN compared to S5C5H8. (b) B2VIH0TN compared to B2C2H0. (c) B2VIH4TN compared to B2C2H4. (d) B2VIH7TN compared to B2VIH7.

with those from only blast loading. In Figure 9.25a, it is observed that the pier suffers extensive spallation on its back surface of the blast-front column and global shear damages in the back column under only blast loading detonated at the deck level of the ship rear where a relatively higher elevation exists than that of the barge (i.e., S5C5H8). With the combination of the ship collision load applied at the piles' cap level with the identical blast loading named S5C5V4TN, the pier endures large global deformations and severe global flexural damages in the piles' head level and the extensive exfoliations and spallation in the front column compared to those under sole blast loading (i.e., S5C5H8). However, the pier still resists total collapse as shown in Figure 9.28a. When the pier is exposed under the explosion with a scaled distance of $Z = 2.61$ kg/m$^{1/3}$ applied at the piles' cap level named B2C2H0, the front column suffers a shear plug damage in its one-third height added to the spallation in its back surface. The severity of this shear plug damage increases at the critical section and it is accumulated with global flexural damages in the piles' head level when the pier is subjected to the combined loading of B2V1H0TN as shown in Figure 9.28b. Despite suffering greater flexural moments at the piles' head level under B2V1H0TN, the bridge withstands total collapse as shown in Figure 9.28a and b. The corresponding residual flexural and axial forces as illustrated in Figure 9.29a and c, respectively, and the vertical displacements of the superstructure cross-section in Figure 9.30a and b demonstrate the resistance of the pier against collapse. In contrast, the whole collapse of the bridge is yielded from progressing a localized shear failure in the front column under the combined loadings of B2V1H4TN and B2V1H7TN as shown in Figure 9.28c and d. The failure of the impacted column resulted from the shear and axial strength losses after about 1.5 s as illustrated in Figure 9.29b and c, which consequently leads to the collapse of the superstructure as shown in Figure 9.30c and d. The failure modes of the pier under combined loading scenarios compared to those under sole impact and blast loads are depicted in Table 9.10.

To determine the predominant failure modes of the bridge pier under different combined loadings, the moment-shear interaction diagrams are plotted for various critical cross-sections of the pier in comparison with the capacity curve that can be obtained using Response-2000 software [59] based on the modified compression field theory [60] as shown in Figure 9.31a–d. It is seen that the pier fails in flexural mode at

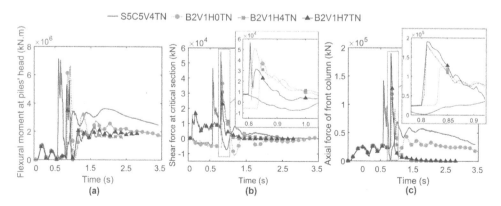

Figure 9.29 Responses of the pier under different combined loadings varying in the vessel types and the loading locations: (a) flexural moment at piles' head, (b) hear force at the critical section, and (c) axial force in the impacted column.

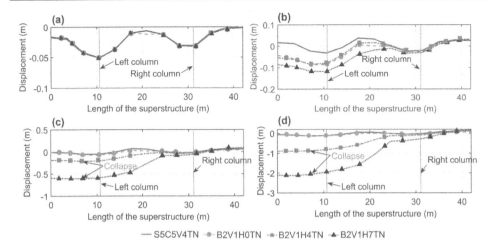

Figure 9.30 Vertical displacements in the cross-section of the bridge superstructure under different combined loadings at: (a) t = 0.5 s, (b) t = 1.0 s, (c) t = 1.5, and (d) t = 2.5 s.

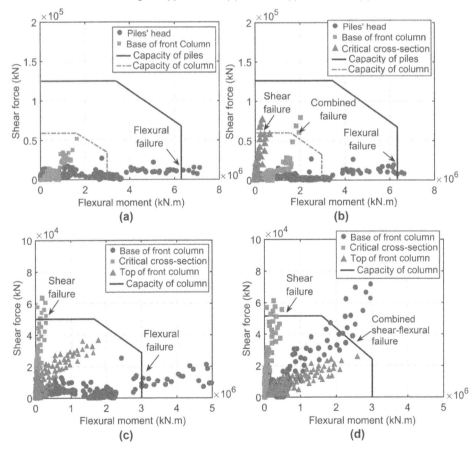

Figure 9.31 Comparison of the moment-shear interaction diagrams plotted for various cross-sections of the pier under different combined loadings of: (a) S5C5V4TN, (b) B2V1H0TN, (c) B2V1H4TN, and (d) B2V1H7TN.

the piles' head level when subjected to vessel collisions on the piles' cap level under S5C5V4TN and B2V1H0TN. Besides, the front column of the pier suffers a combined flexural-shear failure mode at the column base and a shear failure mode at the critical cross-section (i.e., at the middle of shear plug propagated from the impact point with an angle of 45° toward the column base) as shown in Figure 9.31a and b, respectively. In Figure 9.31c and d, it is shown that the failure modes at the base and top zones of the front column tend to a combined failure mode with the increase of the loading height from the column base.

The results from sensitivity assessments of the bridge pier under different combined loadings varying in the loading height from the column base using the proposed damage indices by Equations (9.3)–(9.5) based on the simplified FE-based model are presented in Figure 9.32a and b. It is seen that the damage indices based on the residual axial load ($DI_{(P)}$) and flexural moment ($DI_{(M)}$) increase with increasing loading height from the column base. Moreover, these indices have an almost similar trend of sensitivity to the loading height with a sensitive level of 6.0 m (sudden increase) from the column base under combined loadings. Although this sensitivity level is not notable for the cases under only blast loadings, $DI_{(P)}$ and $DI_{(M)}$ have a more sensitivity level of 10 m under only barge collisions. Therefore, the combination of impact and blast loads significantly decreases the sensitivity level of pier failure to the loading height. However, the damage index based on the residual-shear capacity of the pier column ($DI_{(Q)}$) demonstrates different trends of sensitivity as illustrated in Figure 9.32b. It is seen that the ascending trend of $DI_{(Q)}$ is changed to a descending trend at the loading height of 7.4 m under the only impact and combined loadings. However, this turning level is 4.5 m under sole blast loading. This is because the failure mode of the pier column tends to flexural and flexural-shear failure modes with the increase of the loading height as demonstrated in Figure 9.31a–d.

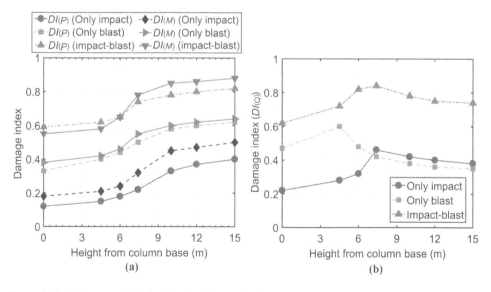

Figure 9.32 Influences of the loading height on the damage indices based on: (a) residual axial load ($DI_{(P)}$) and flexural moment ($DI_{(M)}$) capacities and (b) residual-shear ($DI_{(Q)}$) capacity of the pier column.

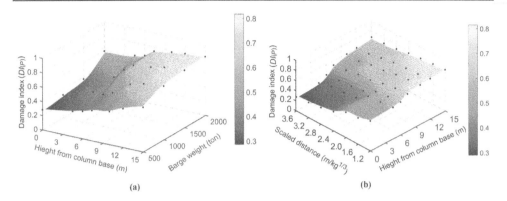

(a)   (b)

Figure 9.33 Sensitivity analysis of pier damage index ($DI_{(P)}$) to the loading height versus: (a) barge impact weight and (b) scaled distance of explosion.

Besides, the sensitivity of the $DI_{(P)}$ to the loading height is also assessed by varying the barge weight and the scaled distance of the blast loading through a parametric study as shown in Figure 9.32a and b. From the damage index results shown in Figure 9.33a, it is seen that the value of $DI_{(P)}$ significantly increases under a combined loading scenario in which the half of cargo capacity of the barge is loaded (i.e., weight of 1,000 tons), and loadings are applied to the height of 7.4 m. Also, it is found that the pier suffers severe damages (for the cases with damage indices higher than 0.6) when it is exposed to combined loadings with scaled distances shorter than 2.4 m/kg$^{1/3}$ for blast loads applied to heights higher than 7.4 m (see Figure 9.33b).

9.5.4 Influence of impact velocity

The influences of the impact velocity on the failure behaviors and responses of the pier are evaluated under the combination of 2,000-DWT barge collisions with an explosion arising from 200-kg TNT applied to the different elevations, including the piles' cap level, and the at height of 7.4 m from the base of columns.

In Figure 9.34a, the local shear plug damages in the impacted column around the impact zone at the height (H_i) of 7.4 m from the column base are formed when the pier is under only barge collision with a high impact velocity (V) of 5 m/s named B2V5H7. In Figure 9.35a, it is seen that the front column of the pier experiences larger shear forces at the critical section (i.e., the middle height of the shear plug propagated from the column base to the impact zone). Besides, from a high-rate barge collision at the piles' cap level, the combination of shear damages and severe flexural damages at the piles' head level is observed as shown in Figure 9.34b. In addition, although the barge collision on the pier column results in a larger value for the first peak impact force, a greater secondary peak impact force is captured when it collides with the piles' cap due to the mobilization of more internal trusses of the barge during the following impulses as shown in Figure 9.35b. Table 9.11 presents a summary of the dynamic responses and failure behaviors of the bridge pier under different loading conditions varying in the impact velocity.

To simulate the combined loading scenarios with high-rate barge collisions, the incidence time of the second peak impact force at 0.43 s for the barge-piles' cap impact scenario and the occurrence time of the peak shear force (which is simultaneous with

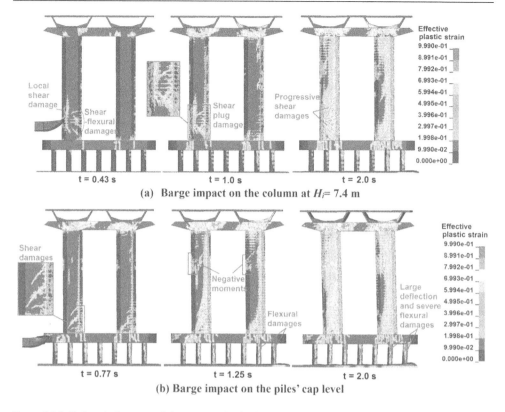

Figure 9.34 Failure behaviors of the pier under high-rate barge collisions with an impact velocity of 5.0 m/s named: (a) Barge impact on the column at H_i=7.4 m (B2V5H7). (b) Barge impact on the piles' cap level (B2V5H0).

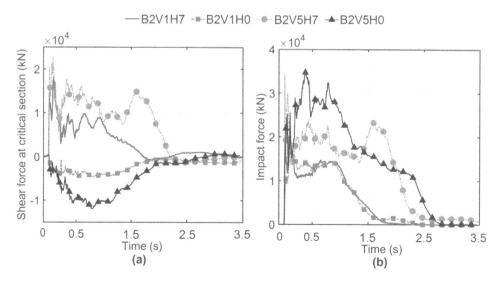

Figure 9.35 Impact responses of the pier under high-rate barge collisions with an impact velocity of 5.0 m/s: (a) shear forces at the critical section and (b) impact forces.

Table 9.11 Comparison of the dynamic responses and failure modes of the pier under different loadings varying in the impact velocity

Loading case ID	Loading type	Maximum displacement at piles' cap (m)	Peak shear force in the front column (MN)	Peak flexural moment at column base (MN m)	Failure	Remarks
B2V1H0	Impact	0.25	4.8	800	Flexural shear	Pier globally suffers combined flexural-shear damages
B2V1H7	Impact	0.22	18	800	Flexural shear, diagonal shear	Minor flexural cracks along with diagonal shear damage
B2V5H0	Impact	0.76	12	980	Diagonal shear	Pier suffers diagonal shear damages at the base both column, and piles' head
B2V5H7	Impact	0.64	23	1,220	Punching shear	Front column suffers a punching shear, and back column endures global flexural-shear damages
B2V1H0TN	Impact-blast	0.6	41	1,325	Moderate spall, and shear punching	Front column suffers shear punching, and back column endures global flexural-shear damages
B2V1H7TN	Impact-blast	0.73	58	1,750	Diagonal shear	Collapse of pier due to diagonal shear failure at loading-front column
B2V5H0TN	Impact-blast	0.88	40	1,430	Shear hinge, and plastic hinges	Pier collapse due to the formation of a shear hinge in the front column and plastic hinges at piles' head
B2V5H7TN	Impact-blast	0.98	42	1,820	Diagonal shear	Collapse of the pier due to a diagonal shear failure in the front column

the occurrence of the final peak impact force at 0.77 s) for the barge-pier column impact scenario as shown in Figure 9.35a and b, respectively, are assumed as the onset times of the sequent detonations.

Figure 9.36a and b demonstrates the failure behaviors of the pier subjected to the combination of high-rate barge collisions (i.e., 2,000-DWT barge) with the impact velocities of 5.0 m/s and the blast loadings of 200-kg TNT with the scaled distances of 2.62 and 1.28 $kg/m^{1/3}$ applied to the piles' cap level named B2V5H0TN, and the height (H_i) of 7.4 m from the column base named B2V5H7TN, respectively. In Figure 9.36a, it is seen that the bridge pier endures an early localized shear failure at the impact zone of the front column under B2V5H7TN, which represents the combined actions of a high-rate barge collision and the explosion loading shortly after the onset of the detonation at 0.43 s. Thus, the progressive collapse of the pier column and the super-structure is observed at the time of 2.0 s. Besides, when the pier is subjected to the combined loading of B2V5H0TN applied on the piles' cap level, the collapse of the bridge pier occurs due to the accumulative effects of the shear failures (shear hinges) in the front column after 1.7 s and global shear-flexural failures (formation of plastic hinges) after about 2.5 s as shown in Figure 9.36b. A shear hinge is formed in the front column due to the accumulation effects of two shear plug damages with opposite directions resulting from each of impact and blast loads at the column's base. Also,

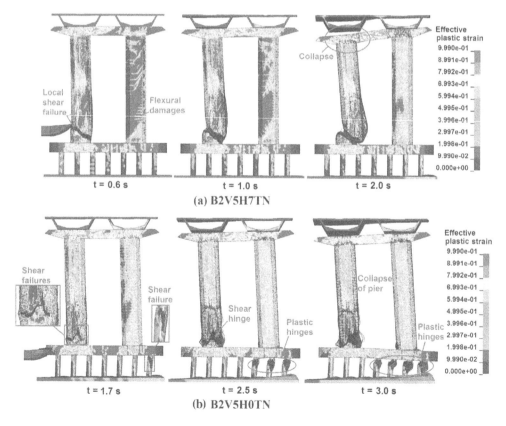

Figure 9.36 Failure behaviors of the pier under combined loadings high-rate barge collisions with a velocity of 5.0 m/s applied to different elevations. (a) B2V5H7TN (b) B2V5H0TN.

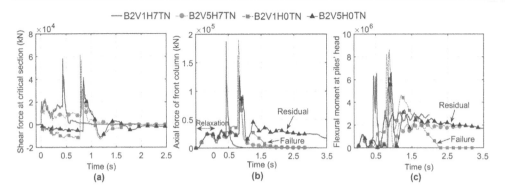

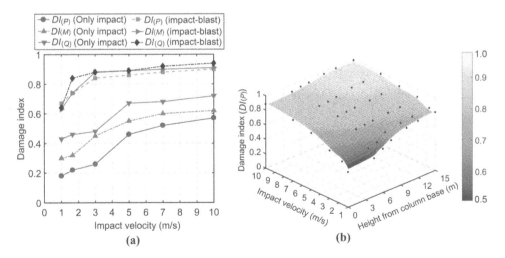

Figure 9.37 Responses of the pier under combined loadings associated with high-rate barge colli-sions with a velocity of 5.0 m/s applied to different elevations: (a) shear forces at the critical section of the impacted column, (b) axial forces of the impacted column, and (c) flexural moments at the piles' head.

losing the shear and axial strengths of the front column under both aforementioned combined loading scenarios can be seen in Figure 9.37a and b, respectively. Also, the flexural strength loss, which indicates the formation of plastic hinges at the piles' head under the combined loading at the piles' cap level under B2V5H0TN, can be observed in Figure 9.36b. The dynamic responses and failure behaviors of the pier under dif-ferent loading conditions varying in the impact velocity are summarized in Table 9.11.

By assessing the influences of the impact velocity on the pier damage indices in Figure 9.38a, it is seen that the magnitudes of all indices significantly increase beyond a velocity of 3.0 m/s when the pier is subjected to sole barge collisions. However, this sensitivity level decreases to a velocity of 1.65 m/s (the recommended velocity by cur-rent design codes for 2,000-DWT barge) when the pier is exposed to the combination of impact and blast loads with a scaled distance of 1.28 $m/kg^{1/3}$. Under this impact

Figure 9.38 (a) Influences of the barge impact velocity on the pier damage indices and (b) sensitivity analysis of the pier damage index $(DI_{(P)})$ to the impact velocity versus the loading height.

velocity, the pier tends to a shear failure since it has a damage index value higher than 0.8, which represents the whole collapse of the pier. Besides, the pier experiences a moderate damage state with a damage index less than 0.6 when the pier is exposed to combined loading with an impact velocity less than 3.0 m/s applied to heights less than 4.5 m from the column base.

9.5.5 Influence of time lag

The influence of the time lags between the initiations of the barge collision and the sequent blast loading is assessed on the failure behaviors and responses of the bridge pier. To do this, three different time stages of the response of the pier at 0.09, 0.8, and 1.62 s are considered the initiation times of the sequent blast loading arising from 200 kg with a scaled distance of 1.28 m/kg$^{1/3}$ combined with a 2,000-DWT barge impact on the pier column at $H_i = 7.4$ m with a low-velocity of 1.65 m/s that are named as the combined loadings of B2V1H7TI, B2V1H7TN, and B2V1H7TD, respectively. These time stages represent the occurrence times of the peak impact force at 0.09 s as shown in Figure 9.16b (simultaneously with the peak shear force at the critical section of the impacted column), the final peak impact force at 0.8 s as shown in Figure 9.16b (which is simultaneous with the first periodic time of the pier), and the peak displacement at the piles' cap level at the simulation time of 1.62 s as shown in Figure 9.17a (which is simultaneous with the free vibration stage of the pier response). In Figure 9.39a and b, it is observed that the pier undergoes larger displacements at the piles' cap and

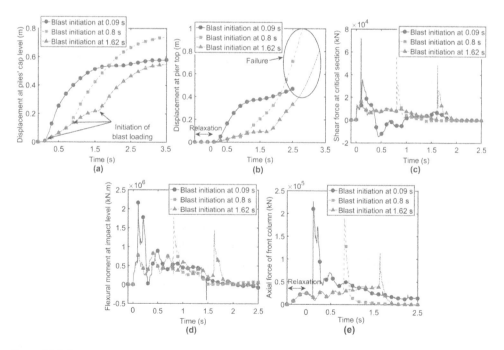

Figure 9.39 Responses of the pier under combined loadings varying in the time lag between the initiations of the loadings: (a) displacement at piles' cap level, (b) displacement at the pier top, (c) shear force at the critical section of the impacted column, (d) flexural moments at impact level, and (e) axial force of the impacted column.

the top elevations under B2V1H7TN in which the sequent blast load is applied at the time of the first natural period at 0.8 s. This is because of the resonance effects of the explosion on the pier response when the following blast load is applied simultaneously with the pier displacement with the most velocity (i.e., the steepest slope of the pier displacement) at 0.8 s. Besides, the pier impacted column undergoes greater shear forces, flexural moments, and axial forces as shown in Figure 9.39c–e, respectively, under B2V1H7TI in which the following blast load is started at the incidence time of the peak impact force, peak shear, and the first peak moment at 0.09 s.

Figure 9.40a and b demonstrates the failure behaviors of the pier under combined loadings of B2V1H7TI and B2V1H7TD in which the sequent detonation is initiated at the occurrence times of the peak impact force and the peak displacement, respectively. It is observed that the pier suffers different failure modes under these combined loads. Such that, the occurrence of global shear failures in the impacted column is observed after 1.5 s, which causes the total collapse of the pier at the following time stage (see Figure 9.40a). However, when the explosion starts simultaneously with the peak displacement (i.e., B2V1H7TD), the collapse of the pier results in due to the combination of the local shear failure (propagated from the column base to the impact point) and the formation of a plastic hinge at the impact level of the column as shown

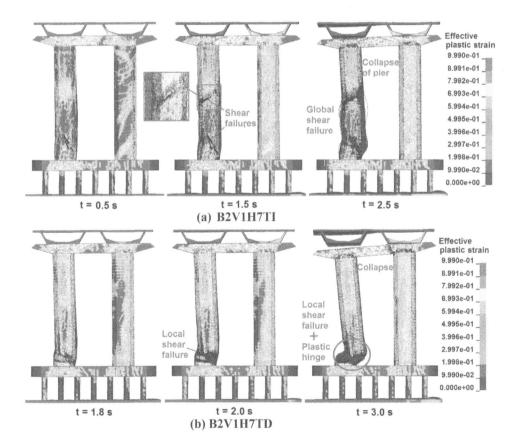

Figure 9.40 Failure behaviors of the pier under combined loadings varying in the time lag between the initiations of the loads. (a) B2V1H7TI (b) B2V1H7TD.

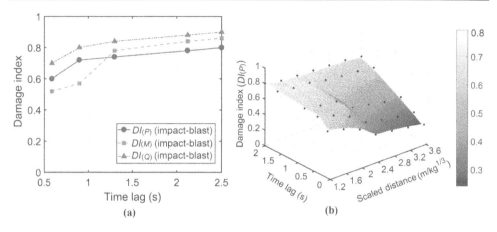

Figure 9.41 (a) Influences of the time lag parameter on the pier damage indices under a combined loading applied at $H_i = 7.4$ m with a velocity of 1.65 m/s and a scaled distance of 1.28 m/kg$^{1/3}$ and (b) sensitivity analysis of the pier damage index ($DI_{(P)}$) to the time lag versus the scaled distance of explosion.

in Figure 9.40b. In Figure 9.41, compared to $DI_{(M)}$ with magnitudes less than 0.6 for the time lags shorter than 0.5 s, $DI_{(Q)}$ results in index values higher than 0.6, which implies the tending of failure modes to shear failures. Beyond this time lag, the pier suffers severe damages composed of shear and flexural failure modes that lead to the whole collapse of the pier since it demonstrates damage index values around and higher than 0.8 for all types of defined indices. Moreover, from a sensitivity analysis on the effectiveness of the time lag versus the variation of the scaled distance of blast loading in Figure 9.39b, the pier suffers severe damages with damage index values higher than 0.6 when it is subjected to blast loadings with scaled distances less than of 2.4 m/kg$^{1/3}$ and the time lags higher than 0.4 s. Table 9.12 presents a summary of the dynamic responses and failure behaviors of the pier under different combined loadings varying in the time lag parameter.

Table 9.12 Comparison of the dynamic responses and failure modes of the pier under different combined loadings varying in the time lag

Loading case ID	Loading type	Maximum displacement at piles' cap (m)	Peak shear force in the front column (MN)	Peak flexural moment at impact level (MN m)	Failure	Remarks
B2VIH7TI	Impact-blast	0.59	72	2,200	Global shear	Collapse of pier due to global shear failure in the front column
B2VIH7TN	Impact-blast	0.73	58	1,750	Diagonal shear	Collapse of pier due to diagonal shear failure at loading-front column
B2VIH7TD	Impact-blast	0.57	47	1,470	Local shear, and plastic hinge	Collapse of pier due to accumulation of local shear failure and plastic hinge in the front column

9.6 Conclusions

The failure behaviors and dynamic responses of an RC girder bridge pier were numerically evaluated subjected to synergetic effects of vessel collisions and blast loadings in comparison with those from sole impact and blast loads. In addition, a simplified model was proposed to assess the vulnerability of the bridge pier to different loading parameters, including the location of applied loads, the impact velocity of vessels, vessel types (varying in terms of impact weight, velocity, and the bow configurations), and the time lag between the initiations of the applied loads through a parametric study. The main conclusions arising from this chapter can be drawn as follows:

- Compared to the collapse resistance of the bridge pier under sole impact and blast loads, the pier experienced more severe damages and larger internal forces under the combined actions of a 2,000-DWT barge collision with a moderate velocity of 1.65 m/s and blast loadings (arising from 200-kg TNT with a standoff distance of 7.5 m) that leads to the whole collapse of the pier. Hence, the current bridge design provisions taking into account blast or vessel collision loads may not be able to sufficiently predict the resistance capacities against the combinations of impact and blast loads.
- The pier suffered more severe localized failures and shear plugs in the impacted column leading to the collapse of the whole pier when both impact and blast loads were applied at the same elevation on the pier column than those arising from scenarios in which loads were applied to the piles' cap level. This is more likely to happen during the barge collision scenarios in real life than container ship collisions. Therefore, more critical combined loading conditions were concluded for barge-pier column collisions than ship-piles' cap collisions. In addition, from a sensitivity analysis, it was found the failure mode of the pier column tends to flexural and flexural-shear failure modes with the increase of the loading height. The pier suffered severe damages with damage index values higher than 0.6 when the loading height is higher than 7.4 m, barge weight is greater than 1,000 tons (the half of barge cargo capacity), and the scaled distance explosion is shorter than 2.4 m/kg$^{1/3}$.
- Different failure modes were observed when the pier was subjected to combined loadings with high-rate barge collisions applied at different elevations. As such, the pier failure mode changed from localized failure modes in the impacted column to global shear-flexural modes at the piles' head reducing the height of applied loads from one-fifth of the column height (i.e., 7.4 m from the column base) to the piles' cap level. From the design point of view, both the structure and substructure components of the pier should be strengthened against combined loadings.
- The pier experienced greater internal forces and deflections when the sequent blast load was applied simultaneously with the occurrence time of the peak impact force. Moreover, the failure mode of the pier and the length of the damaged area significantly depend on the time lag between the initiations of the vessel collision and the following detonation. As such, the failure mode of the pier changed from a global failure (with extensive damage in the pier column) to localized failure mode with increasing the time lag. Moreover, using a sensitivity analysis, it was concluded that suffered severe damages with damage index values higher than 0.6 when it is subjected to blast loadings with scaled distances less than 2.4 m/kg$^{1/3}$ and the time lags higher than 0.4 s.

References

1. Harik IE, Shaaban AM, Gesund H, Valli GYS, Wang ST. United States bridge failures, 1951–1988. *J Perform Constr Fac* 1990;4(4):272–277.
2. Wardhana K, Hadipriono FC. Analysis of recent bridge failures in the United States. *J Perform Constr Fac* 2003;17(3):144–150.
3. Lee GC, Mohan S, Huang C, Fard BN. A study of US bridge failures (1980–2012). Report No. MCEER-13-0008, University at Buffalo, The State University of New York, Buffalo, NY, 2013.
4. Consolazio GR, Cowan DR. Nonlinear analysis of barge crush behavior and its relationship to impact resistant bridge design. *Comput Struct* 2003;81(8):547–557.
5. Kantrales GC, Consolazio GR, Wagner D, Fallaha S. Experimental and analytical study of high-level barge deformation for barge–bridge collision design. *J Bridge Eng* 2015;21(2):04015039.
6. Consolazio GR, Cook RA, McVay MC. Barge impact testing of the St. George Island Causeway Bridge – Phase III: physical testing and data interpretation, structures. Research Report No. BC-354 RPWO-76, Engineering and Industrial Experiment Station, University of Florida, Gainesville, FL, 2006.
7. Consolazio GR, Cowan DR. Numerically efficient dynamic analysis of barge collisions with bridge piers. *J Struct Eng* 2005;131(8):1256–1266.
8. Consolazio GR, Davidson MT. Simplified dynamic barge collision analysis for bridge design. *Transp Res Rec* 2008;2050:13–25.
9. Fan W, Yuan W, Yang Z, Fan Q. Dynamic demand of bridge structure subjected to vessel impact using simplified interaction model. *J Bridge Eng* 2010;16(1):117–126.
10. Davidson MT, Consolazio GR, Getter DJ. Dynamic amplification of pier column internal forces due to barge–bridge collision. *Transp Res Rec* 2010;2172(1):11–22.
11. Gholipour G, Zhang C, Mousavi AA. Analysis of girder bridge pier subjected to barge collision considering the superstructure interactions: the case study of a multiple-pier bridge system. *Struct Infrastruct Eng* 2018;15(3):392–412.
12. Gholipour G, Zhang C, Li M. Effects of soil–pile interaction on the response of bridge pier to barge collision using energy distribution method. *Struct Infrastruct Eng* 2018;14(11):1520–1534.
13. Buth CE, Williams WF, Brackin MS, Lord D, Geedipally SR, Abu-Odeh AY. Analysis of large truck collisions with bridge piers: phase 1. Report of guidelines for designing bridge piers and abutments for vehicle collisions. Report No. 9-4973-1, Texas Transportation Institution, College Station, TX, 2010.
14. Agrawal AK, El-Tawil S, Cao R, Xu X, Chen X, Wong W. A performance based approach for loading definition of heavy vehicle impact events. Report No. FHWA-HIF-18-062, Federal Highway Administration, Office of Research, Development, and Technology, New York, NY, 2018.
15. Do TV, Pham TM, Hao H. Dynamic responses and failure modes of bridge columns under vehicle collision. *Eng Struct* 2018;156:243–259.
16. Gholipour G, Zhang C, Mousavi AA. Reliability analysis of girder bridge piers subjected to barge collisions. *Struct Infrastruct Eng* 2019;15(9):1200–1220.
17. Gholipour G, Zhang C, Mousavi AA. Effects of axial load on nonlinear response of RC columns subjected to lateral impact load: ship-pier collision. *Eng Fail Anal* 2018;91:397–418.
18. Zhang C, Gholipour G, Mousavi AA. State-of-the-art review on responses of RC structures subjected to lateral impact loads. *Arch Comput Method Eng* 2020;28:2477—2507.
19. Wan Y, Zhu L, Fang H, Liu W, Mao Y. Experimental testing and numerical simulations of ship impact on axially loaded reinforced concrete piers. *Int J Impact Eng* 2019;125:246–262.

20. Gholipour G, Zhang C, Mousavi AA. Nonlinear numerical analysis and progressive damage assessment of a cable-stayed bridge pier subjected to ship collision. *Mar Struct* 2020;69:102662.
21. Williamson EB, Bayrak O, Davis C, Daniel Williams G. Performance of bridge columns subjected to blast loads. II: results and recommendations. *J Bridge Eng* 2011;16(6):703–710.
22. Hu ZJ, Wu L, Zhang YF, Sun LZ. Dynamic responses of concrete piers under close-in blast loading. *Int J Damage Mech* 2016;25(8):1235–1254.
23. Williamson EB, Bayrak O, Williams GD, Davis CE, Marchand KA, McKay AE, Kulicki J, Wassef W. Blast-resistant highway bridges: design and detailing guidelines. National Cooperative Highway Research Program (NCHRP Report 645), Transportation Research Board, Washington, DC, 2010.
24. McVay MK. Spall damage of concrete structures. No. WES/TR/SL-88-22, Army Engineer Waterways Experiment Station Vicksburg MS Structures LAB, Vicksburg, MS, 1988.
25. Zhang C, Gholipour G, Mousavi AA. Blast loads induced responses of RC structural members: state-of-the-art review. *Composites, B: Eng* 2020;195:108066.
26. Zhang C, Gholipour G, Mousavi AA. Nonlinear dynamic behavior of simply-supported RC beams subjected to combined impact-blast loading. *Eng Struct* 2019;181:124–142.
27. Gholipour G, Zhang C, Mousavi AA. Loading rate effects on the responses of simply supported RC beams subjected to the combination of impact and blast loads. *Eng Struct* 2019;201:109837.
28. Gholipour G, Zhang C, Mousavi AA. Numerical analysis of axially loaded RC columns subjected to the combination of impact and blast loads. *Eng Struct* 2020;219:110924.
29. Roberts JE, Kulicki JM, Beranek DA, Englot JM, Fisher JW, Hungerbeeler H, Isenberg J, Seible F, Stinson KE, Tang MC, Witt K. *Recommendations for bridge and tunnel security.* American Association of State Highway and Transportation Officials (AASHTO) and Federal Highway Administration (FHWA), Washington, DC, 2003.
30. LS-DYNA 971. Livermore Software Technology Corporation, Livermore, CA, 2015.
31. Fan W, Yuan W. Ship bow force-deformation curves for ship-impact demand of bridges considering effect of pile-cap depth. *Shock Vibr* 2014;2014:1–19.
32. Fan W, Yuan WC. Numerical simulation and analytical modeling of pile-supported structures subjected to ship collisions including soil-structure interaction. *Ocean Eng* 2014;91:11–27.
33. AASHTO. *Guide specifications and commentary for vessel collision design of highway bridges*, 2nd edn. American Association of State Highway and Transportation Officials, Washington, DC, 2009.
34. LSTC. *LS-DYNA keyword user's manual version 971*. Livermore Software Technology Corporation, Livermore, CA, 2016.
35. Yuan P. Modeling, simulation and analysis of multi-barge flotillas impacting bridge piers. Ph.D. Dissertation, University of Kentucky, Lexington, KY, 2005.
36. ASTM. *A36/A36M-08 – Standard specification for carbon structural steel*. ASTM, West Conshohocken, PA, 2008.
37. Getter DJ, Kantrales GC, Consolazio GR, Eudy S, Fallaha S. Strain rate sensitive steel constitutive models for finite element analysis of vessel-structure impacts. *Marine Struct* 2015;44:171–202.
38. Fan W, Xu X, Zhang Z, Shao X. Performance and sensitivity analysis of UHPFRC-strengthened bridge columns subjected to vehicle collisions. *Eng Struct* 2018;173:251–268.
39. Jones N. *Structural impact*. Cambridge University Press, Cambridge, UK, 2011.
40. Symonds PS. Survey of methods of analysis for plastic deformation of structures under dynamic loading. Division of Engineering Report No. BU/NSRDC/1-67, Brown University, Providence, RI, 1967.
41. Manjoine MJ. Influence of rate of strain and temperature on yield stresses of mild steel. *ASME J Appl Mech* 1944;66:A211–A218.

42. ACI 318-14. *Building code requirements for structural concrete and commentary.* American Concrete Institute, Farmington Hills, MI, 2014.

43. El-Tawil S, Severino E, Fonseca P. Vehicle collision with bridge piers. *J Bridge Eng* 2005;10:345–353.

44. Sha Y, Hao H. Nonlinear finite element analysis of barge collision with a single bridge pier. *Eng Struct* 2012;41:63–76.

45. Brumund WF, Leonards GA. Experimental study of static and dynamic friction between sand and typical construction materials. *J Test Eval* 1973;1:162–165.

46. Murray YD. Users' manual for LS-DYNA concrete material model 159. Report No. FHWA-HRT-05-062. Federal Highway Administration, Washington, DC, 2007.

47. Murray YD, Abu-Odeh AY, Bligh RP. Evaluation of LS-DYNA concrete material model 159. Report No. FHWA-HRT-05-063, APTEK, Inc. Colorado Springs, CO, 2007.

48. ASTM. *A615/A615M-05a – Standard specification for deformed and plain carbon-steel bars for concrete reinforcement.* ASTM, West Conshohocken, PA, 2005.

49. Alsos HS, Amdahl J. On the resistance of tanker bottom structures during stranding. *Mar Struct* 2007;20(4):218–237.

50. Pedersen PT, Valsgard S, Olsen D, Spangenberg S. Ship impacts: bow collisions. *Int J Impact Eng* 1993;13(2):163–187.

51. China Ministry of Railways (CMR). *General code for design of railway bridges and culverts (TB10002.1-2005).* China Railway Press, Beijing, China, 2005 (in Chinese).

52. Fujikake K, Li B, Soeun S. Impact response of reinforced concrete beam and its analytical evaluation. *J Struct Eng* 2009;135(8):938–950.

53. Siba F. Near–field explosion effects on reinforced concrete columns: an experimental investigation. Master's Thesis, Department of Civil and Environmental Engineering, Carleton University, Ottawa, Canada, 2014.

54. U.S. Department of Defense. Structures to resist the effects of accidental explosions. UFC 3-340-02. Unified Facilities Criteria (UFC), Washington, DC, 2008.

55. Aviram A, Mackie KR, Stojadinović B. Guidelines for nonlinear analysis of bridge structures in California. Report No. UCB/PEER 2008/03. Pacific Earthquake Engineering Research Center, University of California Berkeley, CA, 2008.

56. Consolazio GR, Getter DJ, Kantrales GC. Validation and implementation of bridge design specifications for barge impact loading (No. 2014/87294). Engineering and Industrial Experiment Station, University of Florida, Gainesville, FL, 2014.

57. Reese LC, Cox WR, Koop FD. Analysis of laterally locked piles in sand. Fifth Annual Offshore Technology Conference, Paper No. OTC 2080, (GESA Report No. D-75-9), Houston, TX, 1974.

58. Matlock H. Correlations for design of laterally loaded piles in soft clay. Offshore Technology in Civil Engineering's Hall of Fame Papers from the Early Years, 1970, pp. 77–94.

59. Bentz E, Collins MP. *Response-2000, V. 1.0. 5.* Toronto University, Toronto, Canada, 2000.

60. Vecchio FJ, Collins MP. The modified compression-field theory for reinforced concrete elements subjected to shear. *ACI J Proc* 1986;83(2):219–31.

Chapter 10

Summary

10.1 Summary of the current work

This book introduces conceptualization and a systematical analytical approach to investigate the vulnerability of reinforced concrete (RC) structures such as beams, columns, and bridge piers subjected to impact and blast loads, and their combinations. In this book, first, the failure behaviors of RC columns and bridge piers under ship collisions have been investigated considering the key dynamic factors such as the strain rate effects of materials, and the inertia of the pier and superstructure. Moreover, the vulnerability of various RC members is investigated subjected to the combination of impact and blast loads varying in terms of two key loading-related parameters, including the loading sequence and the time lag between the initiations of the loads. In addition, since there is no available experimental test in the literature to investigate the failure behaviors of RC members under combined impact-blast loadings, the FE models of the structural members developed using LS-DYNA in the present investigation were separately validated in comparison with the experimental results from impact and blast loading tests conducted by previous studies.

The major contribution and findings of this book can be drawn and summarized as follows:

1 In Chapter 2, from a comprehensive review on the responses and failure behaviors of various types of concrete structures subjected to lateral impact loads, it was found that the impact loads predicted by the current guidelines may be nonconservative due to ignoring the amplification dynamic effects such as inertia and strain rate effects. Besides, most of the previous researches concluded that the axial load parameter has a positive influence on the impact capacities of axially loaded structures when they were exposed to their service levels of axial loads.

2 In Chapter 3, by reviewing the previous works studying the responses and failure behaviors of various types of concrete structures subjected to lateral blast loads, it was concluded that since the vast majority of the current design codes that consider simplified approaches in the prediction of blast loads and the design of structures against explosions, they are not able to efficiently evaluate the brittle damage modes, localized spalling, and shear failure behaviors of concrete structures. Also, no unanimous conclusions were reported on the effectiveness of the axial load when beyond its service level and the column height parameters. In addition, adopting various protective approaches for RC members strengthened

DOI: 10.1201/9781003262343-10

using high-strength composite materials along with high-ratio longitudinal reinforcements that led to diagonal shear failures in most cases.

3 In Chapter 4, the damage progressing process and failure modes of a concrete
cable-stayed bridge pier subjected to a container ship collision was investigated
using FE simulations in LS-DYNA considering the strain rate effects of the materials. The significant influence of the pier concrete strain rate effects on the global
responses of the impacted pier was revealed. Also, the efficient performance of a
proposed simplified ship-pier model using an analytical 2-DOF system in formulating the strain rate effects of the materials was concluded. Also, a key time stage
of 0.5 s was recognized in the determination of the pier's predominant failure
modes by evaluating the damage progression process of the impacted pier from
the minor cracks to the severe damages and cross-sectional fractures. Moreover,
the damage index based on the pier deflection gave a more efficient approach in
describing the damage levels of the pier compared to other damage indices defined
based on the energy, and residual axial load-carrying capacity of the impacted
pier.

4 In-line with Chapter 4, Chapter 5 numerically investigated the influences of the
axial load ratio (ALR), and the location of impact loading on the impact responses
and failure behaviors of RC columns and bridge piers. Significant positive influences of ALR on the impact resistance of RC columns were concluded until a
ratio of 0.5. Also, the failure modes of columns changed from a global failure
mode to local failure in proportion to the increase of ALR. However, the failure
mode of the column changed from a global flexural mode to the shear flexural or
global shear failure modes by decreasing the impact elevation.

5 The nonlinear dynamic responses and failure behaviors of simply supported RC
beams subjected to the combination of impact and blast loads were numerically
investigated varying in terms of various structure- and loading-related parameters in Chapter 6. It was found that the beam suffers more severe spallation in the
depth and experiences larger deformations and internal forces when the impact
load is applied before the sequent blast loading. However, RC beams experienced
more extensive spall damage in their cover zone associated with global shear damages originating from the supports when a uniformly distributed blast load was
applied prior to impact loading. Moreover, by evaluating the influence of time
between the initiations of impact and blast loads, it was obtained that RC beams
suffered more severe spalling damages and the residual plastic deformations
when the blast load was applied during the free vibration stage of the response.
Moreover, when the blast load was applied at the time of the occurrence of the
first peak impact force, larger peak internal forces were generated in the beam.

6 As an extension of Chapter 6, the influences of impact loading rate on the blast
responses of RC beams subjected to the combination of impact and blast loads
were studied. It was revealed that the increase of impact velocity in the combined
loadings where the explosion loads were applied prior to the impact loads will lead
to direct shear failures near the beam supports. In addition, RC beams experienced greater peak moments under combined loadings with low- and middle-rate
impacts in which the following detonation was initiated at the time of the initial
peak of impact force. Besides, the internal forces exhibited greater peak values
under a combined loading with a high-velocity impact in which the following detonation was applied at the time of the initial peak of impact force.

7 In Chapter 8, the vulnerability of a typical RC column commonly used in medium-rise buildings was studied when subjected to combined actions of impact and blast loads with the variability of several loading-related parameters. It was found that the combination of a middle-rate impact loading and a close-in explosion provides more intensive loading conditions when they are applied at the same elevation on the column. However, the combination of an identical impact loading with a far-field detonation leads to more severe failures when they are applied at different elevations. Furthermore, the ALR of 0.5 was accepted as the sensitivity threshold of the column resistance against the collapse under sole blast or impact loading. Besides, the change of the column failure mode from an overall flexural mode to a localized failure mode under combined loadings was sensitive to ALR = 0.3 and beyond. Also, the sensitivity level of the column failure to the impact velocity decreased from a value of 5.0 m/s under only impact loading to 3.0 m/s under combined loadings associated with a close-in explosion applied at the column mid-height.

8 Chapter 9 studied the dynamic responses and failure behaviors of a typical RC girder bridge pier under the combination of vessel collisions and blast loadings. It is concluded that the pier underwent more severe localized failure when both impact and blast loads were applied at the same elevation on the pier column, which is more likely to occur during barge collisions. Different failure modes were observed when the pier was subjected to combined loadings with high-rate barge collisions applied at different elevations. As such, the pier failure mode changed from localized failure modes in the impacted column to global shear-flexural modes at the piles' head by reducing the height of applied loads from one-fifth of the column height (i.e., 7.4 m from the column base) to the piles' cap level. From the design point of view, both the structure and substructure components of the pier should be strengthened against combined loadings. Moreover, the failure mode of the pier changed from global shear failures to a localized failure mode by increasing the time lag and the initiation time of the sequent blast loading. Therefore, strengthening a limited length of the pier column may not be sufficient against the pier failure, which can occur due to global failures rather than localized failure at the impact zone.

10.2 Recommendation for future work

Based on the findings of the completed present investigation and comprehensive review of the previous studies, several aspects as recommendations for future works in the relevant areas can be summarized as follows:

1 Since the theory of the combinations of impact and blast loads was proposed for the first time compared to the literature, the influences of loading-related parameters were mainly investigated against the dynamic behaviors of different structural members. Therefore, it is recommended to comprehensively and uniformly study the effectiveness of different structure-related parameters concerning the responses of various structural members subjected to combined loadings.

2 The presented investigation mainly studied the dynamic behaviors of beams and columns under combined actions of impact and blast loads due to their lower

redundancy compared to slabs and panels. However, it is extremely necessary and recommended to investigate the responses and failure behaviors of RC slabs and panels under such combinations of impact and blast loads for future works.

3 Since the presented work only assessed the vulnerability of RC structural members subjected to the combined actions of impact and blast loads, the need for recognizing the sensitivity thresholds of retrofitted concrete structures strengthened using high-strength composite materials under combined loadings is the topic of importance from the design perspective. Therefore, conducting a series of numerical, analytical, and experimental studies to explore the resistance thresholds of various strengthened strategies on concrete structures using different types of high-strength composites [such as fiber-reinforced polymer (FRP)-based composites] is recommended for future works.

4 Since the currently presented book contents studied the performances of only limited types of structural members as the case studies, it is recommended to provide a series of investigations on benchmark problems for various structural members subjected to combined actions of impact and blast loads.

5 For future works, it is also recommended to carry out reliability analysis and sensitivity assessment of structures and structural members, e.g., bridge piers, subjected to the combinations of vessel collisions and blast loadings varying in terms of various impact- and blast-related parameters.

Appendix A

According to China's bridge design codes, the equivalent static load (F_s) for ship collisions can be calculated as follows:

$$F_s = \gamma V \sqrt{\frac{W}{c_1 + c_2}} \sin\alpha \tag{A-1}$$

where F_s is in kN, γ is the reduction coefficient of kinetic energy (s/m²) which is equal to 0.3 for head-on bow collisions and it is equal to 0.2 for other collision cases, V denotes the ship impact velocity (m/s²), W is the weight of vessel (kN), c_1 and c_2 are the elastic deformation coefficients of the vessel and the impacted bridge, respectively, $c_1 + c_2 =$ 0.0005 m/kN if there are no data to refer, and α denotes the collision angle of the ship.

Index